P9-DZN-272

FROM SILK TO SILICON

ALSO BY JEFFREY E. GARTEN

The Politics of Fortune

The Mind of the CEO

The Big Ten

A Cold Peace

FROM SILK TO SILICON

THE STORY OF GLOBALIZATION THROUGH TEN EXTRAORDINARY LIVES

JEFFREY E. GARTEN

HARPER
An Imprint of HarperCollins*Publishers*

For information, address HarperCollins Publishers, 195 Broadway, New York, NY 10007.

HarperCollins books may be purchased for educational, business, or sales promotional use. For information, please e-mail the Special Markets Department at SPsales@harpercollins.com.

FIRST EDITION

Designed by Janet M. Evans

Maps by Nat Case. © 2015 INCase, LLC. Some material derived from maps created by Abraham Kaleo Parrish, Yale Map Department.

Library of Congress Cataloging-in-Publication Data has been applied for.

ISBN: 978-0-06-240997-3

16 17 18 19 20 OV/RRD 10 9 8 7 6 5 4 3 2 1

For Ina,

Who has been the center of my life for fifty years

CONTENTS

INTRODUCTION

This is the untold story of globalization. It focuses on ten people who made the world smaller and more interconnected. Among those whom you will meet: a desperate teenager who rises from the steppes of central Asia to build the largest land empire in history; a producer of fancy paper products who advances global communication beyond anything achieved in human history; a cognac salesman who engineers the most far-reaching experiment ever attempted to dissolve national borders; a refugee from both the Nazis and the Soviets who leads the computer revolution; and others with similarly remarkable lives. Their accomplishments were not only spectacular in their own eras but continue to shape our world today. In the following chapters I have described who they were, what they did, the improbable journeys they took, and what they had in common. I have also shown how they remain relevant to some of the great global challenges of our times.

Most of us have a basic understanding of globalization, the good and the bad. We've seen how expanding trade can lead to more economic growth, lower prices, greater choice, and new jobs, but also how it undermines existing jobs. Many people have benefited from new investment opportunities in companies and countries around the world, but we've also seen the devastation that comes with international banking crises. We are enriched by cultural and educational exchanges but feel threatened by the spread of terrorism across borders. Every day we experience the ups and downs, the benefits and threats of a more interconnected world. Globalization, however, is anything but a recent phenomenon. It started about sixty thousand years ago, when some 150,000 people walked out of Africa in search of food and security. Over many millennia, these men, women, and children migrated to every part of the world. They intermarried. They traded. They spread and mixed their ideas, religions, and cultures. They fought wars and built empires that brought different populations under political roofs that sometimes spanned whole continents. They created cities that became melting pots of nationalities. They developed technologies and improved communications among themselves. They formulated laws, standards, and treaties governing their growing interdependencies. The story of globalization is no less than the story of human history.

I believe globalization is among the most powerful forces in the world and will become even more so in the decades ahead. It will reshape industries, change the way we work, alter our climate, enrich our cultures, and pose excruciating challenges to governments at all levels—from creating enough good jobs in the face of hypercompetitive trade to dealing with international cyber attacks against our critical infrastructure.

Why read yet another book on the subject? After all, many writers have dealt with globalization from a perspective of sweeping forces such as war, trade, and migration. Globalization has been analyzed by examining international industries such as textiles and oil, and by chronicling specific events such as financial meltdowns and tsunamis. Many books ask whether globalization is beneficial or harmful, and whether it should be encouraged or better controlled. To my knowledge, however, globalization has never been seen through the lens of a small number of people whose heroic deeds gave it a gigantic boost. This is a fundamental omission, for understanding the central personalities of our past constitutes the flesh and blood of history. If we don't focus on critical individuals, we leave out the difference that men and women make when they select one course of action over another. We forfeit the ability to measure contemporary leaders against those who came before them. It would be as if we were studying a war without delving into the motivations, the decisions, the triumphs, and the failures of the top generals. In fact, it is the rich combination of impersonal circumstances *and* human action that makes digging into world history so compelling.

In *From Silk to Silicon* I selected nine men and one woman who met several criteria. First, they had to be transformational leaders. Put it this way: they had to virtually change the world. Many great leaders accomplish something with a big transaction of some kind—they win a big war, they negotiate a major treaty, they persuade a head of state to follow a new course. However, these are not necessarily transformational accomplishments. To achieve that status, leaders have to operate on a more exalted plane, as did the men and the woman I've written about here. Transformational leaders do not exchange one thing for another, nor is their achievement the outcome of a bargain or negotiation, nor did they invent any one thing. Instead, they opened doors to a broad array of possibilities for progress. They changed the prevailing paradigm of how society was organized. They raised the hopes of broad swaths of civilization. They opened highways on which many others could travel.

I also identified people who could be characterized as "first movers," those who initiated or were in on the ground floor of a powerful, fundamental trend or movement that had an outsize impact on the world. In fact, each of my characters can be identified with having ushered in a critical phase of globalization—for example, the exploration of new lands and the search for new treasures, the expansion of governing ever-wider territory, the lowering of barriers to communication and commerce, the spread of new technologies and industrial processes across the world. Another way to think about the individuals in this book is to envision them as the inaugurators of various eras of world history: the Age of Empire, the Age of Exploration, the Age of Colonization, the Age of Global Finance, the Age of Global Communications, the Age of Energy and Industrialization, the Age of Global Philanthropy, the Age of Supranationalism, the Age of Free Markets, the Age of High Technology, and the Age of a Resurgent China. One of my characters led every one of these ages.

My subjects also had to be "doers" and not just thinkers, people who rolled up their sleeves and made something of global significance happen. Thus I stayed away from great philosophers such as Karl Marx, noted scientists such as Marie Curie, or economists such as Adam Smith—important as these people were. We often give too much credit to the power of ideas and not enough recognition to the importance of effectively implementing them on the ground; indeed, generating the purely intellectual breakthroughs is frequently the easy part of great transformations.

None of my characters are saints, to be sure, and several in particular had dark sides and created considerable suffering in their wake. Among the individuals here, you will find some whose efforts to conquer and dominate new lands were brutal if not barbaric, some whose drive to explore and trade involved expanding the heinous institution of slavery, some whose economic and social policies had the unintended effect of tearing apart the fabric of communities and wrecking countless lives. On balance, however, I believe the totality of each person's contributions to the world was decisive in driving globalization to higher plateaus.

I began my search in the twelfth century, when the first great age of comprehensive globalization was dawning, symbolized by the revitalization of the ancient Silk Road. I concluded with the end of the twentieth century, when the third industrial revolution, based so much on the silicon chip, was gathering momentum and when China was just opening to the world—two events that have set the stage for at least the remainder of this century. It's a long time period, to be sure, but I have been influenced by something Winston Churchill was alleged to have said: "The farther backward you can look, the farther forward you can see," which I interpret to mean that looking way back into the past provides much better perspective into the enduring patterns of history.

I'd like to think that individually and collectively the experiences and accomplishments of the ten extraordinary people I have written about provide essential perspective on a number of pressing contemporary challenges. Foremost among them is the future of globalization itself in an era when world economic growth may be slowing, unemployment may be stuck at politically unsustainable levels, terrorism is on the rise, financial crises are recurrent, cyber attacks are growing, and climate change hovers over everything. But *From Silk to Silicon* can provoke thought on other big issues, too—the rise of China, the future of the euro, the possibilities of a world transformed beyond recognition by new technologies.

I have a lot to say in this book about each of the ten extraordinary people, as well as how we may think of them as a group. For now, though, let me indicate only that after having written this book, one overwhelming impression dominates: my protagonists show beyond doubt how much individuals can accomplish against extremely long odds and how transformative and lasting their achievements can be. At a time when we yearn for great leaders in every walk of life, the people I have written about should give us enormous encouragement about what could lie ahead.

November 7, 2015

FROM SILK
TO SILICON

THE COLLECTION OF NATIONAL PALACE MUSEUM

Chapter I

GENGHIS KHAN

The Accidental Empire Builder

1162–1227

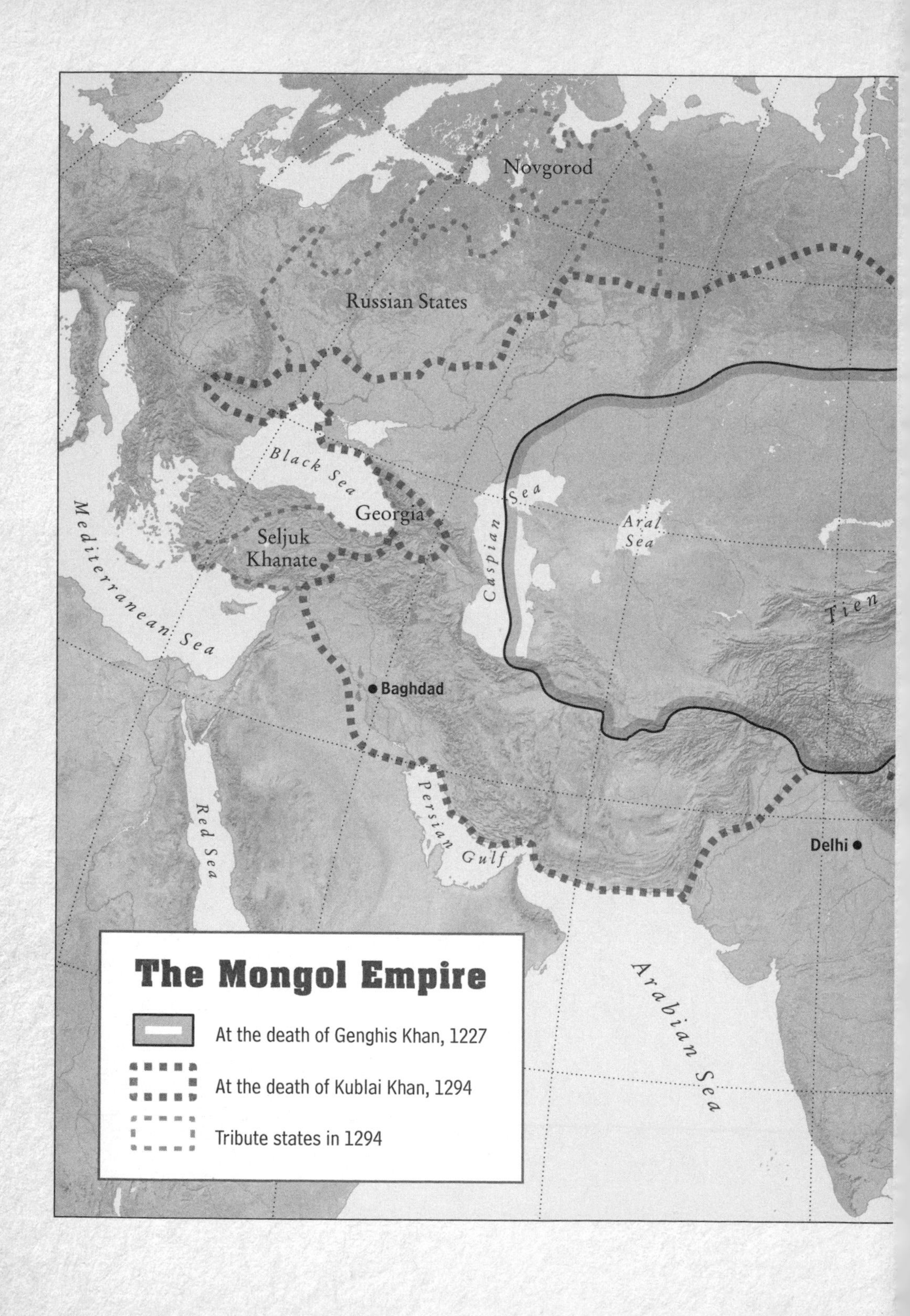
Novgorod
Russian States
Black Sea
Georgia
Seljuk
Khanate
Mediterranean Sea
Caspian Sea
Aral Sea
Tien
Baghdad
Persian Gulf
Red Sea
Delhi
Arabian Sea
The Mongol Empire
At the death of Genghis Khan, 1227
At the death of Kublai Khan, 1294
Tribute states in 1294

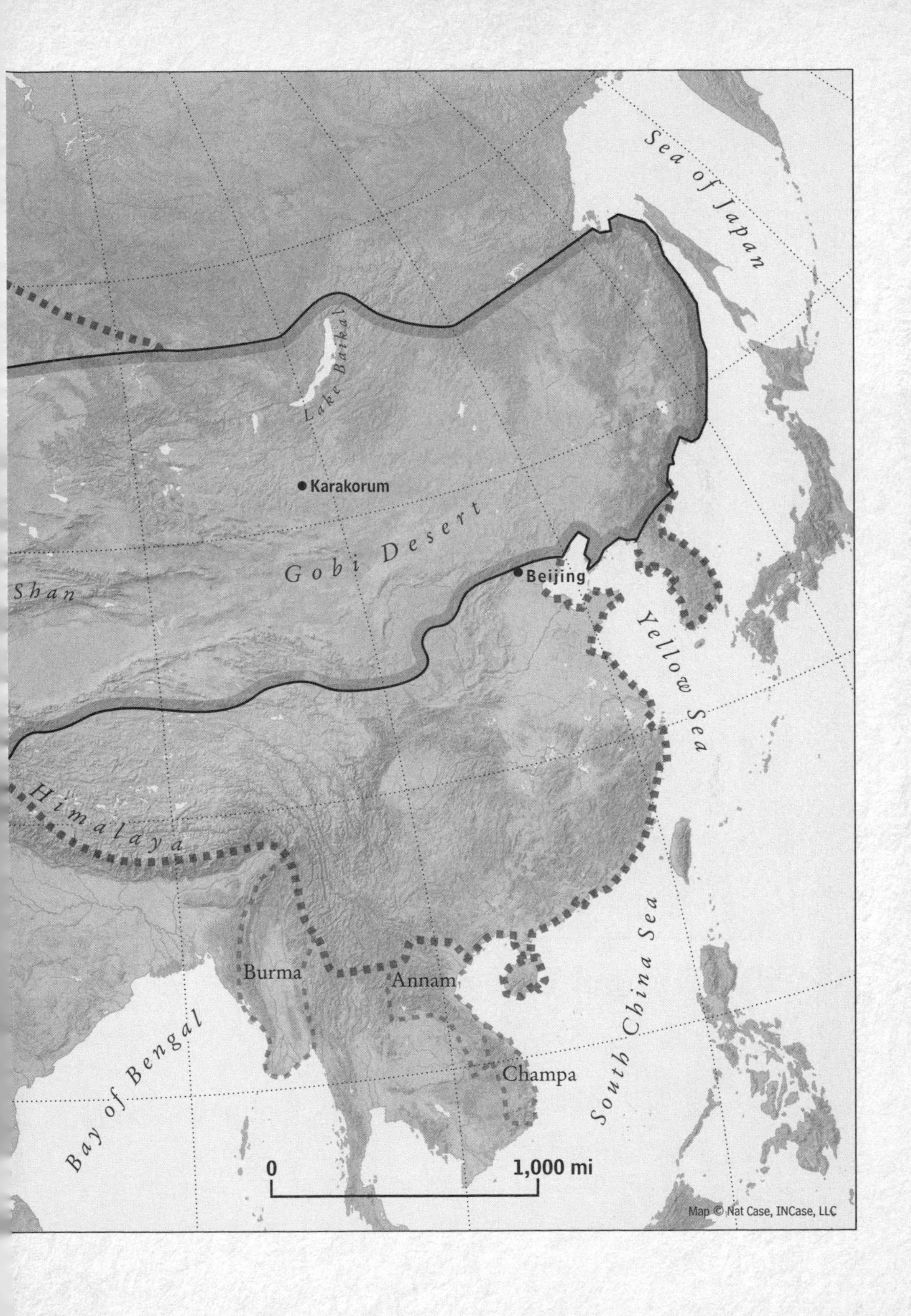

Sea of Japan
Lake Baikal
Karakorum
Gobi Desert
Shan
Beijing
Yellow Sea
Himalaya
Burma
Annam
South China Sea
Bay of Bengal
Champa
0
1,000 mi
Map © Nat Case, INCase, LLC

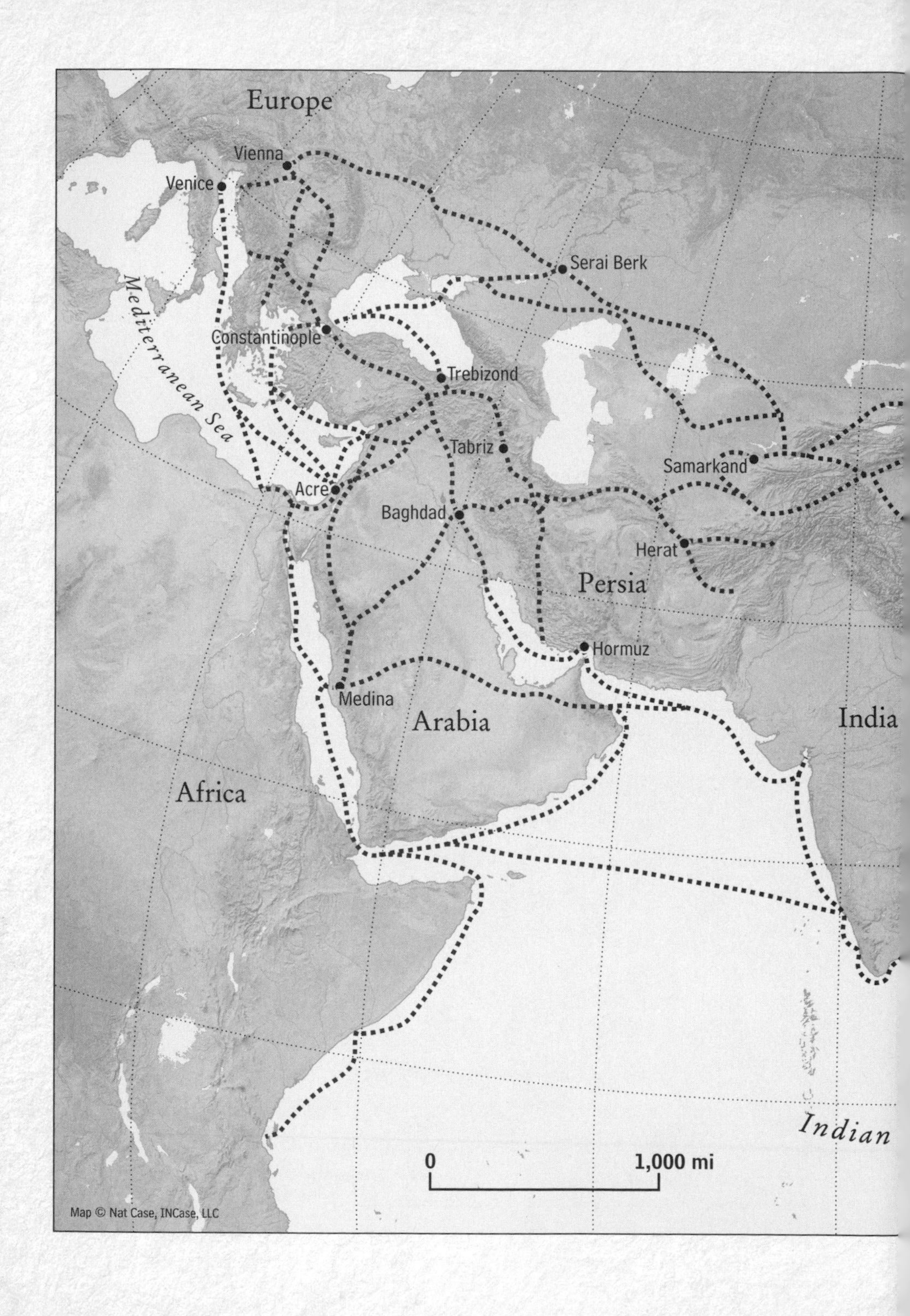
Europe
Vienna
Venice
Serai Berk
Mediterranean Sea
Constantinople
Trebizond
Tabriz
Samarkand
Acre
Baghdad
Herat
Persia
Hormuz
Medina
Arabia
India
Africa
Indian
0
1,000 mi
Map © Nat Case, INCase, LLC

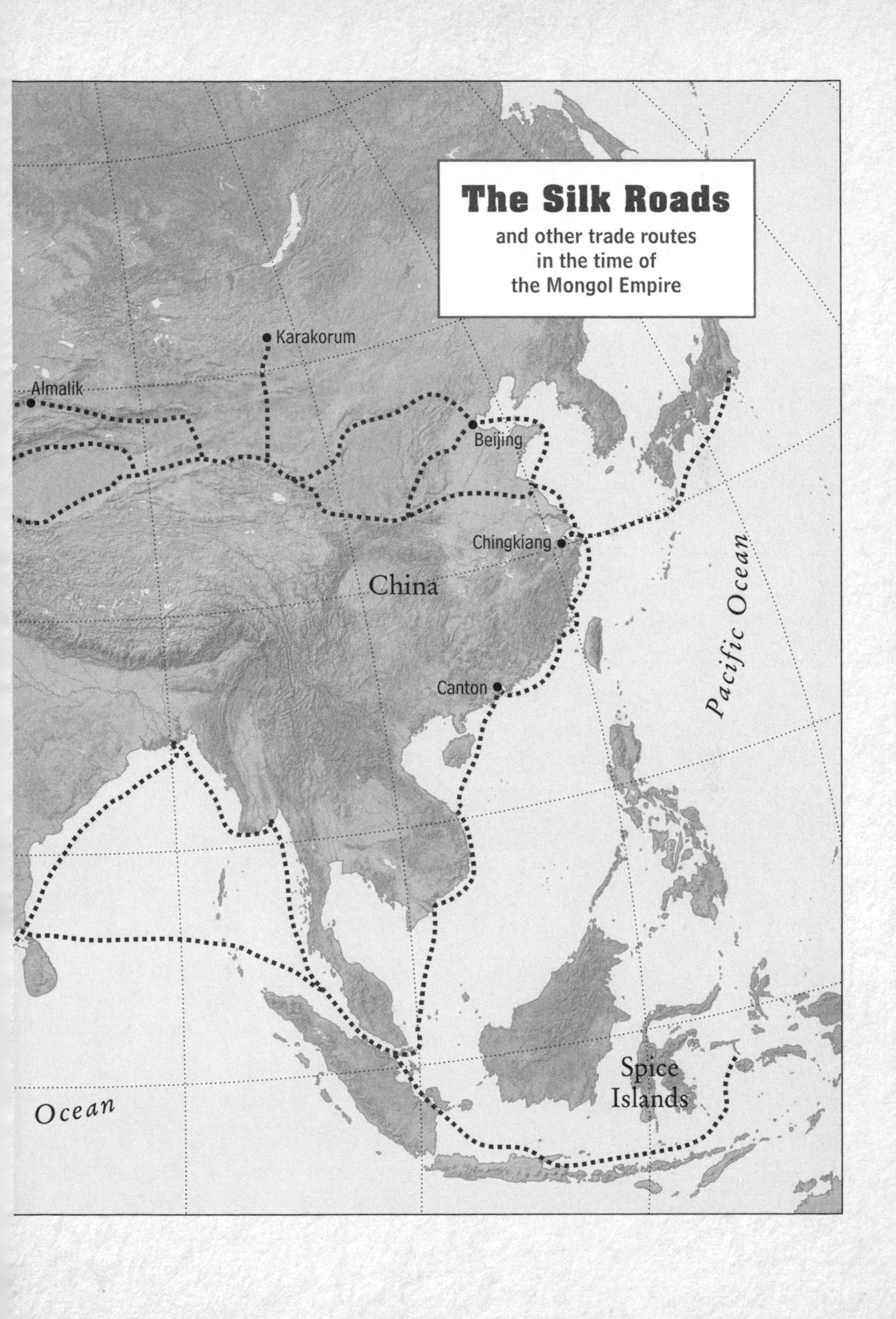

The Silk Roads
and other trade routes
in the time of
the Mongol Empire
Karakorum
Almalik
Beijing
Chingkiang
China
Canton
Pacific Ocean
Ocean
Spice
Islands

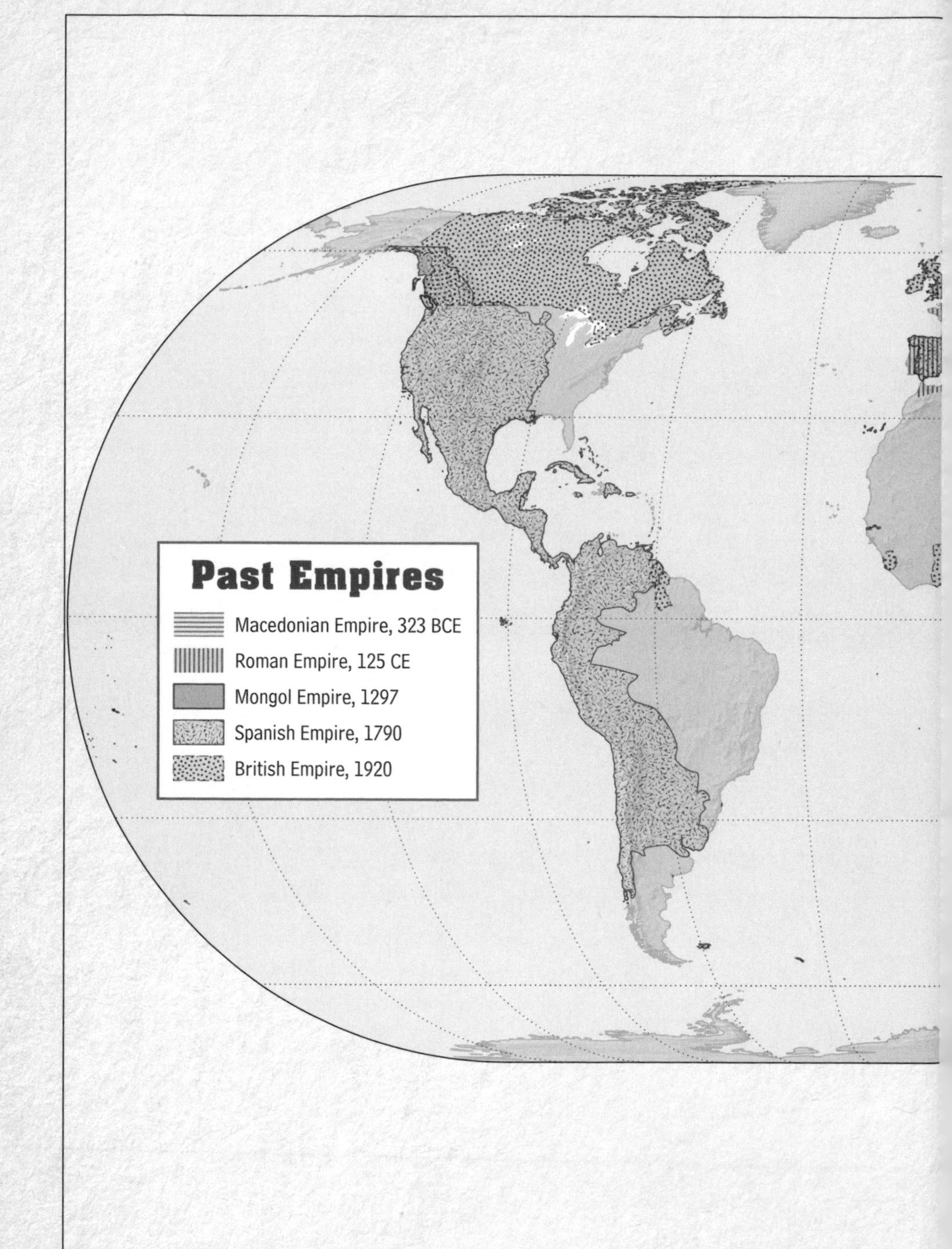
Past Empires
Macedonian Empire, 323 BCE
Roman Empire, 125 CE
Mongol Empire, 1297
Spanish Empire, 1790
British Empire, 1920

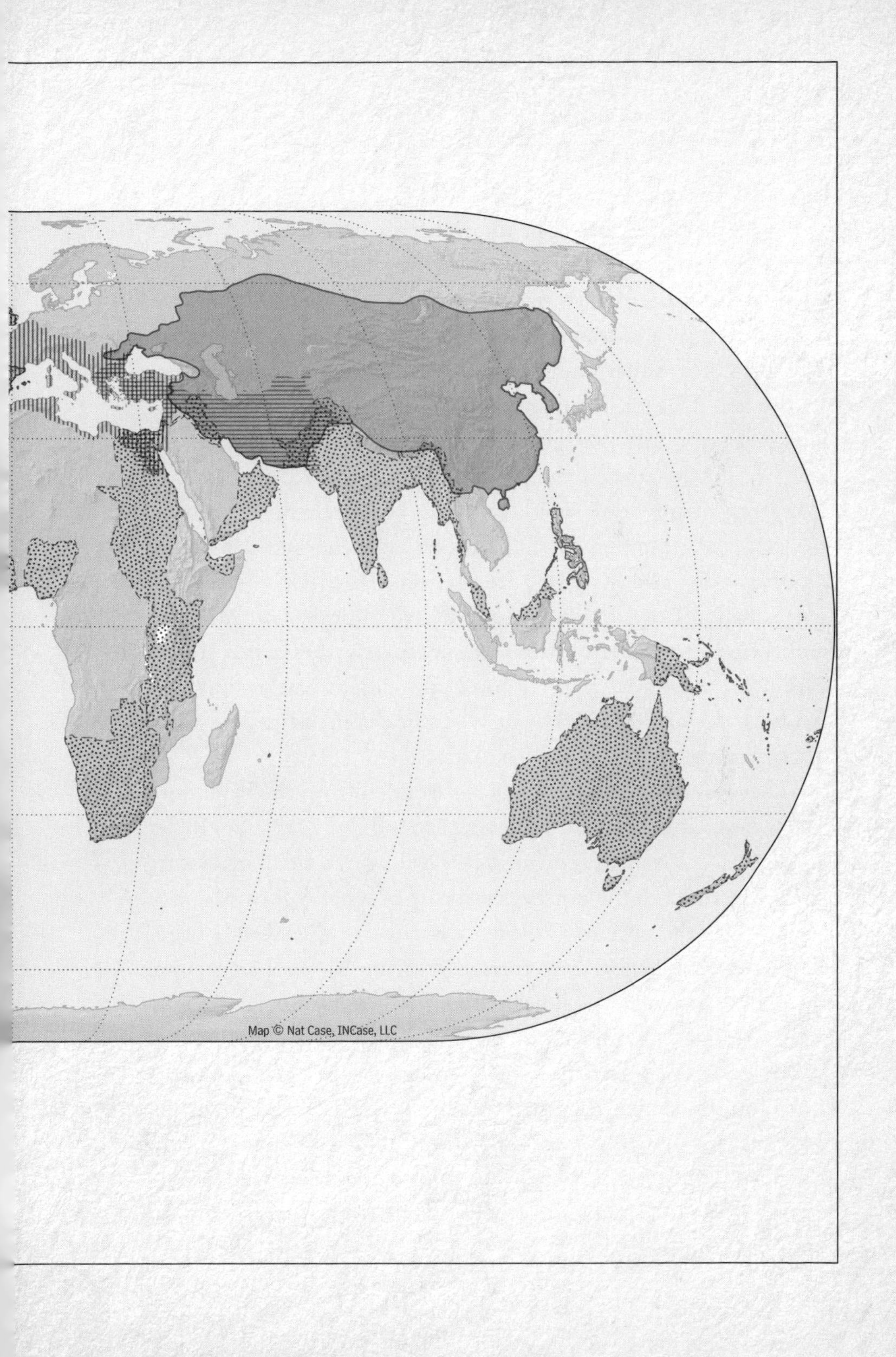

Map © Nat Case, INCase, LLC

Our stories begin in the twelfth century with a leader who conquered lands stretching from the Pacific Rim to what is now eastern Europe, amassing the largest empire in human history. At its height it consisted of twelve million square miles, four times the size of any empire before it and larger in landmass than any that followed it. Genghis Khan not only brought so much territory under one political roof but he started the process, which every other leading character in this book would broaden and deepen, of creating the physical, commercial, and cultural connections that define globalization today. His life reflects the two sides of globalization—the dislocation and destruction that it can inflict and the peace, modernization, and prosperity that it can create. His efforts began with brutal force and violent battles, but also left behind a world united by newly built channels for the spread of commerce and the exchange of ideas.

The legend of Genghis Khan is shrouded from the beginning in darkness. It holds that, sometime around the year AD 1160, a teenage beauty named Hoelun was traveling in a small black cart pulled by an ox in the northeastern part of what is now Mongolia. She was the wife of Chiledu, a warrior of the Merkit tribe. The couple's journey took them across the Mongolian steppe, through grassy plateaus bordered by forests, deserts, and a lake where Yesügei, a leader of the rival Tayichiud tribe, was fishing. When Yesügei spotted the young beauty he resolved immediately to kidnap her, a custom of the times. Together with his brothers, he descended on the caravan and captured Hoelun. Eventually, she gave birth to a son, whom Yesügei named Temüjin, after a rival he had slain in battle. The boy would grow up to become Genghis Khan.

Telling the story of Genghis Khan requires much speculation to fill in the many blanks between the few known facts. Because the Mongols were illiterate, their history was preserved orally. The only known Mongolian documents around Genghis's time are called "The Secret History of the Mongols," written in the mid-thirteenth century in a script borrowed from the Uyghur Turks. No copy of the original survives and the English translation is based on Chinese texts. Many experts view the work as loaded with self-serving political spin. Moreover, because the Mongols couldn't write, the surviving histories are often the work of Chinese and Persian scholars who lived in lands invaded by the Mongols, and display according biases. Still, many of the basics are reasonably clear about who Genghis Khan was, how he rose to power, how he built his world, how he ruled, and certainly about the impact he had.

The Trials of a Future Emperor

When Temüjin was about nine, his father brought him to his mother's original tribe, who were noted for the beauty of their women, in order to find a suitable bride. There Yesügei arranged for Temüjin to marry a ten-year-old girl named Börte, once the two came of age. He left the young boy with Börte's family, expecting him to remain with them for a year in order to get to know his future bride and in-laws. On his way home Yesügei met four men from the Tatar, a tribe that he had recently fought against. Thinking he had not been recognized, Yesügei accepted the Tatars' invitation to share a meal.

When he reached home the next day, he became seriously ill and died, apparently poisoned by his hosts. Hoelun, sensing the imminent threat to her position, sent for her son Temüjin. The young boy, who had looked up to his father as a warrior and leader, was crushed by his sudden death.

Yesügei had been a respected member of the Tayichiud tribe. But several months after he passed away, the tribal leaders, deciding that his family was a costly burden, abandoned them. Temüjin, his mother, and his brother were forced into a harsh exile. Without the protection of a formal community, the family became social outcasts who had to stay on the move to avoid falling prey to bandits. They subsisted on the leaves of wild plants, roots, and boiled millet. In winter they fought temperatures as low as forty below zero; in summer they suffered from oppressive heat.

A turning point for young Temüjin came on the day he found out that his older half brother was hoarding food. Enraged, Temüjin killed him with an arrow. Hoelun came undone. "Destroyer! Destroyer!" she is said to have screamed. "[You are] like an attacking panther, like a lion without control, like a monster swallowing its prey alive." Temüjin was just thirteen years old.

Following the fratricide, leaders of the Tayichiud tribe came looking for Temüjin. On the steppe, the young man was making many friends among fellow outcasts of minor tribes, families that had been abandoned, bandits, and adventurers. The Tayichiud leaders had to wonder whether this bold and violent teenager would one day build a rival tribe to exact revenge for having left him and his family to such a dismal fate. Eventually, they caught Temüjin and put him in jail with his hands tied and his neck bound by a wooden collar, making it painful to move or even to sit. It was a few years before he escaped.

Now about sixteen, Temüjin made a move to reclaim Börte and marry her. As a dowry, Börte's father gave the groom a sable coat. Temüjin took the coat and gave it to an old ally of his father's, Ong, the leader of the Kereyid tribe, who agreed in return to come to Temüjin's aid if ever asked. The opportunity came quickly, for soon after the wedding Börte was kidnapped by the Merkit tribe.

For three days, Temüjin prayed for guidance. While other tribes had embraced Buddhism, Islam, or Christianity, Temüjin was an animist, worshipping the spirits of nature. He prayed to the only god he knew, the Eternal Blue Sky, and then he set out for revenge, backed now by Ong, who marshaled his own tribe as well as others. They routed the Merkits—the first formal military victory for the future Genghis Khan—and Temüjin rescued Börte. Having been abused by her kidnappers, she returned home pregnant with a son and unsure who the father was.

A Mongol Nation-Builder

Temüjin's personal magnetism was attracting more followers. Although still a teenager, he well understood the tentative nature of tribal politics. One moment chieftains would join him and just as suddenly they would sever their ties. To address the challenge of shifting tribal loyalties, Temüjin turned the chaos of plunder into a system of rewards. In each raid he insisted on putting all the booty under his control, a sharp departure from the prevailing custom of allowing each warrior to claim all that he could carry away. By centralizing the process of pillage, Temüjin gave himself a mechanism for distributing the loot to reward loyalty and punish disobedience.

In addition, when he vanquished a tribe, he tended to kill off the political leaders who might challenge him but recruited others who could support his quest for power. He gave no preference to aristocrats or their sons. He also believed in assimilating the tribes he captured into his own, even adopting children of defeated tribes as part of his own family.

Temüjin also compelled loyalty to the emerging Mongolian nation through fear. As he won battle after battle on the steppe, stories of his ferocity spread. In a major campaign against the Tatar tribe, he was reputed to have killed every man above the height of the axle on a cart, one way of making sure to abolish the existing power structure. Years later, after Temüjin became Genghis Khan, and his conquests spread to foreign lands, his reputation for brutality would reach new extremes.

Typically, the Mongols were fighting foes who had superior numbers and weapons. In response, Temüjin turned military agility into a high art. Often he would order his men to surround a tribe and attack from all sides, sowing panic and confusion, then retreat in the many directions from which they had come. Sometimes, in retreat, the Mongols would discard valuable items such as ropes or brass trinkets to slow or distract their pursuers, setting them up for a counterattack. To camouflage the size of his force, he would dispatch his cavalry in waves, and at irregular time intervals, disorienting the enemy. Or when dusk came, he would order every soldier to light five fires, spaced several yards apart, creating the impression of a vast force settling in for the night.

Temüjin was creating a new type of army on the steppe. The old model was one of individuals attacking randomly and chaotically—an "attacking swarm of individuals." In its place, he brought order, a higher degree of control, and a greater variety of tactics. And with every victory, Temüjin sharpened his military skills.

The climactic battle that made Temüjin the sole leader of the Mongolian nation took place in 1204, when he defeated the Naiman tribe. By 1206 his leadership was secure, allowing him to call a convention of chiefs known as a *khuriltai*, which brought together leaders from all over Mongolia. The tents stretched for miles in every direction, and the ceremony lasted for days. Music, wrestling, horse racing, and archery kept everyone in good spirits, while carts distributed delicacies such as boiled horse flesh to the guests. It was at this moment that Temüjin was crowned Genghis Khan.

His followers carried Temüjin on a felt carpet to a throne. The chief shaman gave an invocation, telling Temüjin that "whatever Authority of Power he had given him, was derived from Heaven, and that God would not fail to bless and prosper his Designs if he govern'd his Subjects well and justly; but that, on the contrary he would render himself miserable if he abused that power." Temüjin took the title Chinggis Khan, a name that his future Persian subjects spelled Genghis Khan. The meaning of the name is disputed by historians, but at a minimum it meant Supreme and Universal Ruler. "If you wish me to be your ruler," he asked the chieftains, "are you without exception ready and resolved to fulfill all my behests; to come when I summon you, to go whithersoever I may send you; and to put to death whomsoever I may indicate?" The chiefs roared, "Yes." Genghis replied, "Henceforward, then, my simple word shall be my sword." The chiefs fell to their knees and bowed their heads to the ground four times.

Inventing the Machinery of a Mongol State

While Genghis had proved to be an exceptional military commander, he was also remarkably adept at organizing the Mongolian state, all the more so because, unlike many other conquerors, he had almost nothing on which to build. He was starting with a primitive society comprised of tribes in constant conflict, and in which a principal activity was the confiscation of wives and horses. Though he could neither read nor write, Genghis Khan soon surrounded himself with scholars, scribes, and translators, including many of Chinese or Persian origin, and ordered the creation of a writing system. He set down laws to deal with many habits that triggered feuds and wars, while leaving some room for custom and tradition. Thus he forbade kidnapping, abduction, and enslavement, probably inspired by his own incarceration by the Tayichiuds and the kidnapping of Börte. He declared all children legitimate, whatever their parental status. He made stealing a capital offense. He declared religious freedom for everyone. He instituted a system of taxation but exempted religious leaders, doctors, lawyers, scholars, and other professionals. He created a position of supreme judge to keep track of the laws and punish those who disobeyed them. He decreed that his successors had to be chosen by a *khuriltai*, and said the penalty for any member of his family to claim office without election was death, though this last edict was never honored.

Genghis established a state heavily organized along military lines, but in the process he created order and accountability in place of unruly fragmentation and an every-tribe-for-itself system. He set up a military system based on units of ten—an army of, say, one hundred thousand, broken down into ten regiments, each of which had ten battalions. Each battalion was then comprised of ten companies, and the pattern continued to repeat itself. The new framework brought discipline to previously disorganized tribal forces, with effective communication from top to bottom and easy and rational rotation of command at every level.

Genghis and his mounted army—each of the 150,000 to 200,000 men had multiple horses—could move with exceptional speed, traversing thousands of miles over treacherous mountains and deserts to attack from directions the enemy thought impossibly remote. They traveled with minimum weight, bringing with them their own engineers who would construct what was required on the spot, using local materials rather than dragging a heavy line of equipment behind them. They carried little food and water, preferring to live off the land. Unlike traditional armies that marched in columns, they spread over vast areas with leaders in the middle of concentric circles. Since everyone was illiterate, all orders were verbal and would move from soldier to soldier in the form of a fixed melody to which a new verse was attached, thereby making the instructions easier to remember.

Genghis recruited the sons of his army commanders to form a special regiment that acted as his personal bodyguard. He bought the devotion of the sons with a particularly generous share of the spoils and persuaded them to report on the loyalty of their fathers, ensuring the latter would never rebel lest they put their sons in jeopardy. Day to day, this special guard also provided a range of civil services, from adjudicating legal hearings to controlling entrance to royal tents. They became the emperor's public administration.

A Fragmenting World Is No Match for a Strongman

The Mongol conquest was driven by the same factors that have pushed nomads to attack settled areas throughout history: the weakness of settled civilizations, the desire to seize the goods of societies more sophisticated than their own, and the rise of charismatic leaders.

Well before Temüjin became Genghis Khan, many routes existed to connect China and the Middle East, including seaborne trade in the Indian Ocean. These highways on land and on the oceans came to be known collectively as the Silk Road. The trading cities of Genoa, Baghdad, and Samarkand produced an adventurous merchant class that traveled these routes, spreading achievements in technology and the arts, from the pottery of Song China to the gold- and silver-inlaid furniture of Persia. Traders moved horses, rice, printed books, and all manner of goods along routes that also carried priests and adventurers, ambassadors and entrepreneurs. They spread new technologies for building roads and ships and new ideas and instruments of commerce, such as paper money, new forms of credit and commercial partnerships, and new standards for weight and measurement. The Silk Road was also a transmission belt for the extensive interchange of culture and religion.

The threat to this prospering global crossroads was political: centrifugal forces were breaking apart the power centers of Europe, the Islamic world, and China. Previously undisputed rulers were being challenged by regional and local leaders—who were motivated by their own desire for power and riches—and by clashing cultures and religions. It was into this maelstrom that Genghis Khan was drawn, motivated less by a grand vision than by the lures of plunder and, as we shall see, a desire for personal revenge.

By the first decade of the thirteenth century, while the more sedentary and fragmented societies of China and Islam were preoccupied with repelling invasions, Genghis Khan was bent on offensive expansion, and his militaristic society was superbly organized for that. He also knew that holding his kingdom together would require delivering ever-expanding economic rewards to Mongolian tribal leaders who, fiercely independent by tradition, had been unified not only by his overwhelming military strength but by the prospect of ever-increasing riches. Genghis had created an aggressively ambitious entourage, highly disciplined and driven by a clear mission of attacking and plundering other societies.

Within two years of ascending to the throne, Genghis Khan was not only consolidating his hold on Mongolia but moving into adjacent territory. Of immediate interest to him was the instability next door in China. Long before Genghis, the Mongols had been raiding northern China in pursuit of clothing, furniture, saddles, and cookware—manufactured luxuries that they, fishermen and hunters that they were, could not produce. In response, Chinese leaders tried to play the Mongolian tribes against one another to prevent an attack in unison, and for centuries they had succeeded. In addition, the Tang dynasty, which had once ruled all of the Middle Kingdom, had long since broken into three regional dynasties: the Song in the south, the Jin in northern China and Manchuria, and the Xi Xia in the west.*

That was the situation that Genghis faced to his south. To his west was the Islamic world, vibrant in its economic and cultural life, but existing in a state of political disarray characterized by dynastic rivalries, competing warlords, constant coup d'états, sectarian wars, breakaway regions, and shifting alliances. The sultan himself was at war with his own family, and citizens often distrusted the soldiers in their midst, usually foreign mercenaries. Internal relations were so strained that the Arab caliph in Baghdad was alleged to have secretly petitioned Genghis Khan to launch what would be his first attack on Islam, against the empire of Khwarezm†, an area strategically located between China and Egypt and consisting of parts of modern-day Uzbekistan, Turkmenistan, and Iran.

Resistance and Revenge

Genghis's first strike was against the weakest of the three Chinese dynasties, the Xi Xia. Occupying what today is northern Tibet, the Xi Xia were a highly civilized state with a dozen major cities, a system of official schools, a college with three hundred places to train bureaucrats and scholars, skillful weavers, leatherworkers, builders, and metallurgists—in all, a ripe source of spoils. Within a few years of becoming the supreme Khan, Genghis attacked the walled city of Yinchuan, the Mongols' first assault on a fortified city. At this stage, however, Genghis's army lacked the heavy

---

* The Song dynasty is sometimes spelled as Sung dynasty. The Jin is sometimes referred to as the Jin dynasty, the Chin, or the Jurched Kingdom, sometimes spelled as Jürchen and sometimes referred to as a tribe. The Xi Xia is often referred to as Western Xia or Xia Xia.

† Khwarezm is sometimes spelled as Khwarazm or Khwarizm.

siege weaponry of the Chinese, who had gunpowder bombs and huge double- and triple-bowed siege bows that could fire "arrows" the size of telegraph poles to punch holes in walls half a mile away. Unable to breach the walls, the Mongols tried breaking the dikes to flood Yinchuan. This action also failed to force a surrender, so Genghis's army dug in for a long siege. After nearly a year, during which the Mongol soldiers patiently camped outside the walls of the city, the Xi Xia emperor agreed to a deal. In return for ending the siege, he gave Genghis his daughter as a bride, surrendered camels, falcons, and textiles, and promised an uninterrupted flow of luxury goods back to Mongolia. To the great disappointment of Genghis, the tribute halted as soon as his troops departed. Years later, the Khan and his sons would return to extract retribution for this act of betrayal.

The Mongols' next target in China was the Jin, a better-defended empire with much larger territory than the Xi Xia. Genghis planned the invasion with great precision, deploying some tens of thousands of warriors and thousands of horses, together with camel-drawn carts carrying provisions. He sent these forces down from Mongolia in waves, using messengers on swift horses to coordinate movements, toppling one city after another—Xamba, Wuyuan, Linhe. Genghis pushed the Jin back to Yanjing (now Beijing), which he placed under siege for a year, while his troops ravaged the countryside and seized smaller cities. Eventually they took Yanjing, too, and drove their foes back to Mukden (today Shenyang, the biggest city in Liaoning Province in northeast China). Again unable to crack the strong fortifications of Mukden, Genghis feigned retreat and disappeared. When the delighted citizens opened the gates and held a celebration, the Mongols returned in force to capture the city.

Eventually, the Jin surrendered and the Mongols took control of northeastern China, including what is now the Korean peninsula. They had destroyed scores of cities, massacred thousands of soldiers and civilians, and seized prisoners for use as human shields or guides to their next target. Stories of devastation spread across Central Asia to northern Europe, where one ambassador from the Islamic world reported that the bones of the slaughtered formed mountains, that the soil was greasy with human fat, and that, rather than face the Mongols, sixty thousand girls had committed suicide by jumping from the high walls of Yenking.

Genghis returned to Mongolia, believing the Jin were now a vassal state that was resigned to paying him tribute in perpetuity. Again the booty did not arrive, and again the Mongols would return years later to collect. The cases of the Xi Xia and the Jin appear to have taught Genghis not to rely on promises in the future; he would have to change strategy and start annexing and occupying tributary states.

Attack First, Plan Later

About ten years after his first attack on China, Genghis Khan turned his eyes west to the Islamic world. His original plans did not include seizing and administering new territories; his interest was in trade. In 1217, he sent a delegation of one hundred men, all Muslims, to offer a formal trade treaty to Shah Mohammed of Khwarezm. At that time, the empire of Khwarezm was in a state of constant civil war. When Genghis's trade delegation arrived at the border of Khwarezm, a local governor jailed them as spies. A furious Genghis then sent three messengers to seek the intervention of Shah Mohammed, the highest official in the region. Instead of granting Genghis his request, the shah chose to put the three messengers to death, an act that inspired the local governor to execute the first hundred emissaries as well.

Blinded with rage, Genghis went alone to the top of a hill, bared his head, and again prayed for three days to the Eternal Blue Sky, saying, "I am not the author of this trouble; grant me strength to exact vengeance." Faith in the animist god was central to Genghis's character and a great moral support—or rationalization—for his actions.

Genghis invaded Khwarezm around 1219. By now, the Mongols had adopted Chinese siege technology—huge bows, battering rams, scaling ladders, flame-throwing tubes. They had created a combination of fast cavalry and heavy artillery that could quickly overrun virtually any existing defenses.

Genghis launched this first attack on the Islamic world with no apparent vision of where it would end. It continued all across Eurasia into eastern Europe as looting in one city whetted the appetite of his generals for more. Disappointed in the lack of tribute he was getting from China, he may also have been eager for an excuse to seize new, richer territory to the west. Whatever his motive, he was assaulting a civilization that was as sophisticated as China and, of course, light-years ahead of Mongolia.

The Muslim lands included Arabic, Turkish, and Persian populations. They had replaced papyrus with paper. Many cities had thriving bookshops, and rich families had large libraries of their own. The Arabs translated the Greek and Indian classics, which they used as a foundation for their own advances in virtually every field of scholarship, from architecture to medicine. But intellectual achievement did nothing to unite medieval Islam against the Mongols, who in the five years following the conquest of Khwarezm went on to overrun cities from Afghanistan to the Black Sea, including Gurganj (Turkmenistan, and capital of the Khwarezm empire), Nishapur (Iran), Merv (Turkmenistan), Multan (Pakistan), and Samarkand (Uzbekistan).

The campaign against Islam was unprecedented in its violence, even by the standards of the Mongols. Genghis's men had killed tens of thousands in China, but the death toll in Islamic regions would be many multiples of that. The Mongol army would divide up the local population for either employment in the empire or for death. To prevent political resistance in the future, they would decapitate the social structure by killing off the aristocratic class, mirroring Genghis's previous tactics on the Mongolian steppes. Anyone who owed his position to inheritance or wealth was a target for execution.

Able-bodied but unskilled young men were dragooned into the army as soldiers or were deployed as human shields. When the Mongols found moats surrounding fortified buildings, they would force local men to lie down in the water to become human bridges. It was said he had boiled enemies alive and turned the skulls of his adversaries into silver-coated drinking cups. In one instance, Russian princes who had executed a group of Mongol envoys were captured and forced to lie on the ground. Genghis reputedly put a platform over them and held a banquet on it as the princes slowly suffocated. It is not clear how many of these stories are true, because from the time he was Temüjin, Genghis and his men had helped spread these tales to intimidate their enemies. Nevertheless, Genghis would spare professionals such as clerks, astronomers, engineers, doctors, judges, and skilled workers such as potters, carpenters, and entertainers, earmarking them for employment in other parts of his empire or in his court.

As Genghis advanced deeper into Islamic lands, he sent caravans loaded with plunder back home to Mongolia. The Mongols began to experience luxury that they could never have imagined—including metal tools, Persian porcelain, and sumptuous fabrics and glittering jewelry. The new Mongol royalty and its extended court became conspicuous consumers. Because the Mongols continued to produce almost nothing themselves, because their expectations kept rising, and because they were such accomplished warriors, there was no logical end to the conquests.

The Mongol State Becomes a Cosmopolitan Empire

Still Genghis was eager to oversee his growing empire. He established a central secretariat to record his commands and have them translated and disseminated. Each new conquest magnified the challenge of governing his territory and, seeing the limitations of his fellow Mongolians as administrators, Genghis turned to foreign talent, especially from China and Persia. One example is Yelü Chucai, recruited from the Khitan tribe in northern China, who came to serve as a trusted adviser. Yelü had been trained in astrology, spoke both Mongolian and Chinese, and knew the laws and traditions of the Middle Kingdom. A number of other scholars from the same tribe joined him. Genghis was interested in philosophy, too. Sometime around 1221, for example, he sent for a seventy-year-old Taoist priest from China named Ch'ang-ch'un, someone who had amassed a large following and was widely admired. After nearly three years of travel from China to Genghis's military location deep in western Asia, the elderly Chinese wise man eventually spent years with Genghis, preaching quietism, asceticism, and meditation.

Seeing that every conquest generated a new source of supplies, Genghis recognized that the merchant class could move goods faster than the army. He and his court supported the merchants with financing, business partnerships, large orders of merchandise, and low taxes. It was a primitive but deliberate form of state capitalism that foreshadowed the political economies of Eurasia and East Asia in the centuries to come.

As the empire grew, the Mongols gave priority to logistics. They built roads, bridges, and ferries, as well as new towns to act as supply depots. Most impressively, some ten thousand postal stations were established along the Silk Road and throughout the empire, spaced about a day's walk apart. They usually consisted of outposts where travelers could find fresh horses, food, lodging, and guides. There were even rudimentary passports. Special messenger services were often available to relay news from one outpost to another. The routes were protected from robbers by soldiers and local families.

Genghis and his top leaders carefully blended the mix of goods, ideas, and talents that flowed along their protected continent-wide networks. They gave priority to the military for moving its booty, and the skilled labor it had recruited or commandeered. The Mongols also gave access to select merchants, and all manner of exchanges flourished: cuisines were mixed, physicians and astronomers from China and Persia were exchanged, and literature from different cultures spread. Maps of the empire were produced and refined, dictionaries were translated into multiple languages. Genghis transferred metalsmiths, ceramicists, weavers, bread makers, and cooks from one end of the empire to the other. He even inserted some commanders of defeated Muslim legions into his own army.

Finishing the Job in China

As the campaign for Muslim lands was ending, around 1224, Genghis decided to return to northern China to settle scores with the Xi Xia dynasty, which had not only failed to pay tribute but also had ignored the Khan's request to provide troops for the Islamic campaign. Now in his sixties, Genghis set out to reestablish his dominion in northern China. It took three years, but by 1227 the Xi Xia were ready to capitulate.

At that moment, Genghis Khan died in circumstances that remain a mystery. It is not known whether he was wounded in battle, thrown from his horse, attacked by a deadly virus, poisoned by a mistress, or if his life ended another way. Genghis had left standing orders that all details of his death would be subject to strict secrecy, and so they were. Legend has it that all the soldiers and laborers involved in the burial were killed so they couldn't talk. The practical reason for secrecy was that if Genghis's death became known, it would almost surely stoke rebellion in the Islamic lands and provoke the Xi Xia to reconsider their capitulation. There is no definitive account of whether Genghis, in a final act before his death, or an aide handled the Xi Xia surrender, but it did go through.

The Empire After Genghis

Genghis's heirs continued to expand the Mongol empire. Over the next century they would double its size, adding southern China, all of Iran, most of Turkey, Georgia, Armenia, Azerbaijan, most of habitable Russia, Ukraine, and half of Poland—in all, roughly 20 percent of the world's land area.

Genghis had established not only an empire but a governing foundation that his heirs would build into the *Pax Mongolica*, the period of relative peace and stability that extended over much of Eurasia from 1206 to the mid-fourteenth century. This empire underpinned an era of globalization that was unprecedented. The Mongols revolutionized warfare, which made their conquests possible, but the empire lasted so long because of the expansion of trade, transportation, and communication; the intermixing of people, ideas, and culture; and the unification of administrative procedures. The big advantage that the Mongols had over previous empire builders is that they themselves had nothing in the way of deeply ingrained ideas of politics, economics, or culture to spread abroad. They were not driven by ideology or any messianic impulse—only by the hunger to amass wealth. As a result, they did not impose political ideology or cultural or religious ideas on others but rather created an environment of extreme tolerance, so long as the basic governing regime wasn't fundamentally challenged and so long as the booty flowed smoothly from the far-flung territories to the Mongolian center.

Tolerance for religious freedom was particularly notable and reflected the way Genghis thought. He understood that he had much to gain by showing respect for proud local cultures and influential religious leaders, and he sought out strong relations with them.

Indifferent to controlling matters of religion or culture, the Mongols focused on building commerce and the physical, administrative, and legal infrastructure to help it flow freely. Before the Mongol Empire, for example, few traders were able to travel the entire Silk Road in large part because Arab middlemen in places such as what are today Syria, Iraq, and Lebanon insisted on standing between the buyer and the seller, imposing high taxes on all sides. At the height of *Pax Mongolica*, however, there were no significant trade barriers on the road from the Mediterranean to China, where the big cities became cosmopolitan metropolises. In what today is Beijing, Genghis's grandson Kublai Khan (1260–1294) established special sections for merchants of multiple nationalities from all over the empire, some who came from as far away as Italy, India, and North Africa. The Mongols also encouraged their subjects, particularly the Chinese, to emigrate to foreign trading posts in order to facilitate more commerce.

Genghis's offspring became great transmitters of art and culture through the trading channels they established. Their tastes for colorful clothing, silver jewelry, and images of animals provoked a desire throughout the empire to create items that pleased them, leading to a convergence of styles. Moreover, the interconnection among societies resulted in Chinese textiles and painting becoming even more popular in Persia, where Iranian tile work began to reflect the images of the dragon and phoenix that were popular in the Middle Kingdom. As the number of travelers on the empire's roads rose, some became notable figures. Marco

Polo is the most widely known of these travelers today, but there were many others, including the Muslim jurist Ibn Battuta, the Nestorian Christian Rabban Sauma, the Franciscans John Plano Carpini and William of Rubruck, and the Chinese Confucian Zhou Daguan. Their writings described places from Angkor Wat to Hangzhou to Tabriz to Paris, making these special sites cultural touchstones for elites across the empire.

Under Mongol rule, China's luster as an empire was restored. Starting with Genghis Khan, the Mongol leadership transformed the Middle Kingdom from a civilization torn asunder by civil war and hostile dynasties into a unified nation, capable of outlasting revolts, invasions, and other attempts at foreign domination for six centuries. In western Asia, the Mongols were able to forge unity among competing Islamic chieftains, giving birth to the modern Persian Empire. Although the Mongols never occupied eastern Europe and the Mediterranean countries, they stimulated a revolution in productivity in these regions by exposing the West to specialized tools, new blast furnace technology, new crops that required less work to plant, and to new concepts such as paper money, primacy of state over church, and freedom of religion. In 1620, the English scientist Francis Bacon named three innovations that changed the world, citing the printing press, gunpowder, and the compass. All three came to the West during the heyday of the Mongol Empire.

Historian and journalist Nayan Chanda put it well. "Empires played a key role in the rise of governance as they extended rules and regulations over an expanding territory," he wrote. After pointing to the advances made in the Roman Empire, the Mauryan Empire in India, and the Han Empire in China, he continued:

> The scope of governance grew to a new height under the Mongol Empire. . . . Deep Mongol interest in trading meant

> that the Silk Road across Central Asia emerged as a well-guarded conveyer belt of goods, people, and ideas. The road, with its Mongol sentry points and inns, its postal system, and its rudimentary passport and credit card (paiza) system, provided unprecedented governance for land-based trade and transportation.

The empire's impact on governance in China, in particular, was notable. The administration begun under Genghis would reach its peak under heirs like Kublai Khan, who guaranteed landowners property rights and reduced taxes. Kublai also built an extensive network of schools, professionalized the civil service, and introduced paper money and bankruptcy laws. He established an office for the stimulation of agriculture to improve farmers' lives and crop yields, and a Cotton Promotion Bureau to improve planting, weaving, and textile manufacturing techniques. He promoted the arts and literature, translating Persian and other classics into Chinese. He instituted universal education five hundred years before any ruler in Europe did, and he refused to allow public execution of criminals at a time when they were a popular spectacle in Europe. These achievements occurred well after Genghis Khan, but they evolved from his early attempts at establishing a multicultural society spanning vast territory.

Tracing the Early Roots of Globalization

It is often said that the first golden age of globalization began near the start of the industrial revolution in Europe around 1870 and ended in 1914 when the outbreak of World War I shattered the idea that growing political and economic bonds between nations would guarantee peace. This period had indeed experienced a great boom in global trade, investment, migration, and innovations in communications such as the telegraph. But that period is the wrong starting point. The first golden age was the thirteenth and fourteenth centuries. It was in the era of Genghis and his sons and grandsons when the routes they built and secured opened a new world of possibilities. Europeans were able to buy silk and other rare textiles, exotic spices, even paper-making technology from China. Chinese iron-smelting technology was mixed with Persian engineering skills to build advanced weapons. Medicines from India, China, and Persia were combined to create great advances in pharmacology. Indeed, the story of Genghis Khan is in microcosm a story about globalization itself. It illustrates the force that military conquest has played in furthering the connections among disparate societies. It shows how commerce follows conquest, how commerce and culture intersect, and how transport and communications networks become so important. Together with his sons and grandsons, Genghis also faced the enduring challenge of balancing central administrative control with tolerance for local institutions and culture.

History never repeats itself, as we all know, but we can't ignore some of the parallels. Genghis Khan's world was, of course, far less advanced than ours, but it does presage several of the critical trends of this century. Many Westerners today believe that most innovative ideas and technologies originated in Europe or America. But the Mongol Empire illustrates the pioneering creativity of the East, which is now reemerging with the rise of China and India. Indeed, the growing clout of Asian nations in the global economy in many ways represents a return to the balance of power that began to unfold under Genghis Khan and continued through the mid-eighteenth century. In 1500, for example, the gross domestic product of east Asia was three times that of western Europe and as late as 1820 it was two and a half times bigger. For many centuries before the 1800s, Asia was the world's most dynamic region in terms of technological innovation in such areas as metallurgy, shipbuilding, and farming—led by China and India. Then the positions of Europe and Asia were reversed. Now, however, many of the old trends are reemerging. Asia is again ascending. It now accounts for nearly 50 percent of the global GDP. Its share of production of resources, trade, and investment has also reached levels critical to the global economy. It has the bulk of the world's population. We are coming full circle.

In addition China is once again building connections to Europe via the Persian Gulf and also through South Asia. Its powerful thrusts include the expansion of trade and the construction of oil and gas pipelines, ports, and transnational highways. In 2013, China's president, Xi Jinping, stopped by several of China's neighbors to the west, such as Kazakhstan and Uzbekistan, to emphasize the importance of stronger ties among China and them. At one point in his visit he harkened back to the days of Genghis Khan and said, "I can almost hear the ring of the camel bells and [smell] the wisps of smoke in the desert." Less than two years later he did more than romanticize about the past; he committed to inject $62 billion into three of China's state-owned banks to finance a "New Silk Road," over land and over the oceans, the two being combined in a policy the Chinese call "One Belt, One Road." "One Belt, One Road" also encompasses new rail freight routes through Russia to Hamburg, power stations and manufacturing plants throughout Eurasia, and a variety of financial links. In conjunction with the establishment of a new international bank for infrastructure—the Asian Infrastructure Investment Bank—which China proposed and substantially financed, and which over fifty nations joined in 2015, the modern revival of Genghis Khan's creation could be unfolding.

In world history no group has contributed more to the expansion and deepening of globalization than those who built empires encompassing many millions of people and large swaths of geography. After all, globalization is about connecting on multiple levels and breaking down many of the walls that separate populations of various origins, customs, and beliefs. Globalization also entails developing systems of government that centralize administration and enforce common standards of behavior. The empire that Genghis Khan established in the thirteenth century did both with unprecedented scale and scope. It thus provided one of the most powerful boosts to globalization ever seen. And it is a good bet that a new Silk Road, stretching from the Pacific to the Atlantic Ocean in both spirit and modern physical form, will continue to unfold throughout this century, increasing global interdependence by huge orders of magnitude.

ALBUM/ART RESOURCE, NY

Chapter II

PRINCE HENRY

The Explorer Who Made a Science of Discovery

1394–1460

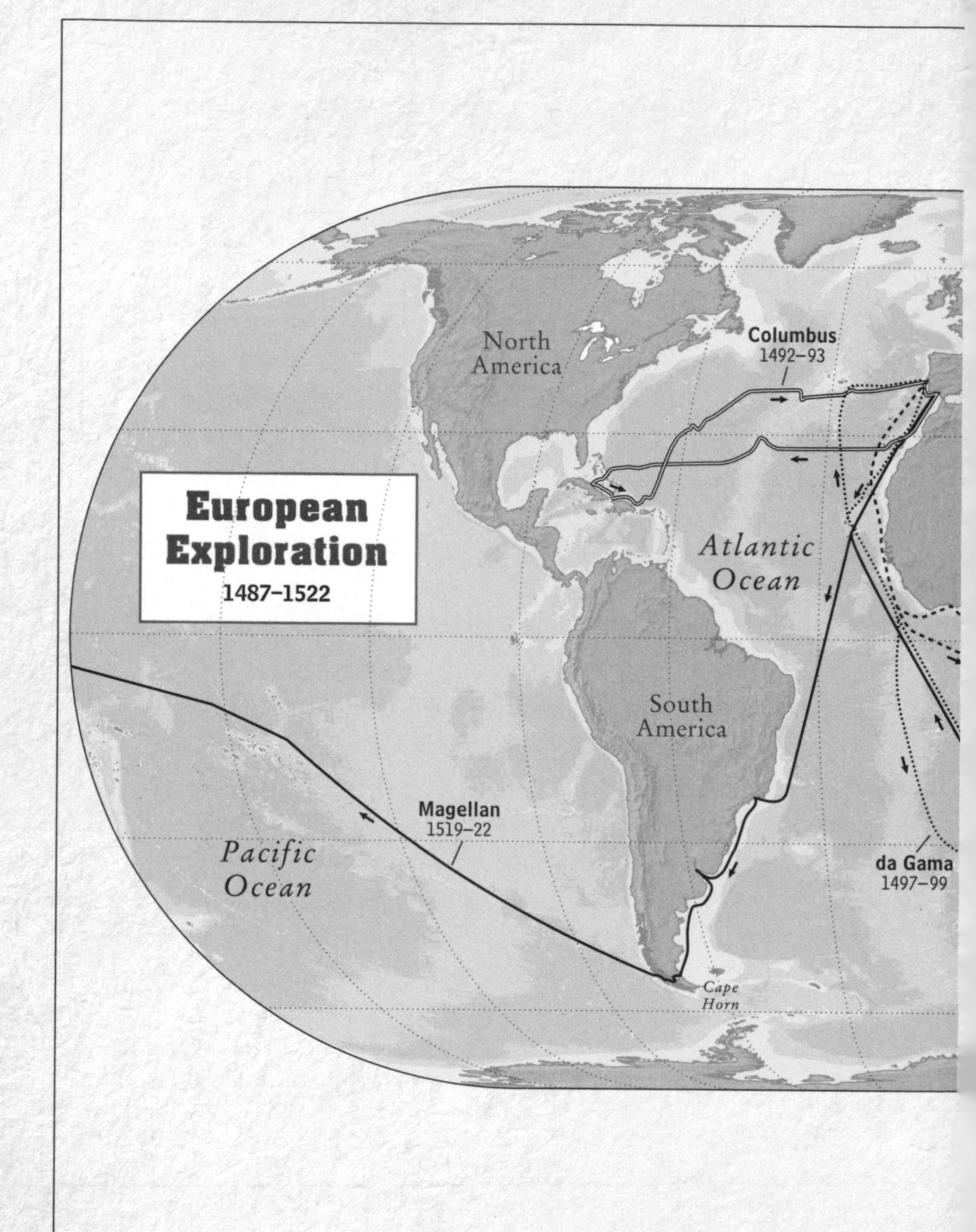
European Exploration
1487–1522
North America
South America
Atlantic Ocean
Pacific Ocean
Columbus
1492–93
Magellan
1519–22
da Gama
1497–99
Cape Horn

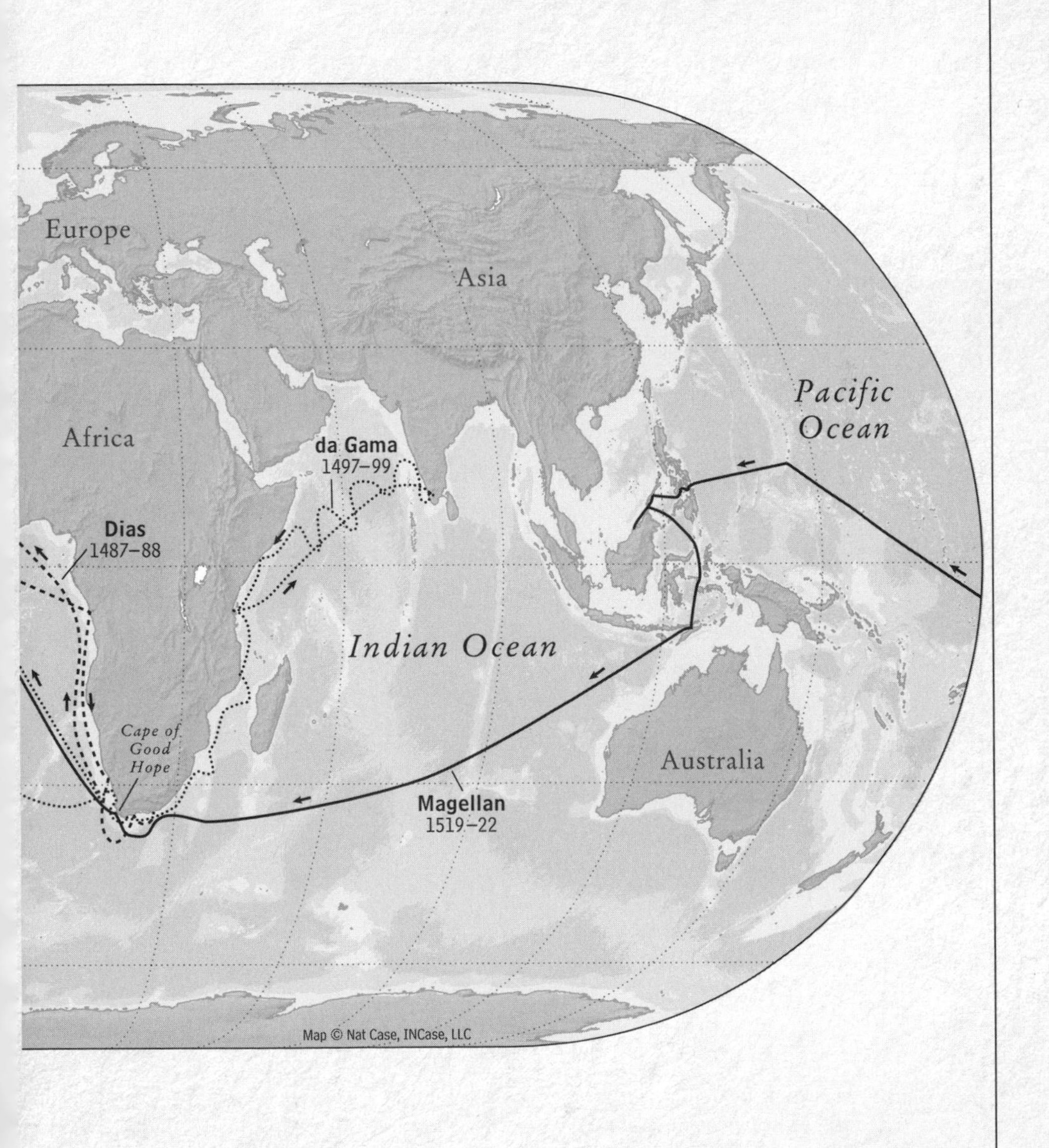

Europe
Asia
Africa
Pacific Ocean
da Gama
1497–99
Dias
1487–88
Indian Ocean
Cape of Good Hope
Australia
Magellan
1519–22
Map © Nat Case, INCase, LLC

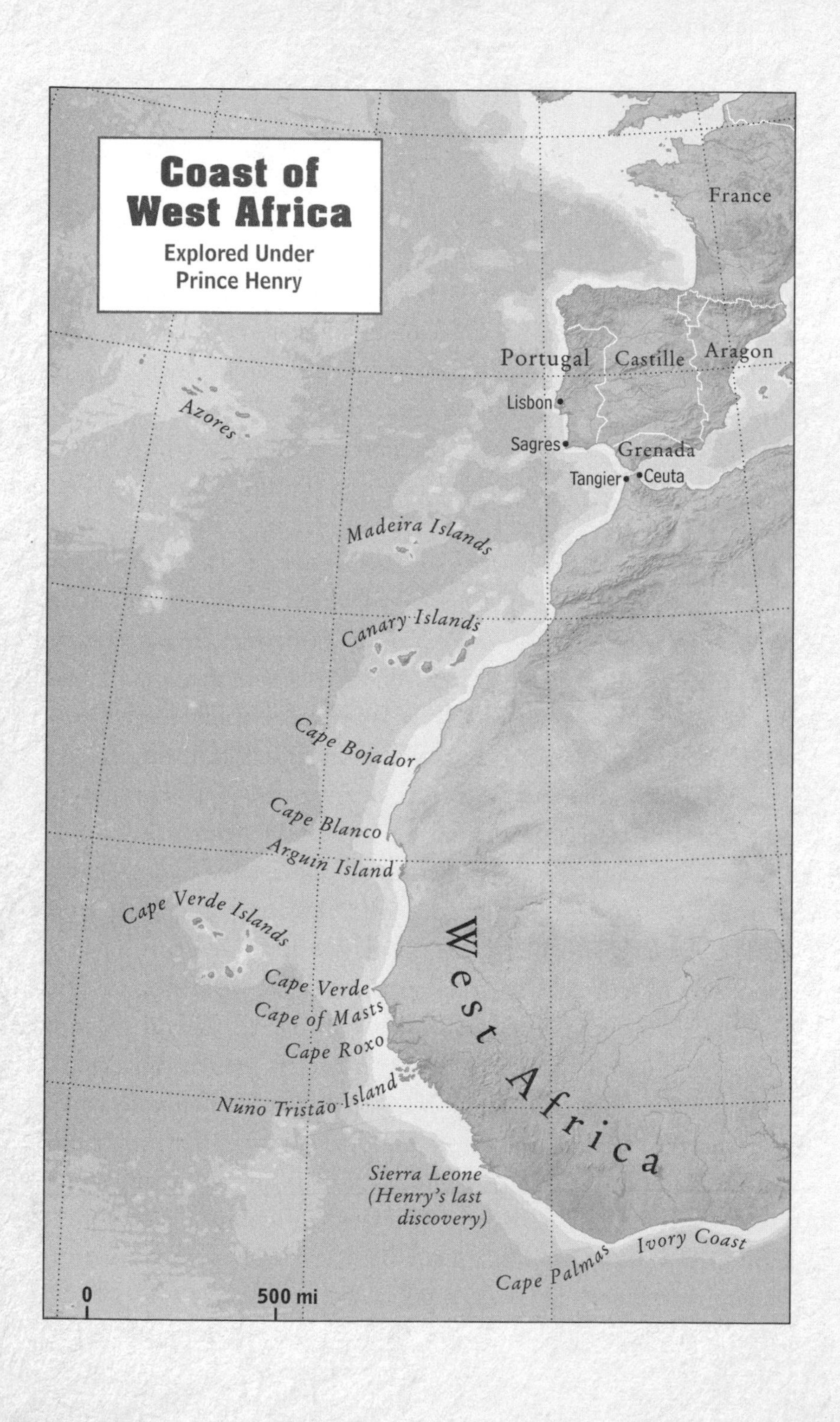

Coast of West Africa
Explored Under Prince Henry
France
Portugal
Castille
Aragon
Lisbon
Sagres
Grenada
Tangier
Ceuta
Azores
Madeira Islands
Canary Islands
Cape Bojador
Cape Blanco
Arguin Island
Cape Verde Islands
Cape Verde
Cape of Masts
Cape Roxo
Nuno Tristão Island
West Africa
Sierra Leone (Henry's last discovery)
Cape Palmas
Ivory Coast
0
500 mi

The ships were built, the soldiers were drilled, the guns were forged, and the provisions were stockpiled—all in top secret. The clamor of an army preparing for war could be heard from one end of Lisbon to another, even in the small villages on its outskirts, but no one knew which war King John I of Portugal was planning for. The citizenry was in a state of high anxiety over the coming military venture, yet ignorant as to its possible motive, or target.

It was 1415 and Portugal was at peace. It had won its independence thirty years before, breaking away from its only neighbor and natural rival, Castile (modern-day Spain). Now relations with Castile were quiet, and the Portuguese princes were idle. In a medieval European culture obsessed by values of chivalry and heroism, the absence of war was misery for the knights and princes of Portugal. With no way to earn the spoils of land, treasure, and honor that could be won only in combat, the king was searching for a foreign target on which to set them loose.

The royal family chose as its potential prize the tiny but heavily fortified port of Ceuta, on the northern coast bordering on what is now Morocco. Located just fourteen miles across the Mediterranean from the Spanish city of Gibraltar, Ceuta served as both the starting point for ships carrying Moroccan wheat to Europe and the northern terminus of the lucrative trans-Saharan caravan trade. From Ceuta, camels and carts took European silver and other products south to the Muslim kingdoms of the Sahel, which stretched from the Atlantic to the Red Sea, and brought back gold, slaves (in very small numbers, at this stage), and ivory for shipment to Europe. Seizing Ceuta would give the king control of this trade and, not incidentally, give Portugal an opportunity to vanquish a Muslim people it regarded as the infidels of Northern Africa.

Many of the king's advisers opposed this risky military adventure. Castile would be leery of Portuguese expansion in its backyard and might intervene on behalf of Ceuta. Portugal was still deep in debt from the war for independence, a strained financial position from which to mobilize enough men and ships for a seaborne expedition in distant foreign lands. Although the Vikings had ventured abroad widely in the late eleventh century, and much exploration had occurred in the Mediterranean, western Europe had yet to send its fleets on such voyages of conquest, at least not in any systematic way. At this point in European history, colonies did not exist.

It was Prince Henry* who argued most passionately in favor of an invasion of Ceuta. As the king's third son, Henry knew he would never inherit the throne and must earn his fame and fortune some other way. In the councils of war, he appealed to his father's sense of destiny, highlighting the glory Portugal would gain throughout Europe by capturing a Muslim seaport, and the commercial benefits of owning a thriving North African trading center. Henry most likely believed the upcoming campaign could promote his stature as a military statesman. Henry convinced his father not only to approve the invasion, but also to entrust him with the battle preparations. Among the Lisbon elite privy to the king's thinking, however, the choice of Henry was not welcome. Many thought him too young, inexperienced, and blinded by ambition to be entrusted with such a critical venture.

The young prince set out to prove himself by preparing exhaustively for the coming invasion. He managed to raise an army of Portuguese knights and mercenaries from across Europe to man the largest flotilla that Portugal had ever assembled. By one estimate, the armada included 59 war galleys, 63 transports, and 120 lighter craft with an invasion force of sailors, rowers, and

---

* His full name is Infante Dom Henrique de Avis.

foot soldiers for a total of around nineteen thousand men. The target of the vast mobilization was kept secret, despite the incessant questioning of Portuguese citizens and anxious foreign ambassadors.

On July 26, 1415, the fleet set sail from Lisbon, heading south with all the fanfare of a great military expedition, the ships flying flags and banners, the sea captains in colorful uniforms, bright awnings over the decks of the galleys. When the fleet reached the southeastern Portuguese port of Lagos, King John announced its target was Ceuta, which soon set off alarms along the Moroccan coast. Ceuta's governor, Sala-ben-Sala, called for military reinforcement from throughout his kingdom. Approaching Ceuta, the Portuguese ships were forced to turn away due to stormy seas, and in one of those moments of confusion on which history so often seems to turn, Sala-ben-Sala decided the threat was gone and dismissed his reinforcements. When the weather cleared, however, the Portuguese returned to mount a fierce attack.

Henry led the assault, but his enthusiasm quickly overtook his skills as a commander. He ran so far in front of his troops that he became separated from them for several hours, long enough to be given up for dead. He nevertheless reemerged to play a lead role in the conquest. The entire battle took just thirteen hours, with heavy casualties among Ceuta's defenders, their bodies piled up in mounds, but only eight Portuguese lives lost—a result of the advantage of surprise and superior weaponry. Within a few days, Henry's men had ritually cleansed the city's mosques so that they could be used by Christians.

The crusade for religion and glory complete, the ruthless plundering began. The Portuguese carted off jewelry, tapestries, silks, brass, spices, gold, silver, and whatever else they could find. It was the first time that Henry saw the riches that lay abroad, and the experience likely whetted his desire to conquer other Islamic lands. King John claimed Ceuta as part of Portugal. He posted twenty-five hundred troops as a permanent garrison and appointed Pedro de Menezes as the Portuguese governor reporting to Henry.

Henry was now in full control of Portugal's first overseas possession, with a new flourish in his title: Prince Henry, Duke of Viseu and Lord of Covilham. A few years later, in 1420, the king would shower another award on his son, asking the pope to make Henry the administrator general of the Order of Christ, a religious-military order in Portugal that was devoted to converting infidels. The order would become a power base for Henry, providing him with an annuity from its extensive land holdings and bolstering his reputation as a religious crusader, not only in Portugal but across Catholic Europe. Whatever he did in the following years, Henry would use these credentials to silence his critics in Lisbon, who would seize on his every slipup to attack him as a rash adventurer who spent too much money for too little return.

The military conquest of Ceuta inaugurated Portuguese expansion abroad, the first step in building Europe's first maritime empire. It demonstrated to other governments that this small coastal backwater of a country, with a population of less than two million people, could be a world power. It marked the first great European military success against Islam on Muslim territory, and it became the first European colonial base in Africa. From this point on, Europeans systematically explored the world, simultaneously penetrating Asia and the Americas. The effort was not without many setbacks, of course. Nor was it without some of mankind's greatest sins, such as Henry's role in expanding the slave trade. But it did lay the essential groundwork for colonization and many other subsequent steps that led to the global society we know today.

Once Portugal assumed control of Ceuta, however, most of the local merchants fled. Without them, the city lost most of its trade ties in Africa and became a Portuguese enclave isolated in a hostile Islamic country. Trying to salvage the situation, Henry became preoccupied with the Arab traders who remained, pumping them for information on trading patterns and commercial networks in Africa and the Middle East. He learned about imported spices that preserved food or could be combined to produce medicines, all of this information being of intense interest in wider Europe. And to make up for the severed supply routes, Henry became intimately familiar with the world of shipping—the acquisition and chartering of commercial vessels, the community of sea captains, the business of purchasing and distributing supplies. He learned how to administer a new colony.

Ceuta was thus a landmark event of Henry's life. He had won his spurs on the battlefield, the recognition he craved from his father, and accolades from the pope, from Henry V of England, and from the king of Castile. He learned about commerce and oversight of a foreign territory. The prince also demonstrated key personal qualities that would shape his future, in particular his relentlessly practical curiosity, and the will to probe systematically for information that could be put to use in future ventures. The Ceuta experience became the foundation for Henry's seminal explorations down the coast of Africa, and for the administration of new lands he and his successors would annex on behalf of Portugal. It would mark the true beginnings of the great European explorations of Africa, Asia, and the new world of the Americas.

An Auspicious Horoscope

The prince was born on March 4, 1394. From his early youth he had a close relationship with his father, a warrior hero who led Portugal to independence from Castile at the Battle of Aljubarrota on August 14, 1385. His mother, Queen Philippa, was a British royal who had grown up in England, and likely passed on to young Henry her tales of English history and the code of chivalry, including stories of brave knights riding gallantly into battle. Like other royals of the time, Henry's schooling would have covered history, literature, biblical and theological studies, astrology, and navigation, with some mysticism and superstition thrown in. Historians agree that Henry was driven by a faith in his own horoscope, which said that he was destined to make "great and noble conquests and to uncover secrets previously hidden from men."

The prince was a product of medieval Europe and the state of Portugal during the late fourteenth and fifteenth centuries. As the Mongol Empire of Genghis Khan began to disintegrate, it was replaced in Eurasia by the reemergence of local kings and strongmen engaged in constant wars, revolts, and thievery. The land passages that once connected China with the frontiers of eastern Europe became more difficult to travel, as each emerging principality established its own borders with new tariffs, middlemen, and commercial customs. Caravan traders yearned for a new route to the Orient that would skirt these troublesome locals.

The best alternative was across the open ocean, but no European nation was fully ready to take to the seas. When it came to maritime capabilities Europe was still way behind China and Persia, and its internal political strife made it difficult to organize national campaigns of any kind, including major seafaring expeditions. Prince Henry's breakthrough in North Africa would greatly boost the emerging sense of Europe's possibilities.

Though the *Pax Mongolica* was fading, Europe had begun absorbing inventions from Asia, including gunpowder and navigational instruments, that would ultimately help shift the balance of influence in Eurasia to Europe. The Continent was rising also on the strength of advances in agricultural and military technology, driven by growing competition among nations, which contrasted sharply with the stagnant, centralized control gripping China. Henry was a product of that competition, but it would be another century before Europe hit full stride, with the establishment of effective nation-states, a spirit of scientific inquiry, and the end of the monopoly of the Catholic Church.

It is no accident that an early pioneer such as Henry emerged from Portugal rather than from the larger European powers. Independence had left Portugal politically quiet and peaceful, while many rival European states felt too threatened by civil struggles or foreign foes to contemplate sending their most daring men on voyages of exploration. Possible rivals farther afield were also steering clear of the high seas. Ming China had been sending Admiral Zheng He on exploratory missions that reached East Africa but was now turning inward and curtailing his voyages. Indian captains were content to limit their travels to the rich opportunities around the Indian Ocean. The Ottomans were constrained by the narrow and often enemy-controlled straits that were their only outlet to the wider world. Only Portugal had a stable and advantageous domestic platform from which to launch explorations across an open sea.

First Discoveries: The Atlantic Islands

After his victory in Ceuta, Prince Henry turned his sights toward the Atlantic.* Seeking to find treasure and convert the heathens, Henry's first choice of targets was the Canary Islands, not far off the African coast. King John would not allow it, however, as the Canaries were claimed by Castile and he was still wary of provoking Portugal's former rulers. Henry thus sent two ships in search of new islands farther west and deeper into the Atlantic.† Blown off course

---

* The precise dates of the Atlantic discoveries are not clear, as different sources contradict one another.

† Prince Henry oversaw the building of the ships, appointed the captains,

by a storm, one of the vessels discovered an island, eventually named Porto Santo, and reported that it was small and uninhabited. The prince sent them back with orders to sail farther. Sometime between 1419 and 1425 (historians differ on the date), they made a larger and more promising find, 30 miles past Porto Santo and 360 miles off the coast of Morocco. Eventually named Madeira, this island was Henry's first major discovery. Although Italian seafarers had seen it decades before and saw no value in it, the Portuguese found timber there that became critical to their shipbuilding industry, as well as the sap of the dragon tree, which would be used to make dyes and medicines. Years later the Portuguese transplanted sugarcane from Sicily and grapes from Crete to the island, turning Madeira into a major exporter of wood, sugar, and wine and a vital cog in Lisbon's imperial economy.

Henry then launched another expedition that unfolded in much the same way as the discovery of Madeira had. After his men came upon a small and worthless island, Henry ordered a longer return trip that found the Azores (sometime in the 1430s; again historical accounts differ). Composed of nine islands on the same latitude as Portugal, and nearly one-third of the way across the Atlantic, the Azores had also been previously sighted and dismissed by other explorers. But the Portuguese encouraged and sometimes coerced citizens to settle there, and they soon built new livestock and fruit industries, creating a way station for explorations across the Atlantic as well as a new laboratory for Portuguese colonial administration. These early successes also gave the prince the credibility in Lisbon that he would need to launch a far more fearsome adventure, the rounding of Cape Bojador on the West African coast.

---

and assembled the crews, but he himself never participated in these or later voyages.

Breakthrough at the Bulging Cape

Cape Bojador means the "bulging cape" in Portuguese, and though the bulge extends no more than twenty-five miles, or less than a day's sail off the coast of what today is Western Sahara, it had limited European exploration for decades. There is some evidence that, centuries earlier, the ancient Mediterranean civilizations of the Phoenicians, the Greeks, and the Carthaginians may all have traveled down the west coast of Africa and past Cape Bojador. But then the many centuries that came to be called the Dark Ages set in, superstitions spread, and by the 1430s it had been generations since anyone had dared pass this point. Legends held that south of Cape Bojador the oceans boiled, sea monsters preyed on ships and people, and the sun was so hot that it would turn white skin black. It was also believed that ships that made it past the cape would be blown by tricky winds onto the treacherous red reefs off the cape. In the popular mind, Bojador was cast as the point of no return on the African coast. If Henry could get an expedition past Bojador, he would break psychological barriers that had held back European exploration for centuries, and liberate Portuguese minds to venture out into the world with an even greater sense of possibility and confidence.

Around 1430 Henry started dispatching his squires to make what may have been as many as fifteen attempts at rounding Bojador. Time and again they failed, turning back mainly because of fear. The squires were representatives of shipowners who oversaw the outfitting and organization of expeditions, but in that capacity they could only do so much to force the hand of skittish captains and crew, tough veterans of many voyages. Henry was convinced that the risks were imaginary. Some historians say this conviction may have been strengthened by fictitious French and Castilian accounts of captains who claimed to have rounded Bojador within the last century. But one way or another, Henry was determined to get a ship past that cape. He had to overcome not only the superstition of the age but the commercial self-interest of captains who saw little to gain in testing themselves at Bojador, when treasures could be won more easily in raids and trade on familiar seas.

In 1434 Prince Henry called in the squire Gil Eannes, who had already failed to round Bojador several times. It is not known exactly what Henry said, whether he appealed to Eannes's patriotism or promised great rewards. Some accounts say that Henry ordered Eannes not to return unless he could describe the lands and seas below the cape, but it is clear that this time the squire set out with courage none had shown before.

A precise description of the fateful Eannes voyage does not exist in any fully reliable form. But one account gives us a picture of what might have transpired. After about a week on the sea, Eannes and company arrived at Cape Not, just north of Bojador. Under the shadow of the sails, the men ate dried fish, salted meat, cheese, and biscuits and drank wine—a feast on which to face the unknown. They proceeded south, with a hot wind blowing off the desert pushing them toward the coastline. When Cape Bojador came into sight, Eannes might have given the order to steer far clear of the dreaded red reefs. They sailed west and then south for a day and a night, then turned east, back toward the coast.

It was a blistering summer day, and the sailors might have gathered on deck and shaded their eyes, perhaps glancing down to check the color of their skin. Land came into sight. It was flat, sandy, and shining under the sun, quite a change from the inhospitable look of the coast north of the cape. The sailors looked north and saw Cape Bojador. They had indeed rounded it. They sailed toward the beach and dropped anchor. A boat was lowered and Eannes rowed ashore. While there were no signs of human life, Eannes did spot some roses—that detail we do know for sure—which he brought home to the prince.

If Ceuta was Henry's most formative experience, pushing past Bojador may have been his most important. It sent a message to all contemporary explorers that the only obstacles left were money and determination. Word quickly spread across the world. The Age of Exploration now unfolded with ever-greater speed, attracting more resources and more men. Year after year, Henry would send out his ships, each going farther and farther down the African coast.*

At this stage, however, Henry had little of concrete value to show for his expensive explorations. Eannes had found no people to convert, no gold or treasure to bring home. To keep going, Henry had to make his adventures pay. So he began looking for a water route across Africa that would serve as a shortcut to the riches of India and China. He hoped to use this same waterway to locate the Christian empire of Prester John, a tribal priest-king said to rule a vast swath of northeast Africa, in the region that is now Ethiopia. Prester John's kingdom was reputed to be rich in gold and silver, with a mighty army of one hundred thousand men, with whom Prince Henry hoped to join in crusades against the Muslim infidels. For lack of time and money, and not least the fact that Prester John was only a legend, Henry would never realize these dreams.

My Kingdom or a Brother

Before the next critical stage of his discoveries, Henry entered the darkest chapter of his life. King John had passed away, replaced on the Portuguese throne by Edward, his eldest son and Henry's

---

* It is possible that it was not the Cape of Bojador that Eannes actually rounded, according to some later accounts. Nevertheless, the impact of breaking the psychological barrier would have been a major achievement. Peter Russell, *Prince Henry "the Navigator": A Life* (New Haven, CT: Yale University Press, 2000), 111.

brother. The prince had for years been eager to conquer Tangier, a coastal city east of Ceuta, and he presented a plan of attack to the new king and his advisers. All were opposed, for Tangier was sure to put up much greater resistance than Ceuta had, and the Portuguese treasury could not afford to raise a navy and army of sufficient scale for this much larger adventure. Once again, Henry's critics said he was too rash, too much the fervid crusader. But as in the debate over Ceuta, Henry proved irresistibly persuasive, arguing that it was Portugal's destiny to convert the Muslims and to unleash the martial spirit of its knights. In the end the royal court agreed to launch another war.

This time, Henry's preparations fell far short. His invasion plans became an open secret, giving Tangier time to prepare its defenses. The prince was able to marshal a force of only eight thousand foot soldiers, half the size of the infantry he unleashed on Ceuta. The invasion of Tangier began on September 9, 1437, with the Portuguese outmanned, outgunned, and outmaneuvered. The defenders repulsed repeated assaults on the city walls. Within days they had surrounded the Portuguese army, forcing Henry to surrender to Sala-ben-Sala, the same governor he had defeated at Ceuta. The governor demanded not only that the Portuguese return Ceuta but also that Henry leave Fernando, his younger brother, hostage as a guarantee that he would deliver on the deal.

To save his army, Henry was forced to accept these terms. The prince was then allowed to retreat to Ceuta, where he remained for a number of months, apparently agonizing over his deal with Sala-ben-Sala. To honor it would mean national and personal humiliation, but to renege meant certain death for Fernando. Henry returned to Portugal to confer with King Edward, and together they decided they would be willing to sacrifice their younger brother if necessary.

For six years they tried to renegotiate a new deal for his release. Jailed like a slave, in isolation and with little food, Fernando finally died in captivity. The botched assault had exposed Henry's arrogance, his bad military judgment, and his unbridled ambition. And it would force him to retrench and focus with even more single-mindedness on what he did well: exploration.

Step by Step

In those days the pope's authority provided a good deal of legal and moral cover for foreign adventures. After the triumph of Bojador, Prince Henry had secured from the pontiff the exclusive rights for Portugal to explore the African coast as far south and as far inland as its expeditions could reach, whether the purpose was trade or war. Henry used the leverage of papal approval to organize missions and to pressure independent merchants to join his ventures, with 20 percent of the revenues of these partnerships going to the prince. The success of Bojador reenergized Henry's spirit of adventure, his zeal for bringing Christianity to the infidels, and his appetite to find riches, especially gold, for himself and Portugal.

It was during this period that Henry began to practice methodical exploration. Between 1441 and 1460, he mounted missions in a controlled manner, each pressing a bit farther south than the last, from Cape Bojador to Cape Blanco, then to Cape Verde, then to Cape Roxo. In this way he moved, step by step, down the African coast.

Henry carefully cataloged every incremental advance in knowledge. Squires, captains, and crews were under orders from Henry to carefully document the coastline, the vegetation, the wildlife, and the culture of any inhabitants. Each specimen and observation his men collected would be brought home and analyzed, creating a base of understanding that the men would build on during subsequent trips. Trips built on one another as sailors gained more and more experience with the navigational possibilities, the winds, and the African people. Henry made it a point for his captains to bring back at least one knowledgeable African native on every trip, as a source of information on the culture and trading environment, and as an interpreter and guide for subsequent trips. The prince encouraged his captains to send emissaries as deep as possible into the interior to better understand the commercial possibilities.

The missions discovered many African tribes. Sometimes these encounters went well and led to new trade ties, but on other occasions they ended in fierce battles. Such friction was not surprising, for though Henry was no longer sending out warships for the purpose of armed conquest, his explorers were instructed to erect a cross at each new destination, claiming the land for Christianity and Portugal. While many of Henry's men sought the renown that came with winning battles against the people of the Sahara coast, an increasing number of the expeditions were financed by merchants, not by Henry or the state, and their interest was purely commercial. The prince had to mediate between the conflicting interests of the warriors and the merchants, and over time and out of necessity, he had to emphasize trade over religious conversion.

The Slave Trade

Soon Henry's missions would discover in Africa a commercial prospect that dwarfed all others in terms of value, one that was heinous in its conception, execution, and implications. Nevertheless, in the eyes of many of his countrymen, the expansion of the slave trade would justify all of Henry's explorations. It was a fact that the roots of this human trafficking went back to the Roman Empire, and that by Henry's time Europeans were buying or bartering for black slaves in the marketplaces of North Africa and putting them to work in Italy, Castile, and also in Portugal. Henry, however, was the first European trader to buy or conscript slaves directly from tribes along the West African coast, establishing the first seaborne operation that systematically and in rapidly increasing numbers exported slaves to Europe.

It started slowly enough with the kidnapping of a few African natives, then progressed to bartering with tribal leaders, exchanging Portuguese linen, silver, and wheat for African men, women, and children. On the morning of August 8, 1444, the first cargo of 235 Africans, taken from what is now Senegal, were delivered to the Portuguese port of Lagos. Historians say this is when modern slavery began. The numbers grew rapidly, and during Henry's life about fifteen to twenty thousand slaves may have been brought to Portugal on his behalf alone.

Some of Henry's sympathetic biographers argue that his main interest in Africans was to convert them to Christianity; that many Africans brought to Lisbon were allowed to integrate into Portuguese society, intermarry, learn commercial skills, and climb the economic ladder; that Henry himself, although entitled to many slaves, took just a few. But such rationalizations cannot hide the violence of this new trade. Henry's men not only bartered linens for humans, they also forcibly seized many Africans, killing those who resisted as a warning to others. They hunted down men, women, and children for capture and shipped them in inhuman conditions, locked inside the hull or lashed to the deck and exposed to the elements throughout the long voyage back to Portugal, where they were sold at auction. Even Zurara, the admiring contemporary chronicler of the prince's life, paints a grim picture of a slave auction in Lisbon.

> What heart could be so hard as not to be pierced with piteous feeling to see that company [of slaves]? Some held their heads low, their faces bathed in tears as they looked at each other; some groaned very piteously looking toward the heavens fixedly and crying out loud, as if they were calling

> on the father of the universe to help them. To increase their anguish still more, those who were in charge of the divisions then arrived and began to separate them from one another so that they formed five equal lots. This made it necessary to separate sons from their fathers and wives from their husbands and brother from brother. . . . And as soon as the children who had been assigned to one group saw their parents in another they jumped up and ran toward them; mothers clasped their children in their arms and lay face downward on the ground, accepting wounds with contempt for the suffering of the flesh rather than let their children be torn from them.

Henry was capable of being a brutal, cold-eyed pragmatist. He had sacrificed his brother for a trading port, and for him discovering African slaves was the equivalent of finding gold. Most of the simmering resentments against Henry in Lisbon—over his manipulations in the royal court, the national humiliation he had brought to Portugal during the Tangier fiasco, and the vast treasure he had spent on unprofitable missions—could be put aside in the face of the prosperity Henry was generating through the slave trade. Whatever he felt about the morality of slavery, he probably saw the trade as a stroke of luck, even divine providence, because it allowed him to continue his Christian mission of exploration, enhancing his status as a national hero.

In 1443 Henry's discovery of Arguin Island proved a landmark event that would give Portugal a firm base in Africa for further expanding the slave trade and extending its influence on the continent. Located just south of Cape Blanco (and today part of the Islamic Republic of Mauritania), the island was about two miles offshore, no longer than three miles wide and half again as long. Its high ground allowed sentries to monitor the strait that separated it from the mainland. It had a freshwater spring and birds, pelicans, and flamingos, which gave visiting seamen the chance to eat meat instead of their normal diet of fish. Arguin was the first and only place within a thousand miles of Portugal where Henry's ships could safely anchor, repair, and get freshwater in Africa.

When the Portuguese found Arguin, it had a dense population of African fishermen. In 1445, twenty-six Portuguese ships arrived to pacify the island and within a decade, ships owned by Henry or his licensed partners were exporting some eight hundred slaves per year out of Arguin. They built a permanent fort out of stone, which became the base for Portugal's slave trading operation and a model for other forts that would soon anchor Portuguese colonies around the world. Soon other European nations would emulate the Portuguese.

Sagres: Henry's Retreat

Prince Henry is known today as "Prince Henry, the Navigator," a title that was conjured up when European biographers started to evaluate in earnest Henry's seafaring accomplishments centuries after he died. It was in some ways a most inapt nickname because, with the exceptions of Ceuta and Tangier, it appears that Henry made no voyages himself. Perhaps he simply did not enjoy being on a ship, perhaps he was afraid of the physical risks, or maybe he saw himself as a merchant prince and pioneer above the seafaring crowd. When he first set out for Africa, exploration was an episodic adventure for merchants and soldiers of fortune. Henry would make it an institution, an ongoing national project. Toward the end of his life he ran this mission out of his fortress in the southern Portuguese city of Sagres, which sat on a high cliff overlooking the Atlantic.* The prince could look out at the

---

* Historians do not agree when he actually established himself at Sagres, but he seemed to have been ensconced there for at least fifteen to twenty years.

ocean and contemplate his next campaign to Christianize the Islamic world or to find riches on the African coast.

Exactly what happened at Sagres and how much those activities advanced maritime science has been a source of controversy among historians. There are those who describe the retreat as a meeting place for maritime experts of all kinds—from shipbuilders to mapmakers to navigational technicians. They would gather and exchange ideas and develop a variety of nautical innovations, with Henry as the driving force. "To Sagres came sailors, travelers, and savants from all over, each adding some new fragment of fact . . . ," wrote historian Daniel Boorstin. "Besides Jews there were Muslims and Arabs, Italians from Genoa and Venice, Germans and Scandinavians, and, as exploring advanced, tribesmen from the west coast of Africa." Others, such as biographer Peter Russell, paint a much less exalted picture, saying there is no evidence that Henry had such an entourage in any formal sense. Whereas Boorstin implies that at Sagres Henry contributed to meaningful advances in the technology of ships and navigational equipment, others such as scholar W. G. L. Randles contend that Henry mounted his explorations without such advances.

Experts seem to agree on two broad propositions, however. First, from the days of Ceuta, the prince was a gatherer of information from an exceptionally wide range of sources: merchants and suppliers, mapmakers and scholars from throughout Europe, and Africans with whom the missions came in contact. He was thus extremely well informed and very likely to have helped his own captains by giving them information they otherwise wouldn't have had. Such activity does not sound startling by today's standards, but it was a pioneering breakthrough in an age when exploratory missions were entirely new and highly irregular, communication was relatively primitive, collaboration was unusual, and the amassing of information and passing it on in a systematic way a genuine novelty. In those respects, Henry was translating raw discovery into useful knowledge for ever more successful ventures for the first time.

The second proposition that most historians would endorse is that Henry's relentless sense of mission was unique, an indefinable blend of religious zeal and lust for gold, with the pious element at least partly a convenient cover for the commercial greed. Still, the religious passion translated into a refusal to fold in the face of setbacks, whether it be criticism in the Lisbon court or humiliation in battle, and into an ability to persuade men to face down any enemy—even their own worst fears—in his name.

Prince Henry died peacefully at Sagres on November 13, 1460. He was sixty-six. Despite multiple streams of income from his lands, royal monopolies on soap and other commodities, joint ventures with merchants for discoveries, and sales of slaves, he ended up deeply in debt.

Exploration after Henry

At the time of his death, Henry still held the exclusive papal right to substantially control all the trade along the Atlantic coast of Africa. In 1469, the African trading concession was transferred to Fernão Gomes, a Lisbon merchant, who continued Henry's pattern of step-by-step exploration in an even more measured way: he would explore three hundred new miles of African coast per year, and in the next five years he would cover what it took Henry's men thirty to achieve. In 1475, the concession reverted to the king of Portugal.

Over the next half century the pace of exploration quickened as intra-European competition to exploit the increasingly profitable opportunities of global trade heated up. In the late fifteenth and early sixteenth centuries, it was often the Portuguese who were first to pass the next barrier, to reach the next sea or continent. In 1488, Bartolomeu Dias rounded the last point, the Cape of Good Hope at the southern tip of Africa, opening a clear ocean path to India. Vasco da Gama took that path to India on his landmark twenty-eight-thousand-mile voyage from 1497 to 1498. In the same decade Christopher Columbus—financed by Spain because the Portuguese king wouldn't support his venture—sailed across the Atlantic to discover the Americas, and Portuguese expeditions would follow. Vasco Núñez de Balboa founded Panama in 1510, and between 1519 and 1522 Ferdinand Magellan's voyage circumnavigated the globe, going westward from Spain. (Magellan himself was killed before the trip ended.) In 1776 Adam Smith, the renowned economist, wrote, "The discovery of America and that of a passage to the East Indies by the Cape of Good Hope are the two most important events recorded in the history of mankind."

If Bojador opened the door to the European discovery of India and America, then the establishment by Henry of early outposts on Ceuta, Madeira, the Azores, and Arguin Island all opened the door to colonization. In the first two decades of the sixteenth century, Portugal established a viceroy in India and captured Ormuz, the gateway to the Persian Gulf, and it occupied Goa, on the coast of India, and Malacca in the Far East. At Portugal's imperial peak its holdings would extend from Brazil in the west to the outer reaches of China in the east.

Portugal's expansion would feed the growing ambitions of the Dutch, Spanish, and British. The first direct challenge came from the Dutch, who attacked Portuguese settlements around the world and vied with Lisbon for control of the trade in spices from Asia, slaves from Africa, and sugar from Brazil. By the end of the sixteenth century, the Portuguese empire was spread thin and fading, giving way to the rising economic might of the Dutch. Facing superior manpower and sea power, the Portuguese bowed out of the global struggle for supremacy.

Legacy

Henry had led the way in turning a nation of peasant farmers and coastal fishermen into an empire built on oceanic discovery and overseas settlements. By the time of his death, his voyages had pushed about fifteen hundred miles down the west coast of Africa to what is now Sierra Leone and had transformed Lisbon into a cosmopolitan hub for the increasingly international culture of seafaring business.

Uniquely for his era, Henry was able to create what historian Daniel Boorstin called a "collaborative national adventure . . . a progressive systematic step-by-step national program for advances through the unknown." He thus set the pattern for future national exploration efforts, even the partnerships with private financiers when state support ebbed. As much as anyone, Henry set in motion the Age of Exploration—the European effort to colonize Africa, Asia, and the Americas—and he helped inaugurate centuries of Western commercial dominance around the globe. This era would no doubt have occurred without Henry. Sooner or later the European drive to acquire Asian products and to spread Christianity would have produced another transformative figure. But it was Henry who seized the moment.

In the century after Prince Henry, continuous trade among all key regions of the earth was established for the first time, and the world was becoming a single network. An ever-expanding array of commodities and finished products were being interchanged—sugar, pepper, nutmeg, cloves, cinnamon, grain, wine, timber, gold, silver, silks, textiles, porcelain, horses, to name but a few. Ships were carrying crops such as tobacco and potatoes, or animals such as horses and pigs, to regions where they were previously unknown. At the same time these vessels carried parasites and insects that would transplant malaria and smallpox from

Eurasia to the Americas, and syphilis from the Western Hemisphere to Eurasia. If the Mongol era opened by Genghis Khan was the first golden age of globalization, the European era opened by Henry was surely the second.

In 1975, one of the architects of space exploration, Wernher von Braun, wrote that Henry's seaside castle at Sagres "was the closest precedent to what the space community is trying to accomplish in our time." He went on to compare men such as Bartolomeu Dias, Ferdinand Magellan, and Vasco da Gama as the people whom the prince trained, and he described the launching of the historic voyage of Christopher Columbus as being part of the exploratory environment Henry created. In 1980, author Carl Sagan described the *Voyager* space capsules as "lineal descendants" of Henry's ships. Clearly, Prince Henry has been an inspiration for centuries.

Today we are still exploring new territory. In outer space, for example, we have landed on asteroids, sent men into space for well over a year, discovered new planets, and even sent space capsules beyond our own solar system. An increasing number of countries and companies are involved, pooling talent, money, research, and crowdsourced knowledge. Entire industries are emerging, driven by commercial incentives, and educated speculation exists that small colonies could be established on the moon or Mars by the middle of this century. No one can be sure how all this will evolve, any more than Prince Henry could foresee the future, but as in the early stages of European exploration, the direction and the powerful momentum seem assured. The headline of an editorial in the usually staid *Financial Times* in 2014 reflected the reality. "Our Manifest Destiny Is to Move Beyond Earth," it said. With regard to the oceans, we are in the early stages of cataloging the extraordinary amount of biodiversity and natural resources that lie deep below the earth's surface. In all these pursuits we can see the patterns set by Prince Henry—the relentless quest for new information, the systemic approach to building on the lessons of the last mission, the exhilaration that human beings experience by breaking through new geographical barriers, and the relentless drive to go farther and farther even when the payoff is uncertain. Without such efforts, globalization would have never become as wide and as deep as it is today. And with ongoing attempts to push the boundaries of knowledge and experience, we can be sure that our world will continue to get smaller and more interconnected for as long as human beings survive.

NATIONAL TRUST PHOTO LIBRARY/ART RESOURCE, NY

Chapter III

ROBERT CLIVE

The Rogue Who Captured India for the British Empire

1725–1774

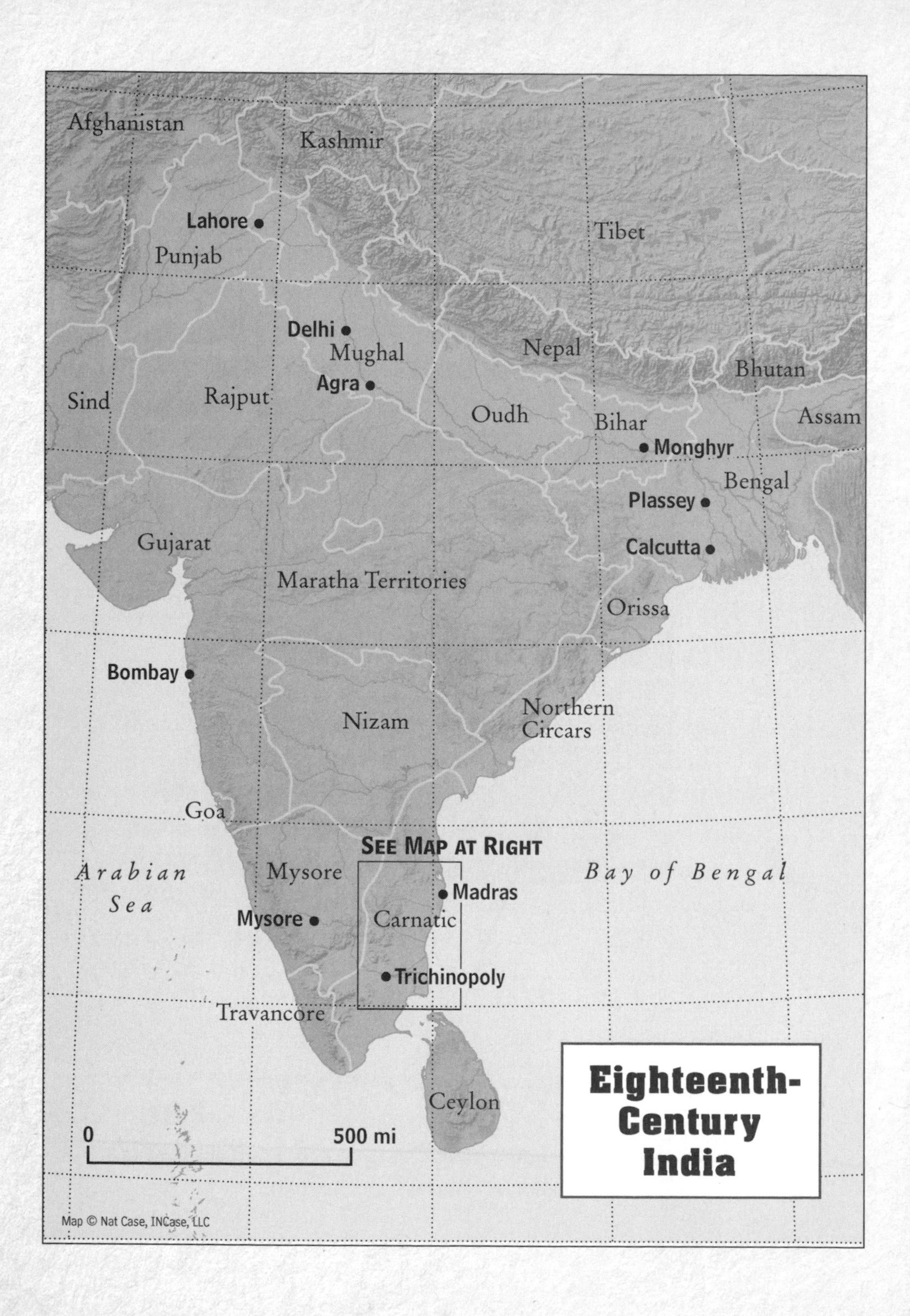
Afghanistan
Kashmir
Lahore
Punjab
Tibet
Delhi
Mughal
Nepal
Agra
Bhutan
Sind
Rajput
Oudh
Bihar
Assam
Monghyr
Bengal
Plassey
Gujarat
Calcutta
Maratha Territories
Orissa
Bombay
Northern
Circars
Nizam
Goa
SEE MAP AT RIGHT
Arabian
Sea
Mysore
Bay of Bengal
Madras
Mysore
Carnatic
Trichinopoly
Travancore
Ceylon
0
500 mi
Eighteenth-
Century
India
Map © Nat Case, INCase, LLC

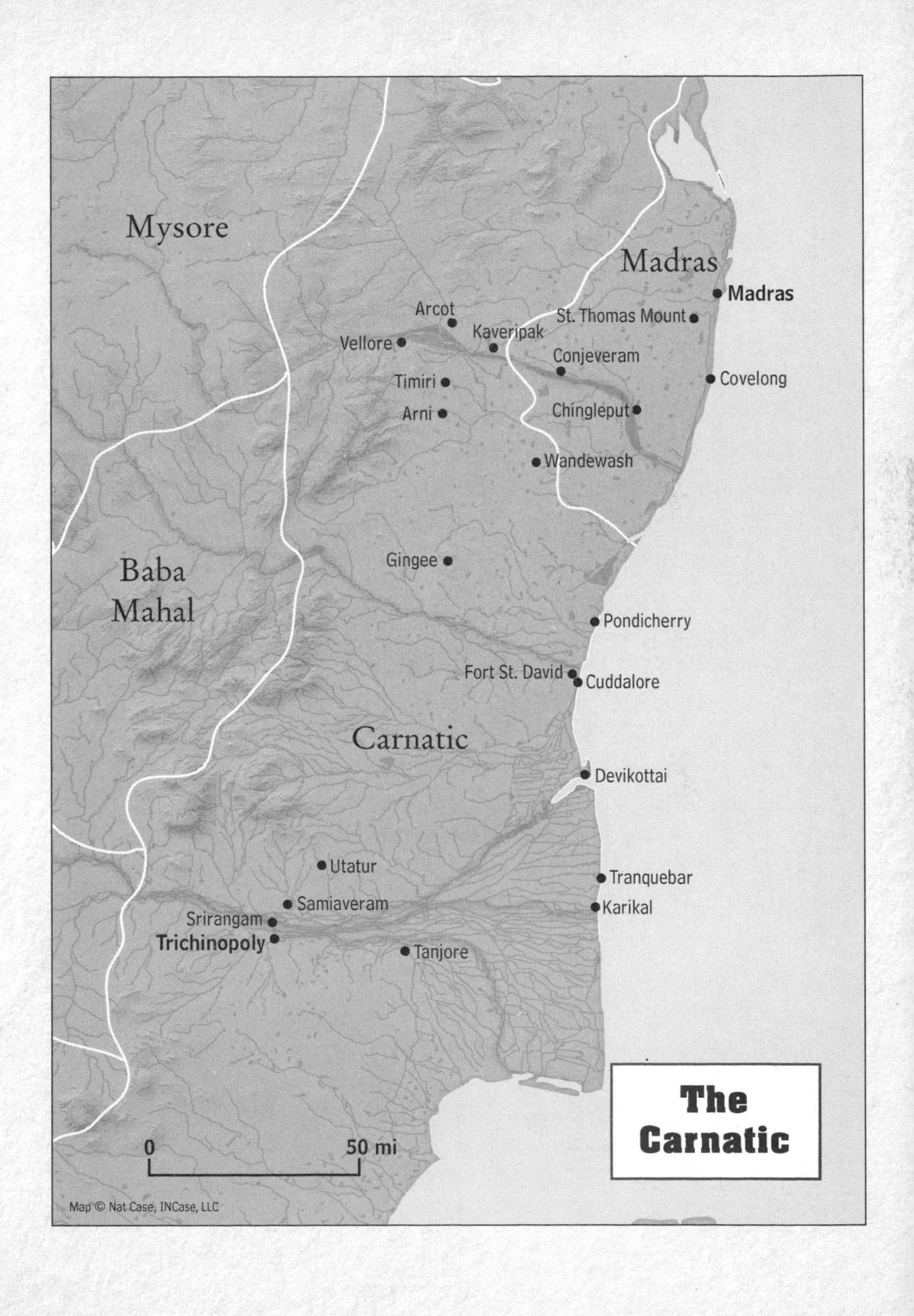

Mysore
Madras
Madras
Arcot
St. Thomas Mount
Vellore
Kaveripak
Conjeveram
Covelong
Timiri
Arni
Chingleput
Wandewash
Baba
Mahal
Gingee
Pondicherry
Fort St. David
Cuddalore
Carnatic
Devikottai
Utatur
Tranquebar
Samiaveram
Karikal
Srirangam
Trichinopoly
Tanjore
0
50 mi
The Carnatic
Map © Nat Case, INCase, LLC

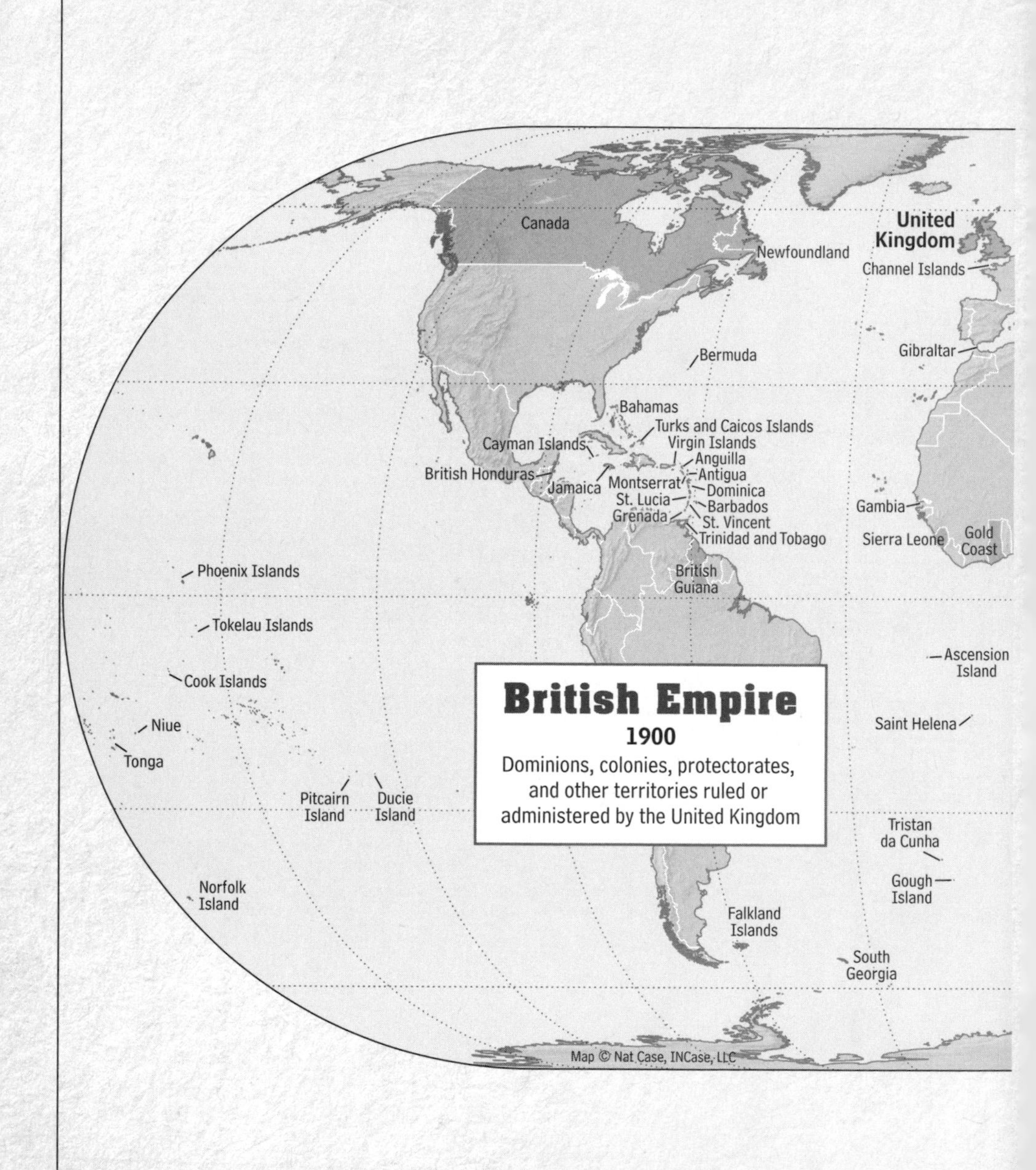

British Empire
1900
Dominions, colonies, protectorates, and other territories ruled or administered by the United Kingdom
Canada
Newfoundland
United Kingdom
Channel Islands
Gibraltar
Bermuda
Bahamas
Turks and Caicos Islands
Virgin Islands
Cayman Islands
Anguilla
Antigua
British Honduras
Jamaica
Montserrat
Dominica
St. Lucia
Barbados
Grenada
St. Vincent
Trinidad and Tobago
Gambia
Sierra Leone
Gold Coast
British Guiana
Phoenix Islands
Tokelau Islands
Cook Islands
Niue
Tonga
Pitcairn Island
Ducie Island
Ascension Island
Saint Helena
Tristan da Cunha
Gough Island
Norfolk Island
Falkland Islands
South Georgia
Map © Nat Case, INCase, LLC

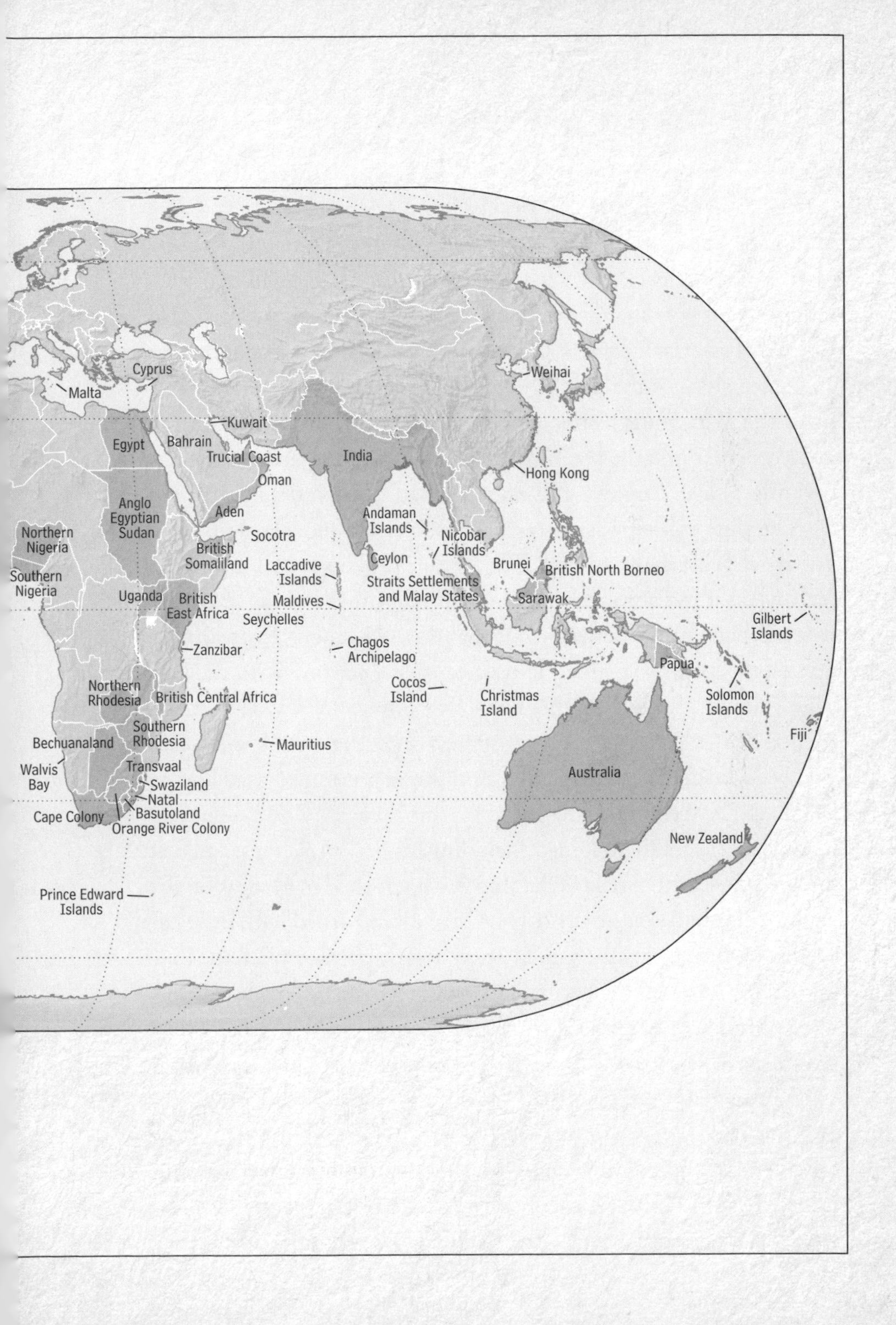

Cyprus
Malta
Weihai
Kuwait
Egypt
Bahrain
Trucial Coast
India
Hong Kong
Oman
Anglo
Egyptian
Sudan
Aden
Andaman
Islands
Northern
Nigeria
Socotra
Nicobar
Islands
British
Somaliland
Ceylon
Laccadive
Islands
Brunei
British North Borneo
Southern
Nigeria
Straits Settlements
and Malay States
Uganda
British
East Africa
Maldives
Sarawak
Seychelles
Gilbert
Islands
Zanzibar
Chagos
Archipelago
Papua
Cocos
Island
Christmas
Island
Northern
Rhodesia
British Central Africa
Solomon
Islands
Southern
Rhodesia
Fiji
Bechuanaland
Mauritius
Walvis
Bay
Transvaal
Australia
Swaziland
Natal
Basutoland
Cape Colony
Orange River Colony
New Zealand
Prince Edward
Islands

On March 10, 1743, seventeen-year-old Robert Clive stood on a London dock with all his worldly possessions packed into a few small suitcases, ready to seek his fortune in India. The teenager was about to board the *Winchester*, which at 500 tons was one of the biggest and most advanced ships of the day. A combined merchant vessel and warship, the *Winchester* reflected the dual nature of its owner, the East India Company, the most important trading arm of the expanding British Empire, and one that had its own defense force.

The ship was supposed to reach India in six months, but it did not arrive for well over a year. The voyage took its first unfortunate turn when it ran aground on a hidden reef off the coast of Brazil, where it had sailed to catch winds that would propel it eastward. To lighten the load, the captain jettisoned cargo, starting with the luggage of lower-ranking officials like Clive. Soon after, a violent storm rocked the ship, tossing Clive himself into the sea. The captain grabbed a bucket, tied a rope to it, and threw this improvised lifeline to Clive. The teenager managed to scramble back on board but lost many of his sparse belongings, including his hat, wig, shoes, and belt buckle. After further delays the ship finally docked in Madras on the east coast of India on June 1, 1744. Clive disembarked with little left but the clothes on his back and had to borrow money from the captain at an exorbitant interest rate in order to replenish his wardrobe.

Accounts of Madras during this period suggest what Clive would have seen as the *Winchester* came into the harbor at twilight. Warm air drifted off the land, carrying the whiff of spices and burnt dung, as the new arrivals rode toward the twinkling lights of the tiny colonial settlement on boats built from coconut fiber. They likely were carried by Indian boatmen the last few yards to terra firma, where British visitors were instantly struck by the dramatic juxtapositions: of Hindu peasants in filthy loincloths and Muslim traders in long silk robes, both carrying bundles on their heads; of wild dogs rummaging in garbage pits near food stalls covered in flies, not far from the broad thoroughfares and white palatial residences where the richest British traders lived, protected by uniformed guards.

Located in the southeastern coastal province of the Carnatic, home to all of the most important European trading centers in India, Madras was the centerpiece in a far-flung string of British enclaves that included footholds in Calcutta and Bombay. At the center of the settlement in Madras was Fort St. George, which included a small compound with a mansion for the governor, barracks for soldiers, and warehouses and residences for the Company. Only Europeans were allowed to live in the compound, although a number of wealthy Western merchants also lived in estates outside. Clive's home would be two spartan rooms in the Company hostel within the fort.

Coming of Age inside the Company

The East India Company had four classes of employees: senior merchants, junior merchants, factors (traders), and the lowly clerks who were called writers. Clive belonged to the latter rung. Typically, a young writer would need about five years to be promoted to factor, another three to become a junior merchant, and another three to senior merchant. Clive immediately fell into a numbing work routine. Day after day, he stood behind a high desk with about twenty other clerks, tracking bills, receipts, and inventory levels, marking ledgers and accounting for East India Company trade in products such as indigo, saltpeter, and cotton. Every morning the writers were awakened by the firing of a gun, then attended church before breakfast. Business was conducted until noon, followed by lunch, a nap, and a return to work at 4:00

p.m. Clive spent a number of evenings in the library located in the governor's mansion. Though never a serious student, he delved into a wide range of authors—Bacon, Descartes, Confucius, Erasmus, Hobbes, Machiavelli, Shakespeare, and others. It seems to have been the only period of his life when Clive showed any intellectual curiosity and lasted about a year.

Clive was born into a lower-middle-class family and experienced few real joys as a child. His father, a lawyer and businessman, sent Clive at the age of three to live with an aunt and uncle after something—it is not clear what—went awry with the family finances. Clive was a mercurial child, his moods alternating between good cheer and bad temper. His uncle wrote to his parents that Clive was "fierce and imperious." He was in and out of several schools, not because of academic problems but because he had trouble getting along with fellow students or submitting to discipline. Something of a daredevil and ruffian, he once climbed the steeple of the neighborhood church and sat there threatening to fall off, creating a panic below. As a teenager, he led a gang that extorted protection money from shopkeepers, threatening to smash their windows if they refused.

When Clive turned seventeen, his father arranged an interview with a friend in the East India Company, which had a stellar reputation as an avenue to adventure and wealth, affording an unusual opportunity for young men no matter what their education. The Company paid low salaries but allowed employees the opportunity to grow rich—sometimes astoundingly so—through various forms of trading for their own account or through accepting gifts and bounties for military victories.

Clive kept his distance from the Company's attempts to recreate a social scene that resembled typical life in England, with dinner parties, dancing, after-dinner drinking, and card games. He made few friends, despite the close quarters in which everyone lived and worked. He wrote many letters home and pined for replies, waiting for ships to bring some news from his family, but virtually no communication came his way. "I have not enjoyed one happy day since I left my native country," he wrote to a cousin. What worried Clive most were the debts he owed the ship captain who brought him to India. Clive admired his father and craved his acceptance, and was mortified when he had to write home asking for money to repay the debts—doubly so when he continued to get no reply.

Less than a year after he arrived in Madras, Clive loaded a flintlock pistol, placed the muzzle against his temple, and pressed the trigger. Click. He tried a second time. Again, only a click. A friend walked in and, hearing from Clive what had happened, took the gun, pointed it in a safe direction, and pressed the trigger. A loud bang shattered the silence. The story may be apocryphal, but most of Clive's biographers believe it is credible because Clive had suffered fits of depression all his life. Now the pressure of his debts was driving Clive to rock bottom in a harsh work environment—made oppressive by the sweltering heat and the ever-present threats of diarrhea, fevers, giant ants, and dust storms—that drove an unnatural number of Company employees to opium, heavy drinking, and an early grave.

The central question, then, is how did Robert Clive skyrocket in the next fifteen years to become the equivalent of the chief executive officer and the commander in chief of the East India Company, not to mention one of England's richest men? A suicidal youth, a teenage clerk with no pedigree in class-ruled England, a young man deeply in debt and with few friends, no political connections, and no commercial experience or military training in a company that demanded all of the above—how did he do it? How did he play the central role in building what had been a small trading company—with scattered outposts, consisting all together of a few square miles in a few Indian cities and trading only with England—into a multinational company with its own army that controlled over a third of the Indian subcontinent? Most important, how could he, being such an unimpressive young man, have laid the foundation for an empire—the British Empire—that was such a powerful force in the history of globalization, spreading to all corners of the world the concepts of market economics, democratic ideals, and rule of law?

National Rivalries Open the Door for Clive

Clive came to India just as Britain was entering the first industrial revolution and taking off as an imperial nation. England was becoming a parliamentary democracy, and as political power shifted from the monarchy to the legislature, economic influence was filtering down from high society to a new mass-consumption society, built on the growing wealth of the middle class. A financial revolution was importing, mainly from the Dutch, techniques such as the sale of public bonds in order to finance a larger navy and supporting infrastructure for imperial expansion. No European country, with the possible exception of Holland, was as politically and socially progressive or as eager to satisfy the appetites of a growing consumer base through global commerce.

To be the first nation to conquer all of India, Britain would have to compete with France. Though Britain was the greatest naval power in the world, France had the most powerful army in Europe, and the two were in a state of nearly constant conflict throughout the eighteenth century. When Clive arrived in Madras in 1744, it was an open question as to whether France or England would conquer the subcontinent. This contest was the international backdrop for all of Clive's subsequent victories.

It was a contest that could also not be understood without reference to internal politics in India itself. Clive was entering a world that was larger and more sophisticated than most British citizens could fathom. The population of India was at least twenty times that of the United Kingdom. At the time India probably accounted for a quarter of the world's gross economic product, compared to about 3 percent for Great Britain, and it was home to one of the world's most advanced civilizations in terms of its literature, art, and architecture. In an essay he wrote about Clive, Lord Macaulay, a leading member of Parliament and noted historian, was indignant that Britain was paying more attention to its colonies in the Americas than to India, an empire as highly civilized as that of Spain, which had conquered the Americas. Macaulay wrote that the Indians had cities fairer than Toledo, buildings more costly than the cathedral of Seville, "bankers richer than the richest firms of Barcelona or Cádiz." It would have required an incredible stretch of the imagination to think that Britain would rule India one day.

Nevertheless, India had long before reached its zenith under the Mughal emperor Aurangzeb* in the late seventeenth century. When he died in 1707, the forces of political disintegration had set in, sowing unrest in a nation where the Mughal rulers were Muslim but two-thirds of the population was Hindu. By the time Clive arrived, central authorities were losing ground to autonomous and semiautonomous rulers in the provinces, who raised their own taxes, built their own armies, concluded alliances, and went to war with one another. In this highly fluid political environment, the British were represented by the East India Company, which traded goods and raw materials from a few outposts on both Indian coasts. The Company's officials had to bow before the Indian emperor and his deputies, asking for permission to do business. They had to negotiate complex treaties, pay bribes, and vie for princely favors—such as territorial concessions and tax relief—in competition with the French and the Dutch, who had their own versions of the East India Company.

---

* His full name is Muhy-ud-din Muhammad Aurangzeb.

The Indian princes and European companies were scrambling to build alliances, which were often broken as soon as a better offer came along. Often, two Indian princes would unite against a third, each side backed by the British or French, making the India market a surrogate field for European rivalry. As commercial competition accelerated, France and England began to expand their protective forces around their companies. They took to the battlefield against each other and used their advanced military capabilities against the far larger but less disciplined Indian armies, with their antiquated weaponry. In this fragmenting empire, an ambitious and cunning man like Clive could succeed wildly.

"Whosoever Commands the Trade of the World, Commands the Riches"

Clive would turn the East India Company into his vehicle for dominating India. Founded by a group of British entrepreneurs and established as "the Company of Merchants of London Trading into the East Indies," it would be known colloquially as "The Company." Officially established by Queen Elizabeth in 1600, the original charter created a strictly commercial enterprise with a fifteen-year monopoly (over other British firms) on trade anywhere in the world outside of Europe. Britain began cutting deals with the Indian emperor to create trading posts on the subcontinent as early as 1612 in Surat and had extended these footholds to Madras and Bombay by 1670, when King Charles II gave the Company rights to do more than trade; it was now authorized to acquire territory, mint money, raise armies for protection, and exercise civil and criminal jurisdiction over its personnel. Like its French and Dutch rivals, the Company was becoming the embodiment of European mercantilism, a doctrine that justified the use of any measures, commercial or military, necessary to dominate global economic competition. Sometime around 1600, Sir Walter Raleigh encapsulated the mercantile view that would drive European leaders when he said, "Whosoever commands the trade of the world, commands the riches of the world and consequently the world itself."

Until Clive entered the picture, however, the Company remained first and foremost a fragmented trading operation. Its outposts in Madras, Calcutta, Bombay, and elsewhere all reported separately to the headquarters in London. There were few connections among the various branches, which competed with one another. It was only under Clive, and against the backdrop of the unraveling political fabric of India, that the Company began to consolidate its influence and exercise its political and military powers in full. The more territory the Company controlled, the more leverage it had to demand more concessions from the princes, and the more it dealt directly—as opposed to dealing through middlemen—with Indian merchants. By the time Clive finally left India in 1767, he had completed the transformation of the Company from trading enterprise to territorial power.

Clive built an impressive military force. Under him, a small, informal security force designed to protect the Company's trading operations grew into an army of over one hundred thousand men whose purpose was indistinguishable from the Company's mission—to conquer new markets any way it could. Under him, the Company became the largest and most complex trading operation in India, with territorial powers over Madras, Calcutta, and Bombay, and flourishing commercial operations. Exports to Asia increased 100 percent. The Company came to account for 20 percent of all imports into Britain, where it became synonymous with financial stability. In the four decades between 1709 and 1748, the Company failed only twice to pay dividends to shareholders.

During Clive's lifetime, the Company became increasingly intertwined with the financial and political life of Britain. The political interests who drew their clout or wealth in some significant measure from the Company included its wide network of suppliers and its shareholders, who ranged from the rising middle class to the upper class and into the halls of Westminster. Almost a third of the members of Parliament held Company stock, and thirteen of the Company's twenty-four directors as well as twelve of its former military and civilian leaders had seats in Parliament. The many Company employees who returned to England to buy estates and businesses also constituted a formidable lobby. No other entity except the Bank of England was a creditor to the government, and no other institution could float such volumes of bonds to the public. The Company had established a gold-plated reputation and occupied a special position in London's financial world—by virtue of the kind of investment it became for so many British citizens and the periodic financial surpluses it could lend out.

The Company helped elevate Britain from second-tier to major power status in Europe. No other agency of the British government flew the Union Jack in so many exotic locations. It was Robert Clive who made the Company into the force that it would become.

From Writer to Warrior

Here is how it all happened. For two years after his attempted suicide, Clive continued to suffer from boredom, loneliness, and depression. Then on September 7, 1746, the European battle for supremacy in India began in earnest, with the French invasion of Fort St. George. Clive awoke to the boom of cannons and looked out to see French ships offshore and French soldiers loading into small boats that would carry them to land. All around him was the commotion of women and children being ushered into shelters to wait out what looked certain to be a debacle for the British.

The fort's defense was in the hands of a ragtag protection force, ill trained, poorly equipped, incompetently led, and outnumbered three to one. When a French shell hit the British supply depot, Company soldiers looted their own stores. Some got drunk on stolen liquor, while others left the fort to loot nearby houses vacated by fleeing civilians. The British cannon fell apart after discharging just a few rounds. Within twenty-four hours, Nicholas Morse, the British governor, surrendered. Clive was captured and the French paraded him and other humiliated British captives through the town square. The British could have their freedom, said the French, if they swore not to take up arms against the French again and if they paid a bounty to the victors. Those who refused were sequestered at the fort. Clive was among them, but a few nights later, with the help of some locals, he and two others under arrest escaped by donning Indian clothes and applying dark paint to their faces in order to impersonate local men.

The three men traveled fifty miles south to Fort St. David, another British outpost, startling the guards when they threw off their costumes and revealed that they were British. Clive decided to enlist as an ensign. Just twenty-one, he likely knew that soldiers could amass wealth as easily as merchants, and in addition he could not bear to let the French victory stand. Over the next year, Clive would learn soldiering under the able commander, Major Stringer Lawrence, who had already repelled two French attempts to storm the fort. Lawrence taught Clive how to deceive an enemy with feigned retreat, how to surprise him with all-night marches, and how to wage war guerrilla-style.

Under Lawrence, Clive began to gain military experience in a series of small battles. He exhibited all the traits of a formidable military leader, including a flare for the kind of unconventional warfare that was necessary in eighteenth-century India. During one of the French assaults, Clive led a platoon of thirty men in holding off a counterattack by the entire French force in driving rain. He was soon promoted to the rank of lieutenant, with the role of quartermaster, a job in which he could make a commission on all the supplies delivered to Fort St. David. This was still entirely legal under Company rules, and it put Clive in a position—call it a merchant soldier—that he would occupy for many years. As quartermaster, Clive had to negotiate for everything from cattle and camels to rice and curry, gaining an intimate knowledge of how to think like the locals, knowledge that would soon become invaluable on the battlefield. He had garnered respect, he cherished the camaraderie of soldiers, and he sensed limitless possibilities to acquire wealth.

For most of the next decade, Clive was engaged in one battle after another. Some were important skirmishes with French forces, some were big battles with Indian troops supported by the French, all were part of a long continuum of combat. Clive proved himself to be increasingly skillful in employing deception, cunning, and ruthlessness; in other words, in doing exactly what wasn't expected by the enemy—not unlike Genghis Khan. The Indians, who had grown accustomed to the more conventional tactics of the Portuguese, Dutch, and French, faced a different kind of adversary in Clive. Macaulay put it this way: "This man, in the other parts of his life, an honorable English gentleman and a soldier, was no sooner matched against an Indian intriguer, than he became himself an Indian intriguer, and descended, without scruple" to their level.

Clive engendered enormous loyalty and respect from his troops, in part because he showed no fear of combat himself. He understood the mind-set of the Indian troops who were employed by the British and fought alongside them—they were called sepoys—and won their allegiance, too. By 1757, Clive emerged victorious over the French and the Indians. Two of the epic battles were at Arcot, where he captured a good swath of southern India, and the Battle of Plassey, where he won for England the riches of the provinces in the north.

The Battle of Arcot and Southern India

In November 1751, Clive became the central figure in the main battle for the Carnatic, the most important province on the southeast coast, where the French were headquartered at Pondicherry and the British at Madras. Two Indian princes were battling for the post of nawab (governor) of the Carnatic, with the British backing Muhammad Ali and the French supporting his rival, Chanda Sahib. With Ali surrounded in the town of Trichinopoly, the British came to his rescue under the command of Captain Rudolph de Guingens, with Clive at his side.

After marching for nearly two months to reach the outskirts of Trichinopoly, de Guingens decided to pull back in the face of vastly superior French and Indian forces. Clive was furious. He had not forgotten the humiliating surrender at Fort St. George and quickly secured approval from the governor at Fort St. David to lead a return assault on Trichinopoly, where he slipped past the heavy defenses to meet secretly with Muhammad Ali. Together he and Ali devised an unconventional counterattack. Rather than try to break the enemy line at Trichinopoly, Clive would surprise everyone and attack Chanda Sahib at his base in Arcot, the provincial capital.

It was the first of Clive's many big gambles. Now twenty-six, he had been promoted to captain, and, in preparation for the upcoming sneak attack, he gathered under his command most of the East India Company's troops in the region. A prolonged battle at Arcot would expose all of the British forts in the Carnatic to French assault, since all of the English soldiers would be traveling with Clive. On the other hand, a British victory could draw every

Indian who hated either the French or Chanda Sahib to the English side.

Through blistering heat Clive marched his five hundred men toward Arcot at breakneck speed, learning on the way that Sahib had more than one thousand soldiers. As he reached the city, a violent electrical storm opened up the skies, turning the ground to mud as the British slogged toward the walls of Arcot, ready for the worst. Instead they found the capital empty. The Indian commanders had received word that Clive's men were marching through the monsoon, an act that they believed defied the very gods themselves. As a result, they had fled before Clive arrived, just as they would have done in the face of a supernatural event. Clive raised the flag of Muhammad Ali over Arcot and appointed officials to administer the capital in Ali's name, but everyone in the region saw Clive as the real power in the capital.

A few miles away, Chanda Sahib was massing for a counterattack, his force swelled to three thousand now, with French reinforcements on the way. Clive, meanwhile, was holed up in a large fort with a battalion that had shrunk to 320 men (of whom one-third were Europeans). That was too few to man the milelong walls of the fort, particularly given that the moat was dry in places and some of the towers too feeble to support cannon. In another surprise move, Clive decided instead to attack, launching an early-morning guerrilla raid that not only scattered the startled enemy but also provoked a series of counterattacks by them. Eventually Clive faced ten thousand men.

Battered by the heat, by dysentery, by enemy cannon and sniper fire, the British defenders were nonetheless holding their ground when word spread that six thousand Marathas—mercenary guerrillas—were willing to come to Clive's support—for a price. Realizing that the Marathas could turn the tide, Chanda Sahib knew that he had to move fast and sent word informing Clive that he had to make a choice: surrender or face immediate assault.

Clive refused to surrender and Chanda Sahib's assault came quickly, with Sahib's elephants battering down the fortress walls on several sides. Clive's men met the attack in lines, the front ranks firing and passing the spent rifles back for reloaded ones. Clive ran from tower to tower, issuing instructions, propping up morale, dodging bullets, and inspiring his men with his courage. Two moves then turned the battle: Clive's men shot at the elephants, who turned and ran, sowing chaos in Sahib's ranks.* Then the British shot a prominent Indian commander, which to local fighters of the time constituted a signal from above of imminent defeat. Sahib's army fled, ending a fifty-day siege that left heavy Indian casualties, while Clive lost just ninety men.

Arcot turned the tide in India toward the British. It created an aura of British invincibility in the minds of the Indian princes and gave the Company's forces the momentum they needed to go from victory to victory. Clive became a legend as a commander who could win against any odds and as a staunch ally who had saved Muhammad Ali. He had won the admiration of Indian leaders and soldiers, who volunteered in droves to serve with him. Later, the French would accuse Clive of mounting abuses—bombing a French civilian camp, for example, or firing on French troops who had raised the white flag. His fame mixed with notoriety, tingeing his reputation with the image of the bare-knuckle street fighter. But Arcot remained the foundation from which Clive would ascend to rule India.

After Arcot, Clive marched back to Trichinopoly, where Ali was still holed up, and liberated the city. Throughout the entire episode, he held on to his post as quartermaster, amassing a small

---

* Whether Clive ordered his men to shoot the animals or whether this just happened in the heat of battle is not clear.

fortune by charging a commission on every sack of rice, every load of firewood, and every delivery of swordfish in his territory. He had settled his debts and began sending large sums home to his family, who were, not surprisingly, now more willing to correspond with him. On the strength of his son's growing fame, his father began to ascend in British society. As for Clive himself, he had long since left the writers' rooms where he started out. He moved into more splendid quarters, married a young Englishwoman, Margaret Maskelyne, and tried to settle into a normal family routine. But normality would always elude him. Racked by regular attacks of fever and abdominal pain that became intertwined with deep depression, Clive took opium to relieve the stress. By 1753, he and Margaret concluded that they had had enough of India.

The Money Runs Out

Clive arrived back in London to a hero's welcome and the warm embrace of his parents. The directors of the East India Company sought his advice on matters of high policy, but Clive was a political novice, and England was strange territory for him. He faced open doors everywhere but, much as he tried, he didn't seem to fit in anywhere—not in upper-class society, not in the British political power structure. He ran for Parliament and won but never actually occupied his seat because the election was almost immediately contested for technical reasons unrelated to him. His money began to run out, too.

When the Company offered Clive the rank of lieutenant colonel, with a mission to dislodge the French from central India and to serve as understudy to the governor of Madras, Clive accepted. In April 1755, after less than two years in England, he and his wife boarded the *Stretham* back to India. The dock was filled with well-wishers and Clive was given a nine-gun salute, a far different send-off from the one he had received as a teenager twelve years earlier. His second tour would multiply his wealth beyond imagination and ensure his enduring fame as one of the founders of the British Empire.

The Black Hole of Calcutta

When Clive reentered India at the end of 1755, the sole British stronghold in the Bengal region—the richest province in India—was Calcutta. Its crown prince was young Siraj-ud-Daula, who had recently ascended to a throne that encompassed the provinces of Orissa and Bihar as well as Bengal. The prince resented the Company's growing empire, its arrogance, its control of his premier city. Not only was the Company refusing to pay tribute to him (as the Dutch and French had been doing), it was defying agreed limits on how much its employees could trade tax-free with local merchants and on how far it could expand Fort William, the main British base in Calcutta. Adding to the prince's fury, the Company was also protecting an Indian local who had betrayed him.

Siraj-ud-Daula decided to make an example out of the Company. On June 20, 1756, he seized Fort William and ordered his men to imprison the entire garrison in one of their own cells. Siraj-ud-Daula then went to sleep, and his subordinates refused to wake him as the prisoners began to scream for relief. The prisoners numbered between 50 and 150—historians offer different numbers—but by all accounts they were packed into an underground cell that measured just fourteen feet by eighteen feet, with small barred windows on the ceiling serving as both the entranceway and breathing hole. By morning a third of the prisoners had suffocated to death. Thus did "the Black Hole of Calcutta," with all its angry, racist connotations, enter the English language.

Top officials of the East India Company, seeing no choice but to exact a painful retribution, assigned Clive to lead this mission. In early June 1757, Clive set out with a force of twelve ships and twenty-five hundred men bound for Fort William. When they arrived, they found it deserted. Continuing upriver toward the prince's residence, Clive was well aware that he was sailing toward what could become a conquest of the wealthiest territory in India. He wrote to his father of how this "grandest" of his undertakings "may enable me to do great things," as the future would indeed show. To the Company chairman in London, Laurence Sullivan, Clive wrote that "this rich and flourishing kingdom may be totally subdued" by a mere two thousand Europeans, as "the Indians are indolent, luxuriant, ignorant and cowardly. . . . They attempt everything by treachery rather than force."

Clive saw India as a morass of intrigue and corruption. He assumed every foreigner—European or Indian—was double-dealing, and he responded in kind. When Prince Siraj-ud-Daula heard Clive was coming after him, he offered a truce that would restore all British rights in Calcutta. Clive accepted but then kept right on moving toward him, sure that the prince would break the deal as soon as betrayal served his purposes.

Clive and Siraj-ud-Daula were headed for war, and both knew it. Once again facing vastly superior forces, Clive learned through informers that one of the prince's generals, Mir Jafer, was ready to defect to the British in return for a written guarantee that he would be the next prince of Calcutta. At the same time, however, a well-informed merchant named Omichand was threatening to expose Mir Jafer's treachery unless he, Omichand, was guaranteed a major share of Calcutta after its fall. He also wanted a written guarantee—signed by both Clive and the ranking British admiral, Charles Watson. Figuring he could not promise Calcutta to both Mir Jafer and Omichand, Clive prepared two postwar treaties, one for each. Admiral Watson did not want to be a party to this scheme, but somehow Clive secured his signature anyway. Historians have debated whether Watson authorized someone to sign for him or whether Clive crafted or ordered up a forgery. Watson died soon after, leaving no clear record of what happened.

"Like Caesar, He Made His Decision and There Was No Turning Back"

The big battle came on June 23, 1757, at the small town of Plassey, which stood between two branches of the Hooghly River. With the monsoon coming, Clive worried that if he crossed the river, it might swell with the rains and block his only avenue of retreat. He was dependent on timely backup from the potentially duplicitous Mir Jafer, who was responding vaguely, if at all, to his letters. Torn between the desire to attack and the need for dependable reinforcements, Clive chose to attack. "This was his Rubicon, his burning of the boats," wrote biographer Robert Harvey. "Like Caesar and Cortés, he had made his decision and there was no turning back."

Marching his men through the torrential rain and deepening mud, Clive reached Plassey in the dead of night and awoke to see a force of Indians he estimated to be about fifty thousand strong arrayed on the field against him, with French reinforcements reportedly on the way. Clive had just three thousand men and told a colleague he planned to put up a daytime fight before slipping back to Calcutta the next night.

The day broke, arguably the day in which the fate of India would be sealed, although no one could have known that at the time. The heavy rains had taken their toll. The battle started with the strangely sporadic booming of the Indian artillery because the monsoon silenced many of the guns and left others too wet to fire consistently. Some of the cannoneers were ordered not to fire by Indian commanders who hated the prince they were there to defend. Meanwhile English artillerymen, trained to protect their ammunition from the rain with tarpaulins, were firing with great effect. Several of the senior officers of Siraj-ud-Daula fell, and soon there was chaos in the Indian ranks. The French hung back as well, perhaps also unwilling to commit to a prince who inspired such patchy loyalty in his own men. Mir Jafer never did take to the field. Siraj-ud-Daula retreated to his palace. In short order the battle was over and Clive had won.

Emperor Behind the Curtain

Later critics would say Clive was lucky at Plassey because the enemy was exceptionally incompetent. But as great military strategists would acknowledge, often battles are won by a combination of luck and what happens before soldiers even take the field. Such was the case here. Clive's political deal with Mir Jafer, imperfect as it was, may have kept the general on the sidelines. And the superior training of Clive's troops put them in a position to exploit the monsoon, at the moment when it was paralyzing the enemy. By defeating one of the great viceroys of the old Moghul Empire, the British emerged as the acknowledged rulers of Bengal, with Clive as the emperor behind the scenes in the largest and wealthiest provinces of India, constituting a kingdom of forty million people.

Choosing to overlook Mir Jafer's failure to help in the battle, Clive pronounced him nawab of Bengal and ruler of its three provinces. (Omichand reportedly had a mental breakdown upon discovering that his copy of the treaty was the bogus one and that he would not be ruler of Bengal.) A few days later, Clive entered the provincial capital at Murshidabad, walking in the middle of a personal entourage of hundreds and wading through a cheering crowd of thousands. Settling into expansive guest quarters, he found that the city vaults had been looted and held only a tenth of the treasure he had expected. Nonetheless, the city leaders lavished gifts of jewels and bullion on Clive and arranged payments to the Company, its directors in Calcutta and London, and its military commanders.

In the subsequent years Clive's influence in Bengal kept expanding. His presence, reputation, and troops enforced Mir Jafer's authority. He installed officials who would be beholden to the British for their positions and would cater to the Company's interests. He arbitrated between feuding political factions. He developed personal ties with princes of neighboring provinces.

The Company's directors in London named Clive governor of the Calcutta branch of the East India Company, which gave him three overlapping roles: governor of Calcutta, political proconsul of Bengal, and supreme military commander over both. Meanwhile, tributes were being paid to him in London. In a speech on the floor of Parliament, Britain's foreign minister and minister of war, William Pitt, said, "We had lost our glory, honor and reputation everywhere but in India. There the country had a heaven-born general who . . . was not afraid to attack a numerous army with a handful of men, and overcome them." He then compared Clive to Alexander the Great.

Clive soon received another honor, which would make him both fabulously rich and hugely controversial. The emperor of India, a figurehead in Delhi, awarded him an imperial title—"Flower of the Empire, Defender of the Country, the Brave, Firm in War"—that entitled him to a *jagir*, a lifetime stream of revenue. Mir Jafer awarded Clive the stream of rent paid by the East India Company on land in Calcutta, making Clive, in effect, landlord of the Company that employed him.

Clive's behavior became increasingly ostentatious. He moved only in large convoys of soldiers, elephants, and servants. He gave lavish parties. He wrote his father that the job of running India was too big for the Company and that the British government should take it over, making him, Robert Clive, the first governor-general of India, reporting to the Crown. The first part of this thought—the government takeover of the East India Company—would in fact come to pass, although long after Clive's time in India. As for the second part, some highly placed Company officials got wind of Clive's ambitions and thought he was succumbing to megalomania.

An Alien in England

Clive decided it was once again time to capitalize on his fame by entering British politics. On February 21, 1760, he boarded the *Royal George* with his wife, this time thinking he was leaving India for good and returning to a life of wealth, privilege, and influence in his home country.

He arrived in London as the hero of Plassey and the conqueror of Bengal. He received an audience with King George II, along with many accolades from the East India Company and an honorary degree at Oxford. His stupendous wealth easily enabled him to win a seat in the House of Commons. But he was still seen as an outsider, a garish arriviste, identified only with his exploits in India. The attacks, in part, reflected the jealousy and hypocrisy of the old elite in London, who seemed to resent the use of personal fortunes in the pursuit of political power—but only if the fortune was new money, like Clive's.

It was now becoming clear, even to Clive, that "Clive of India," as he would be dubbed years later, could not realize his burning ambition to become "Clive of England." Harshly criticized for accepting as a gift-for-life the rent that the Company was paying to Calcutta, he spent much of his time fighting Company attempts to rescind his *jagir*. At the same time, the empire he had set up in Bengal was unraveling. Mir Jafer was chafing under the control of his new Company bosses, who replaced him with his corrupt son-in-law. The latter was soon deposed as well. At the same time, the Company's military and administrative structure fell into disarray, and the long tradition of private dealing became increasingly disreputable, with employees forcing Indians to sell to them at steep discounts and buy at inflated prices.

"See What an Augean Stable There Is to Be Cleansed"

On March 12, 1764, the Company's directors proposed that Clive return to India for a third time to clean things up, and he readily agreed. Now nearly forty, he was given all that he asked for, including near-dictatorial powers over all of the Company's operations in India as well as more troops. In May 1765, he arrived in India, determined to be a zealous reformer, to bring good government to the corrupt empire he had done so much to create, and one he now condemned for its lack of discipline and purpose. On arrival in India, Clive wrote to a friend, "See what an Augean Stable there is to be cleansed. The confusion we behold, what does it arise from? Rapacity and luxury, the unreasonable desire of so many to acquire in an instant what only a few can, or ought to, possess . . . in short the evils, civil and military, are enormous but they shall be rooted out."

Clive took on his new challenges with the same energy and restlessness that he showed on the battlefield. In order to curtail the freelance moneymaking activities of the Company's officials, Clive realized that they needed to earn enough income so that they need not pursue dubious side deals. He thus raised middle-management salaries to much higher levels. In order to raise pay, Clive had to tap into the revenues of the Company's monopoly on salt, a highly controversial measure that reduced the Company's profits. Clive established that from then on, anyone serving as governor-general—the post he held—could not engage in any outside moneymaking activities and that their compensation would be tied directly to the revenues of the Company in India. This may sound like an early echo of today's campaigns to restrain the high rate of increases in CEO pay, but Clive was attempting something even more difficult; he was pushing through permanent cuts in executive compensation, considering the fortunes that could be made by accepting gifts and bribes or cutting private deals.

Next, Clive implored the Company in London to recruit and send men with higher levels of skills and integrity, aiming to professionalize the Company's workforce. In fact, his goal was to create a new governing class of civil servants, and they would become the embryo of the salaried, independent civil service that was to be the model of how Britain ruled India and many other colonies. Meritocracies may be a normal thing today, but they were a novelty in the class-ridden society of eighteenth-century Britain. Clive also established a postal system that linked all the branches of the Company in India, facilitating the exchange of information and closer coordination.

Clive next attacked the administration of the British military in India. In place of the unified but unwieldy organization he found, he restructured the armed forces into three separate brigades. He clarified the tangled lines of command between the British soldiers and the Indians who fought under them. He also prohibited commanders from receiving special allowances from Indian officials. This prohibition was so controversial that the entire British military establishment in India tried to mutiny, but Clive prevailed.

Most important, Clive clarified the murky balance of power between British and Indian leaders by gaining more legitimacy for the de facto rule of the Company. He approached the figurehead emperor in Delhi and received his formal consent for an arrangement that would underpin British rule for the next century. The Company would have the legal right to administer all of Bengal. It would collect all the taxes, apportion the money, and make all the major administrative appointments. In return it would pay an exorbitant royalty to the ruling princes and institute arrangements to allow the Indians the veneer of governing. Clive knew, however, that it would be politically explosive if the Company itself were seen to be collecting revenues. So he set up an Indian administrative structure that would report to the Company and be controlled by it. For the first time, Britain had the legal authority to exercise direct political and commercial control over India, bringing colonial rule out of the informal shadows.

Clive's achievements were particularly remarkable given the rogue he had been before he became a reformer, and the context of the times. First, the industrial revolution was in its early years, and the discussion of how to best manage large enterprises was only just beginning. It most certainly had not reached India yet, so Clive was advocating standards there that no one had even pondered before. Moreover, Clive was trying to change shady habits of self-enrichment that had been building up inside the Company for decades and from which he himself had earned his own fortune. It is a wonder that this ex-warrior and profiteer is the man who put the whole idea of a British empire on a more orderly and civilized path.

Clive's actions are even more impressive since he had only eighteen months, as this stay in India would be cut short by deteriorating health. Many writers have marveled at what he accomplished. "After securing a continent by the time he was only 35, [he] had set down the administrative foundations of the empire that was to last two centuries," wrote Robert Harvey. Macaulay would praise Clive for "one of the most extensive, difficult, and salutary reforms that ever was accomplished by any statesman."

Clive returned to London during the summer of 1767. As soon as he arrived, trouble hit Bengal in the form of a massive famine. Many of Clive's reforms had not yet taken full hold, and employees were already breaking the new ban on private dealing, hoarding grain and selling it at exorbitant prices when supplies ran low. Critics attacked the Company for famine profiteering, and hungry locals attacked it in violent skirmishes. The Company looked like it was spinning out of control.

"Mr. Chairman, I Stand Astonished at My Moderation"

Meanwhile in England, the Company had become ever more controversial. Its supporters pointed to the benefits that accrued to its thousands of shareholders, to the tax revenues that flowed to the British Treasury, and to the global influence that Britain was acquiring from its commercial activities. Its detractors attacked the whole conception of a commercial organization that could embody near-sovereign powers and challenge the authority of the state. They criticized business executives who cut political deals with Indian princes and conducted themselves in ways that would never be tolerated at home. The attacks were enflamed by the many Company employees who were coming home and flaunting their stupendous wealth—just as Clive had done—engendering deep resentment among Englishmen of all classes. Now hated by both reformers and by the officials whose dirty deals he had tried to stop, Clive's many enemies set out to destroy his reputation and claw back his fortune.

In 1772, Parliament held a wide-ranging inquest on Company practices in India, focused in particular on Clive. He came under attack for his double-dealing treaties and the admiral's allegedly forged signature at Plassey, not to mention the way he amassed his fortune. Unrepentant, Clive revealed a mind-set utterly divorced from the British public, out of touch with mainstream values, and indignant at being challenged. He told the committee:

> Consider the situation in which the victory at Plassey had placed me. . . . A great prince was dependent on my pleasure. An opulent city lay at my mercy; its richest bankers bid against each other for my smiles; I walked through vaults which were thrown open to me alone, piled on either hand with gold and jewels! Mr. Chairman, at the moment I stand astonished at my moderation.

In the end, Clive was exonerated, but his reputation was ruined and his political career was finished. It was a stupendous comedown, and his melancholy returned in full force. At the same time, he was afflicted by tropical diseases he had contracted over the years and was taking liberal amounts of opium. On November 22, 1774, the man who did so much to create the foundations of the British Empire took his own life. Accounts vary as to how he did it—a penknife to the throat, an overdose of opium? His wife and family never said, and there was no postmortem, no inquest, and no official explanation. He was buried in an unmarked grave within the yard of an obscure church in Moreton Say, a small village in eastern Shropshire, where he was born. Years later a plaque was installed with the words PREMIS IN INDIS [First in India].

The Clives Who Followed Clive

Robert Clive was an adventurer, a soldier of fortune, a conqueror, an occupier, an administrator. He belonged to a genre of men who arise within all empires and who have often wielded great influence over large swaths of geography and large numbers of people. All of the Company's rivals had their own Clive. The Dutch East India Company had one in Jan Pieterszoon Coen during the seventeenth century; the French East India Company had one in Joseph-François Dupleix, who clashed with Clive in India. They often represented the unsavory balance between good and evil that characterized many powerful men and women who pushed the envelope of globalization.

Clive laid the foundation for the East India Company to become an industrial powerhouse and for the Company to be the instrument with which Britain conquered India. As Oxford professor John Darwin writes, the British Empire was a global system connected by extensive links on all continents fostered by the British navy, which controlled the seas and was available to protect Britain's colonies; by extensive diplomatic ties among the same participants; by deep cultural and educational exchanges; and, of course, by commercial and financial interdependencies—all intersecting in the government corridors in London. This world system was strongly supported by a complex network of mail services, telegraph wires, undersea cables, passenger steamers, and eventually imperial air routes. India was the jewel in the crown. With regard to what India itself was to become, here is how Darwin puts it:

> Imperial India was more than the countries of modern 'South Asia.' It was 'Greater India': a 'sub-empire' ruled from Calcutta, extending from Aden to Burma, and with its own sphere of influence in the Persian Gulf, Southwest Iran, Afghanistan. . . . 'Greater India' might even include coastal East Africa, whose metropole was Bombay.

Of course, Clive wasn't solely responsible for building the East India Company or for conquering India. Many other people played vital roles in shaping the Company's business strategies, building its ties to India and China, expanding and training its commercial and military workforce. What Clive did, above all else, was to create the space, set the direction, and stimulate the momentum that allowed the Company and British control of India to grow. At Arcot and Plassey, he bent the river of history and turned the tide. "[Arcot] was the first great triumph of British arms in the history of India," said Robert Harvey. Plassey, as we have seen, led to British supremacy, via an Indian surrogate, over the wealthiest provinces of India. The Bengali-English scholar Nirad Chaudhuri summed it up well: "It has to be noted that, if he was the founder of the British Empire in India, he was only this in a limited sense—that is to say, he was the founder of British power, which could no longer be successfully challenged: the process had gone so far [under him] that it could only go forward and never be erased."

And so the East India Company would eventually build its own ships in England and India; it would run massive warehousing and foundries; it would acquire the power to print its own currency and annex territory by force. It would construct and manage wet and dry docks and manage and adjudicate law in the areas it controlled. Not surprisingly it was England's largest single employer. By the nineteenth century, it had become synonymous with Britain's global interests and global reach. At its height—some twenty-five years after Clive died—the Company had established control of over five hundred thousand square miles of Indian territory, an area more than three times the size of California, and containing almost ninety-four million people. Its army consisted of 250,000 men, 45,000 of whom were Europeans and the rest Indian recruits. It had its own military academy, its own diplomats, its own intelligence service replete with Indian spies and informants. It established trading and military dominance over Malaya, Hong Kong, and Singapore and engaged in both military and commercial operations in Burma, China, and Afghanistan.

By the 1850s, however, the Company had become too intertwined with Britain's global interests to be run as a commercial company. London took away the Company's administrative authority in India and, in 1874, shut it down. From then on it was the British government directly pulling all the military strings, and many different companies conducting commerce.

The Company had advanced globalization in a number of ways. It responded to the massive increase in consumer demand in England by importing goods such as pepper, silk, tea, and opium from all over the world. It also exported goods from England.

The Company invested heavily in boats, ports, warehouses, and roads in dozens of its colonies, and in new trade and financial arrangements, such as insurance, that allowed trade to flourish among its territorial possessions. In this regard it not only facilitated trade to and from England but it made deals between other countries, as well—between China and India, for example.

Its growing ranks of investors and employees constituted a powerful source of political support for global trade and investment, and a force for the intermingling of British, Indian, and other foreign cultures.

By importing fine Asian goods such as embroidered silk and lacquered furniture, the Company inspired British entrepreneurs to push innovation at home, accelerating the industrial revolution.

Furthermore, the East India Company showed how government and commercial enterprises of the same nation can make common cause in expanding commerce and culture across borders. Indeed, it demonstrated that when it comes to globalization, the line between the state and its companies can be thin or invisible. History in the aftermath of the Company demonstrated this phenomenon. In the three decades following World War II, American companies such as United Fruit and ITT and Washington worked hand in hand in both commercial and political arenas in foreign countries. Today, corporations such as Europe's Airbus Industrie have symbiotic relations with their governments.

The closest modern equivalents to the Company, however, are the giant state-owned companies of China, such as the Sinopec Group, the petroleum behemoth. As Beijing extends its influence across East Asia and works to secure supplies of oil and other raw materials from Africa to Latin America, its large state firms are and will continue to be critical vehicles of this expansion.

In Joseph Conrad's epic 1899 novel, *The Heart of Darkness*, the narrator, Charles Marlow, becomes obsessed with a Mr. Kurtz, the head of a remote colonial outpost deep in the Belgian Congo. Kurtz represents various facets of the European culture of imperialism, a culture that derives from the clash between the ruler and the ruled in foreign settings, one that produces megalomaniacal and violent tendencies, and an absence of moral bearings bordering on madness. We could be forgiven for seeing some of these qualities in Robert Clive, the embodiment of British imperialism. Writer and historian William Dalrymple even described Clive as "an unstable sociopath." Nevertheless, a guileful operator perfectly adapted to the shifting terrain he came to dominate, Clive is a fine example of the serendipitous fit between man and historical circumstances. It is true that in various phases of his life, he embodied the ruthlessness that eventually pushed British rule to forty-three colonies on five continents, linking London to East Asia, South Asia, North America, and Latin America, controlling a quarter of the world's landmass and one-fifth of its population. He did all this with intense determination, cunning, brute force, and ostentatious greed. Whereas Genghis Khan and Prince Henry expanded the physical horizons of mankind, Clive and others like him penetrated the lands that had been previously discovered by implanting ideas that became, for better or worse, more common across broad geographical boundaries—all this to a much greater extent than the Mongols ever did.

Clive embodied the reforming zeal that transplanted British ideas of effective democratic governance, rule of law, market-oriented economics, political pluralism, and modern education systems across that vast territory. The empire was a conduit for the political and economic liberalization that characterized so much of the twentieth century and the first decade of the twenty-first. Indeed, it was the British Empire that, in tandem with the American democratic capitalist system, created the global economy as we know it, based as it is on consumer-driven markets, rule of law, and the ideal—at least in North America, Europe, and a growing number of emerging nations—of free and open societies.

Especially in its first several decades, the British Empire did not expand according to any strategic blueprint designed in London. Instead, it was the result of constant improvisation by relentless men with personal agendas that included glory for England, missionary zeal for its values, personal adventure, and the search for fortune. Clive was the epitome of such men.

There are no known authentic pictures of Mayer Amschel Rothschild. This painting, by Moritz Daniel Oppenheim, was commissioned by Rothschild's sons after the Napoleonic Wars and shows Prince William, the landgrave, entrusting Mayer Amschel Rothschild with his treasures.

Chapter IV

MAYER AMSCHEL ROTHSCHILD

The Godfather of Global Banking

1744–1812

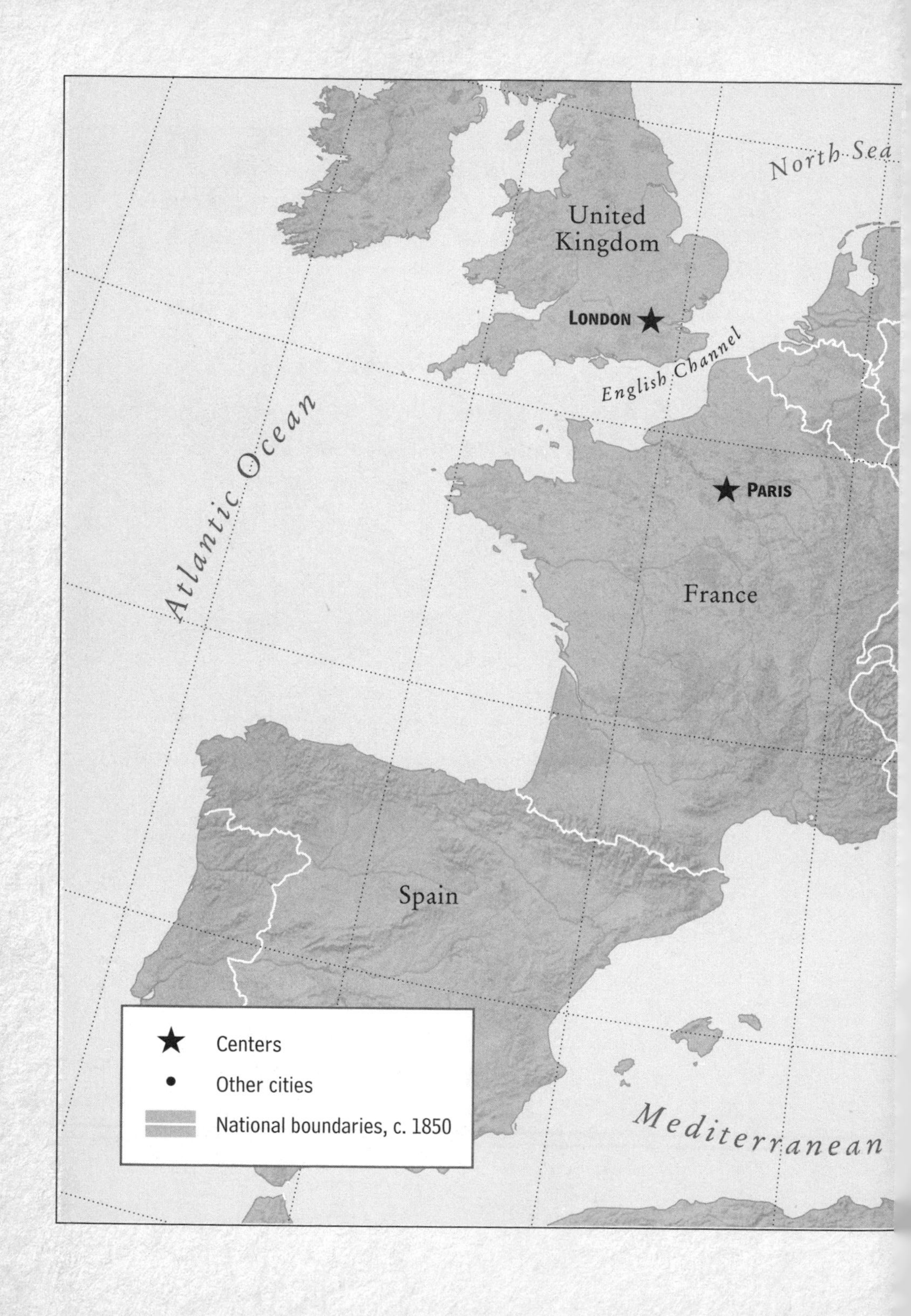

North Sea
United Kingdom
LONDON
English Channel
Atlantic Ocean
PARIS
France
Spain
Mediterranean
Centers
Other cities
National boundaries, c. 1850

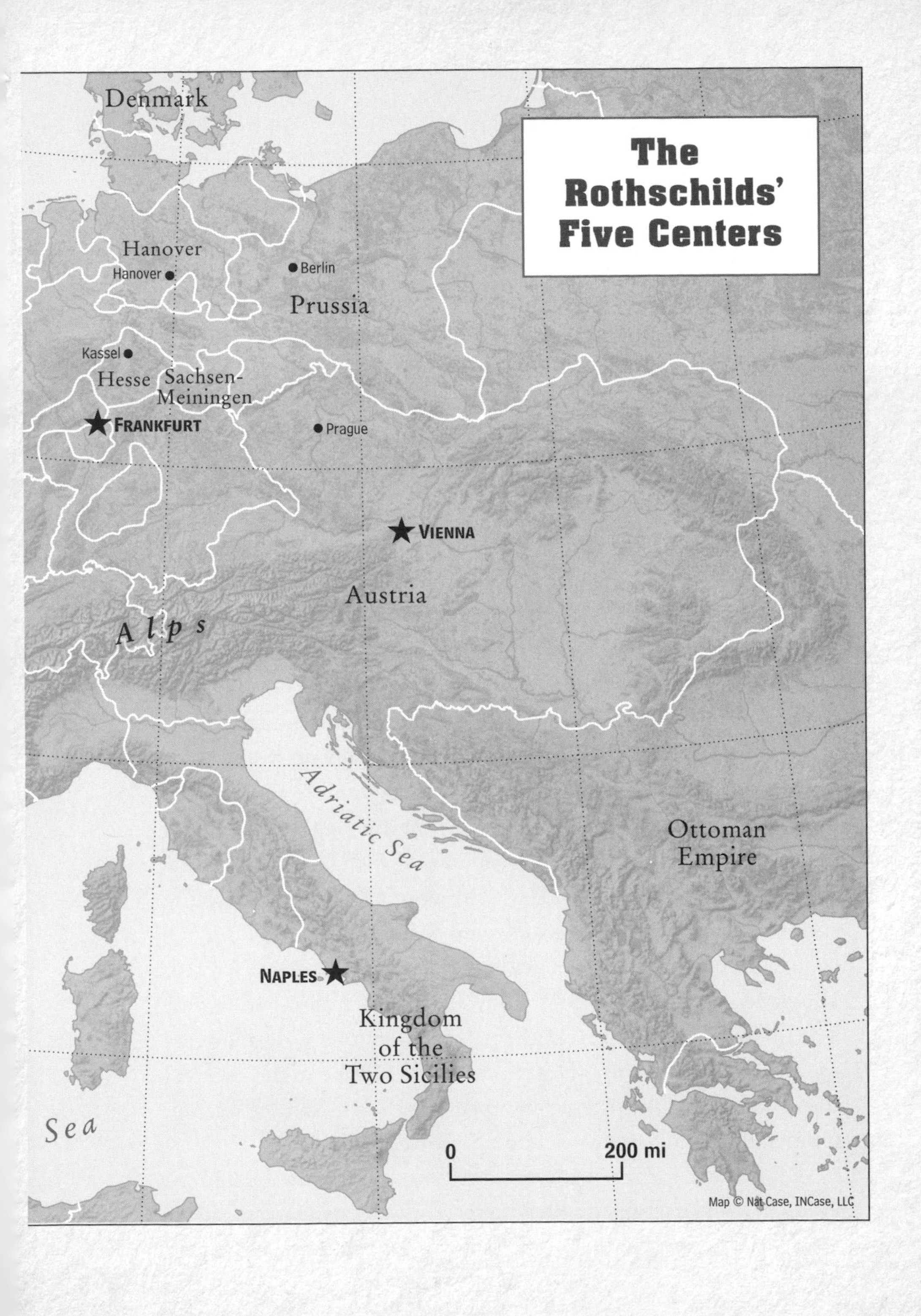
The Rothschilds' Five Centers
Denmark
Hanover
Hanover
Berlin
Prussia
Kassel
Hesse
Sachsen-Meiningen
FRANKFURT
Prague
VIENNA
Austria
Alps
Adriatic Sea
Ottoman Empire
NAPLES
Kingdom of the Two Sicilies
Sea
0
200 mi
Map © Nat Case, INCase, LLC

Mayer Amschel Rothschild was the son of a lower-middle-class coin and silk trader in the Judengasse, the Jewish ghetto of eighteenth-century Frankfurt. The ghetto contained not one tree or one patch of grass, and has been described by historians as "slum-like," "soulless," and "foul smelling." An anonymous traveler of Rothschild's day wrote that if you were coming to Frankfurt, you need not ask directions to find the Judengasse. All you needed was to follow the stench.

To the Jews, however, the ghetto was a vibrant community, and Mayer never moved away, even after he became the most powerful banker in the world. The neighborhood had its own government administration, its own courts, and charitable institutions that helped widows and provided health care for the poor. It contained a medical clinic, a public bath, a theater, a communal bakery, and four synagogues that were the center of a tight-knit society.

Mayer's street was a quarter mile long, and so narrow that horse-drawn wagons had a difficult time turning around. As a boy Mayer often walked in the middle of the road because raw sewage flowed down the sides. The youngest of ten surviving children, he was one of five sons who shared a tiny bedroom in a narrow house, just ten feet wide and nine hundred square feet altogether, which the Rothschilds shared with another family. The house had a view across the narrow alley to a wall of houses much like it, many with colorful names that spoke to the cheerful possibilities within these cramped quarters—White Turnip, Golden Well, and, in Rothschild's case, Red Shield.

Every day, Mayer walked from his house several hundred yards to his synagogue, where he received an elementary educational typical for the early 1750s in the Judengasse. He studied the Bible, Jewish scripture, and scholarly commentary on Jewish literature and tradition. Despite what looked to outsiders like a life of misery, this ghetto produced not only rabbis and cantors but also doctors, mathematicians, actors, musicians, teachers, and highly entrepreneurial businessmen, though none rose to the heights reached by Mayer Amschel Rothschild.

The sense of community was strengthened by the prejudices of the Frankfurt government, which for three centuries confined Jews to the ghetto, drastically curtailed their civil rights, and periodically mounted pogroms against them. An edict that limited the number of houses in the ghetto led to severe overcrowding. The wooden buildings were vulnerable to fires, which burned large parts of the Judengasse to the ground in 1711, 1721, and 1774. The windows of the houses that overlooked the city had to be boarded up every night so that Jews couldn't peer into Christian houses. Jews needed special passes to leave the ghetto in the evenings, or on weekends and Christian religious holidays, and they were prohibited from walking in the parks or near religious monuments.

Jews did not often venture outside the ghetto, but when they did they could not walk more than two abreast, and they were often roughed up by arrogant locals. They were forced to make extra payments for protection of their property and themselves. They were prohibited from acquiring land or farming, or from trading in spices and most other commodities. The attitude of the Frankfurt establishment was typified by a sign that its government had painted on one of the entrances to the city. It depicted a fat sow holding up its tail. Next to the animal was a Jew with his tongue hanging out to lick the excrement. Other Jews were shown sucking the sow's teats. Overlooking the whole scene was the Devil, an approving look on his face.

Who could have predicted that a child of this ghetto would become the founder and leader of the most powerful bank the world had yet seen? Indeed, the House of Rothschild would rival in its power and influence the major sovereign governments of the day. Many of them came to the Rothschilds to underwrite the cost of their military engagements; others needed the bank to refinance their debts. Newly independent nations required the Rothschilds' help to launch themselves. Major industries needed their financing to expand. The Rothschilds were leading pioneers in financial globalization, working out of the unlikely confines of the Frankfurt ghetto to expand the reach of banks across borders, connecting borrowers and creditors of different nationalities with increasing ease, channeling capital from one part of Europe to another, or from Europe to Asia or Europe to Latin America, in order to finance global trade and investment in all manner of industrial ventures. Over the centuries the financial industry has been a key pillar of globalization. Among the distinctive contributions of the House of Rothschild to the industry's global reach were its organizational skills, its cohesion as an international entity, and the innovation it brought to markets, particularly the critical international bond market.

Germany as a nation did not exist when Mayer was still in school. The German-speaking peoples were spread among some three hundred kingdoms, states, principalities, and duchies, most of which printed their own money (offering foreign exchange experts a lot of profit-making opportunity). Within this patchwork, the "city-state" of Frankfurt occupied a vital role as a major port on the river Main, which flowed into the Rhine, the most important waterway in Europe. Frankfurt was known throughout the Continent for semiannual trade fairs that brought together buyers and sellers from far-off lands. It was a famous meeting place for bankers and traders, and it was the de facto financial center for princes, dukes, and others who governed the numerous territories surrounding it. The region and the city were nominally ruled from Vienna by the Holy Roman

Empire, itself a vast conglomerate of northern and central European principalities.

Mayer's first exposure to this complex, fragmenting empire came in occasional apprenticing tasks for his father, such as sorting out coins at the Frankfurt trade fairs. Tragically, when Mayer was twelve, both of his parents died in one of the smallpox epidemics that swept through Europe. He moved in with relatives, who soon decided to send him to Hanover to apprentice with Wolf Jakob Oppenheimer, a prominent Jewish banking house where Mayer's late father probably had some connections. It was with Oppenheimer that the young Rothschild received his first glimpse of how foreign trade was financed, and it was here that he learned about rare, historic coins from places such as ancient Rome, Persia, and the Byzantine Empire. At the Oppenheimer bank, Mayer came into contact with wealthy princes and other collectors, and he gained an understanding of what it was like to be a Jew in the service of powerful men. There is an expression for this in Jewish history: Court Jew.

The Court Jew

Through the ages, Islamic and Christian rulers often hired Jews to advise them on financial matters. In royal courts seething with political intrigues, Jews were seen as fringe characters with no political future and no personal agendas. Since the elites looked down on commerce, they left financial matters to Jews, who were often granted an exception to the laws against charging interest on loans. After centuries of persecution and exile, Jews sometimes had access to considerable knowledge of other lands and cultures, plus a web of family connections in the Jewish Diaspora across Europe and the Islamic world—connections that could prove invaluable to kings and ministers with more limited networks. Court Jews gained a reputation for adeptly trading on gossip, skillfully parlaying one important relationship into another, and marrying well to expand their wealth and influence. In one way or another, Mayer Amschel Rothschild would embody all of these characteristics.

Whether it was the Abbasid caliphate in the ninth century, the Syrian and Egyptian courts in the twelfth and thirteenth centuries, Catholic Spain in the fourteenth century, the Ottoman Empire of the fifteenth century—whatever the period, Jews were found counseling potentates. They were major figures in the management of state finances, and they were also called on to raise funds for the armies and navies of the day. Nevertheless, they were subject to the whims of their masters. "Lengthy terms of imprisonment, cruel torture, and the expulsion of wives and children were fates suffered by not a few [of them]," wrote historian Michael Graetz. In the late seventeenth century, for example, Samuel Oppenheimer had served Emperor Leopold for three decades as supplier to the Austrian imperial army and grew powerful enough to personally finance the debts of the state, but he landed in jail after major snafus in supplying the troops. His subsequent bankruptcy caused financial convulsions throughout Europe. Shortly before Rothschild's time, Jud Süss Oppenheimer (1698–1738) had risen to become administrator of the far-flung Württemberg domains but was hanged for treason when the political winds shifted against him.

When Mayer was coming of age as a financier, the Holy Roman Empire was well behind England and Holland in financial sophistication, and its fractured Germanic region had no central institution like the Bank of England or the Bank of Amsterdam to standardize or stabilize currencies. This made it difficult for Germanic rulers to raise funds and finance armies, forcing them to turn to private entrepreneurs and financiers—often Jews. In the broader region, Jewish families became prominent financiers, including the Oppenheims of Cologne, the Bambergers of Mainz, the Habers of Karlsruhe, and the Warburgs of Hamburg. In Frankfurt, the Sterns, Speyers, Erlangers, and Seligmans stood out.

Of course, not all the great banking families were Jewish, and what was to become the House of Rothschild was but the latest of a long, grand tradition of family-owned banks in general. Some of the prototypes can be found in the thirteenth- or fourteenth-century Italian cities of Pisa, Florence, Venice, and Genoa, where banking families such as the Ricciardis, Bardis, and Peruzzis financed kings, trade, and war; or later in the fifteenth through seventeenth centuries with the renowned Medicis. These Italian families also developed letters of credit, deposit accounts, and new financial instruments such as bills of exchange that allowed people to trade precious metals using slips of paper rather than the metals themselves.

Mayer Amschel Rothschild followed the pattern of the court Jews and of previous great banking families, except that the company he created would grow bigger, more powerful, more closely coordinated throughout Europe, and more durable than any before or since. At a time when capitalism was replacing the old feudal economic order and providing seemingly unlimited opportunities for clever financiers, the Rothschilds would tower over their many competitors for most of the nineteenth century.

The Coin Dealer Gets a Break

Mayer returned from his apprenticeship with Oppenheimer before his twentieth birthday. Operating out of his crowded house, and partnering with Kalman, his invalid brother, Mayer set up a business trading rare coins, medals, precious stones, carved figures, and antiques. He sent out the first of the elegant leather-bound catalogs that he would distribute for the next twenty years, bringing him into contact with many wealthy customers. These catalogs were early evidence of his extensive due diligence on the history of the artifacts he was selling, as well as his meticulous attention to the art of presentation and to the precise requirements of his rich clientele. On the days when he could freely leave the ghetto and on others when he obtained a pass, he was in constant motion, traveling by stagecoach to see customers in Darmstadt, Mainz, Wiesbaden, and other capitals of surrounding principalities. He was building an impressive customer base that eventually included the king of Bavaria and the Duke of Weimar. A big breakthrough—maybe *the* big breakthrough of his life—came early. It stemmed from his ties to General von Estorff, who had been impressed by Mayer when he was a teenage apprentice at Oppenheimer and had followed his career since. Estorff was close to Prince William, the son of King Frederick II of Hesse-Kassel, who held the title of "landgrave" and was one of the wealthiest men in Europe.* In 1764, King Frederick appointed Prince William ruler of the tiny principality of Hanau-Münzenberg, where the prince was eager to establish a major library and coin collection. He turned to Estorff, who set up a highly unusual meeting between a royal figure and a collector from the Frankfurt ghetto, Mayer Amschel Rothschild. The Jewish merchant won a role as the prince's occasional buying agent at the Frankfurt trade fairs (where royalty would not stoop to handling cash transactions). The deals remained small but increased in frequency.

In 1769, after several years of diligent service, Rothschild persuaded Prince William to grant him the title of court agent, an honorific designation that gave Mayer added prestige in royal circles. Rothschild wasted no time posting the official emblem on the door of his house. While he was expanding his transactions

---

* "Landgrave" was a title of a nobleman who exercised sovereign power over his domain and reported directly to the Holy Roman Empire, not through a bishop or duke.

with William, Mayer met and married Gutle Schnapper, who lived just a few houses away. It was an arranged union, as was the custom. He was twenty-six years old; she was seventeen. Gutle came from a prominent family in the Judengasse and brought with her a good cash dowry and her father's extensive business connections. She would eventually bear nineteen children, ten of whom survived, five of whom were the sons who would constitute the founding partners of the future House of Rothschild.

Mayer's next big step was to become a key intermediary between the German prince and British banks. The roots of this deal were planted when Prince William started renting out conscripts from his principality as mercenaries to the British as they were fighting the American Revolution. London paid well for William's soldiers, but it did so in British notes that had to be redeemed at a British bank. At that time British banks had no branches outside of England, so William dispatched German merchants to redeem the notes for cash in London, then exchange them for local German currencies. These middlemen kept a small percentage (the "spread") as a fee for delivering the cash to William.

It took Mayer some time to persuade William to grant him a role in redeeming British war notes. He succeeded by first winning the trust of Carl Friedrich Buderus, a rising star in Prince William's Treasury. Mayer helped Buderus to invest his own funds and was successful enough that Buderus became an ardent supporter, lobbying the prince on Mayer's behalf. Befriending and enriching top aides to royalty became an indispensable element in the rise of the Rothschilds, and in their ability to remain so prominent for so long.

Before granting this Jewish banker a role in the critical war bond trade, William checked Mayer's track record with clients, all of whom praised his attentiveness and his extreme reliability. What puzzled William was that Mayer did not seem to live in a manner that reflected the financial success his references described. Either he was hiding his earnings or reinvesting everything in the business. He was in fact reinvesting, another hallmark of the Rothschild formula for the approaching nineteenth century.

Once William gave Mayer a share in redeeming British notes, Mayer pressed constantly for more, offering to take a lower commission in return for more volume. This approach would guide the family from Mayer through his sons: accept thin margins to squeeze out rivals, profit on volume, then raise prices when the competition was eliminated. There was, of course, nothing original in this time-worn formula, except that Rothschild protected his monopoly position through exceptionally deep personal relationships with customers.

When King Frederick II died of a stroke in 1785, the throne of Hesse-Kassel passed to Prince William along with the landgrave title, and Rothschild found himself connected to one of Europe's greatest fortunes. The prince's treasure included English bonds, art, and land, and Rothschild leaned even harder on Buderus to win more business from the new landgrave.

Becoming International Bankers

Mayer was expanding in other directions as well. He had become a general merchant buying and selling English wools and cottons, coffee, sugar, rabbit skins, and other merchandise. He extended loans to shopkeepers and provided credit to some of his coin customers. By the early 1780s he was able to buy a bigger house in the Judengasse that had two hidden cellars, one accessible by a trapdoor through the vestibule, the second through a door concealed by a false wall—features that would one day prove critical to Rothschild's survival. Still, living conditions were extremely cramped for his growing family. Closets and cupboards everywhere were jammed with dishes, clothes, and business papers.

Behind the house was a nine-foot-square room that became Rothschild's first business office.

The European political scene was in constant turmoil. During the early decades of the nineteenth century, European leaders were constantly seeking to redraw national borders through war and diplomatic maneuver. Napoleon was reasserting imperial control over France in the wake of the 1789 revolution and working to extend his power across all of Europe. His campaigns turned the entire Continent into a theater for his wars with England, Austria, Russia, and Prussia, the entire scene being characterized by shifting alliances and counter-alliances. The Rothschilds saw in this chaos almost limitless opportunity to help governments raise money for armies, to refinance war debts, and to help nations circumvent trade embargoes.

Rothschild was entering the emerging world of international merchant banking. His rivals in this new field included Hope and Co. of the Netherlands and Baring Brothers of the UK. These merchant-banking pioneers competed to develop international connections that could provide the best access to information on the changing politics, customs, laws, and economic rhythms of a continent in upheaval. Mayer and his five sons would navigate this world of wartime intrigue and danger by creating a finely honed multinational team with unmatched financial sophistication and the secretive culture of a smuggling operation. In this setting the Rothschild clan would first work with their original Prussian client, Prince William, and his allies in Austria and England. But they would also find a way to play all sides, taking on the French as a client at times, too.

Backing All Sides in the Napoleonic Wars

Napoleon started to move against France's neighbors in 1792, when Mayer was forty-eight. To protect his home city from the French, Rothschild secured funding and supplies, including food, uniforms, and packhorses for an Austrian force to defend Frankfurt. He also made sure that Austria's soldiers received their wages, all of which earned him substantial commissions.

Meanwhile, Prince William had decided to support England against France, in return for a subsidy of 100,000 British pounds, which London paid in paper IOUs. William wanted a local intermediary who would buy these IOUs, charging the smallest possible fee. Mayer won a good deal of this business, further binding him to his wealthy German client and deepening his ties to English financial institutions that had to redeem the IOUs.

As he maneuvered to profit from the fallout of the French Revolution, Mayer was also staying one step ahead of the industrial revolution then accelerating in England. The English wool and textile industry had become the envy of Europe. Rothschild became a major importer of English cloth, eventually eliminating the English middlemen and their large commissions by sending his third son, Nathan, to establish a Rothschild-owned operation to buy textiles in Manchester, England, and to transport them directly to the family operations in Frankfurt.

Nathan had never been to England and at first spoke no English, but he would build the first and most important branch of the Rothschild network in a nation that was the foremost military and financial power of the day. Working closely with his father and other family members back in Frankfurt, Nathan not only purchased cloth but had it customized for different buyers in Germany, as well as in Switzerland, Austria, and Russia. The Rothschild family arranged for sea and land transport and for trade financing, and soon expanded to include trade in goods from the British colonies, including indigo, tea, dried fruit, sugar, and coffee. It was a private, family-owned company, to be sure, but in its scope it had shades of the East India Company.

Nathan had learned from his father: He would do nearly anything to acquire a good client and demonstrated the entrepreneurial energy and logistical capability to build a network of suppliers all over the UK. He would beat out his rivals on price and make profits on large volume. Like Mayer, he was unusually secretive in this work, with a knack for making important people trust in his reliability.

Early on Nathan attracted the attention of Levy Barent Cohen, one of England's richest Jews, who would serve as his mentor. Within several years Nathan married Cohen's daughter, Hannah, gaining in both wealth and access to key British officials. Tripling his capital in just a few years, Nathan gradually became a merchant-prince, handling a sizable part of London's trade.

The Bond Market Pioneers

Even as the family expanded its roles in the European cloth trade and the British war effort, Mayer opened another line of business. Working with the government of Denmark, he and his eldest son, Amschel, who also worked out of Frankfurt, designed a complex bond issue. Rothschild purchased all of the available Danish bonds for immediate resale to Prince William, who remained an anonymous buyer. William's appetite for sovereign bonds seemed unlimited, and Mayer managed a series of similar deals on behalf of other governments for William and other buyers, including for the German princedom of Hesse-Darmstadt. Mayer was transforming his trading operation and foreign exchange house into a multinational bank of significant scope and scale. The size, complexity, and proliferation of these bond issues were startling. By 1810, Mayer had begun making loans to French entities in the German region—Britain's enemies—figuring that the safest way to navigate the political fault lines of Europe was to have allies on all sides.

During the war between France and the Austrian alliance, Prince William was caught in the middle and tried to play both sides by claiming neutrality. Nevertheless, Napoleon accused him of disloyalty and in 1806 ordered French troops to occupy Hesse-Kassel, William's kingdom. William was forced to flee, leaving his fortune, including his gold, silver, rare coins and metals, bonds, titles, deeds, and contracts, plus all his financial records in his home, which was vulnerable to confiscation by the French occupiers. Moving from location to location to avoid capture by Napoleon's men, William increasingly relied on his adviser Buderus and, through him, the Rothschilds, to manage and protect his holdings and to collect the payments on loans he had made across Europe. Mayer often assigned Amschel to travel with the prince in order to help keep funds flowing to the prince's moving entourage.

The Rothschilds were so successful in building William's fortune under these trying circumstances that he eventually made the family his exclusive financial agent. Mayer transferred much of William's assets to London, where Nathan would invest them in British securities. Already a big player in London, Nathan became a market mover. The Rothschilds also became major creditors, negotiating loans all over the Continent, using William's assets while protecting his anonymity and dazzling the prince with their ability to account for every cent of his mobile fortune.

In the midst of a war with constantly shifting alliances, Mayer kept a low profile and conducted business in private, almost always face-to-face with his clients. While most bankers expected customers to come to them, the Rothschilds went to the clients, traveling in the primitive wagons of the day. Mayer was in Hesse-Kassel to meet with Buderus and William so often that Jewish merchants and bankers there protested that he should have to pay the same taxes they did (which Mayer refused to do). Eventually the family started to travel in stagecoaches built with secret compartments to hide documents. They employed codes in their extensive correspondence to one another, occasionally writing German words using Hebrew letters in ways only they could understand.

As the French noose continued to tighten around William, the risks grew for the Rothschilds. It was inevitable that General Lagrange, the new French governor of now-occupied Hesse-Kassel, would find at least some of the wealth the prince had hidden in various castles. In one incident, the French came across several chests of William's jewels and securities in the basement of an abandoned castle, but Rothschild was able to bribe Lagrange to take only a small percentage of the loot and allow the rest to escape. It is not clear why Lagrange would settle, but Mayer may have guaranteed Lagrange bigger benefits down the road. In another case, the Rothschilds persuaded debtors to continue making payments to William, even after Napoleon offered these same debtors a reward to send payments to the French Treasury instead.

Infuriated with William's advisers for dodging their efforts to confiscate his fortune, the French arrested Buderus. They squeezed him for information on the whereabouts of William and the Rothschilds, but Buderus refused to talk. French pressure scared William into moving to Vienna, where he could avail himself of the protection of the Austrian emperor, who welcomed him but, knowing of his fortune, also wanted a loan. The Rothschilds arranged a meeting, at which the parties could not come to terms on a loan deal. However, the direct ties established between the Rothschilds and the Austrian court would prove crucial to the family in the years ahead. In 1800, Mayer and Amschel were made official agents of the Austrian court. Mayer Amschel Rothschild was by now perhaps the wealthiest man in the Judengasse, with German customers from Hamburg to Berlin, and international customers from London to Paris, and Amsterdam to Vienna.

The Blockade-Runners

The war saw power shift from France to its rivals and back. In 1806, Napoleon was ascendant in the German regions, establishing a French Confederation of the Rhine that included occupied Frankfurt. The French set up a blockade against English goods entering the confederation, and Frankfurt merchants responded by organizing smuggling operations. None of these blockade-runners were more enterprising than the Rothschild family.

Nathan mobilized his own fleet of ships to transport British goods to the coasts of Germany and Holland, where his brothers stood ready to move the merchandise onto the Continent. An elaborate clandestine messenger service linked all the brothers to their father in Frankfurt.

In fact, by 1808 Mayer and his sons were taking huge risks in moving contraband and dealing with William behind Napoleon's back—with all this activity directed by Mayer from his house in the Judengasse. France had at this time appointed Baron Karl von Dalberg, a German prince-bishop who had collaborated with Napoleon, to be the governor-general of Frankfurt. Dalberg, a calculating pragmatist, had limited personal financial resources, a void that Mayer seems to have filled with loans and investment advice, cementing a critical new relationship in high places.

Dalberg would help save Rothschild. The French authorities had been watching Mayer and his sons for some time, intercepting their correspondence and keeping tabs on visitors to their house. But the Rothschilds were traveling so often, each in a different direction, that it became nearly impossible for the French authorities to keep track of them. Mayer and his sons had also become masters of deception, using codes, decoys, and informers to alert them when French inspectors were nearby.

In June 1808, the investigators arrived at Rothschild's house. Mayer claimed to be bedridden and unable to talk to them, so they grilled the rest of the family for an exhaustive week, with each family member, including Gutle, under house arrest in a different room. Because of the advance warning, Mayer and his sons were able to hide in their cellars all the records and deeds of ownership, as well as several trunks of valuables belonging to William. Eventually, Dalberg intervened to call off the investigation, which discovered no real evidence. Then Mayer moved to ingratiate himself more closely with the French. In 1811, he arranged a generous subsidy for Dalberg. And the grateful governor, in turn, arranged a French passport complete with all the right introductions for Mayer's son Jacob, who set up a new branch of the family bank in Paris.

Mayer now had foreign branches on both sides of the running war between England and France and had become one of the wealthiest Jews not just in the ghetto but in the bustling commercial hub of Frankfurt. He was managing Prince William's vast portfolio on the fly and was instrumental in helping England to finance its armies and its allies. Nathan, in particular, was carving out a very profitable niche in smuggling currencies and gold past the French, gaining a reputation as a mastermind of undercover operations.

In 1814, with British forces stranded behind enemy lines on the Iberian Peninsula without supplies or paychecks, Nathan stepped in and offered to smuggle gold to the English troops. The government in London, lacking any other plan, agreed so long as Nathan took all the risk. The Rothschilds pushed ahead, buying gold on the open market, much of it from the British East India Company, and shipping it to ports at Boulogne, Dunkirk, and Gravelines, where Rothschild's sons took personal responsibility for transporting the gold to Frankfurt or Vienna, then on to General Arthur Wellesley (later to become the Duke of Wellington), the British commander in Spain. Some of the gold was deposited in local Spanish and Portuguese banks with which the Rothschilds had built close relations, and on which Wellesley could then draw funds in local currencies. This complex operation combining clandestine logistics and furtive foreign exchange transactions was the kind of business at which the Rothschilds excelled. It would provide the template for future services to London, in which the Rothschilds helped England deliver huge financial subsidies to its allies, and would help carve the Rothschild legend in stone.

Creating a Global Market

At this time Mayer was also laying the basis for major advances in the global financial markets. Money had already been moving across borders in growing volumes and at increasing speeds, but the House of Rothschild built on this incremental process of change in revolutionary ways.

It is difficult today to imagine how tough it was to transfer funds across borders in the early 1800s, when the job had to be done by couriers carrying heavy loads of gold, silver, or currency, a means of exchange that was awkward, slow, and vulnerable to bandits and pirates. Even national financial markets were exposed to these risks, and the idea of an integrated international market was still far off. It would be a stretch to suggest that the financial techniques developed by Mayer Amschel Rothschild and his sons were completely original. Financial innovation had been occurring for centuries all over the world, including in China, Persia, Venice, Genoa, the Dutch market, and in England itself. Moreover, markets are close-knit networks of buyers and sellers, and they are highly adaptive precisely because so many players have an incentive to continuously push innovation. What the Rothschilds did at a minimum was to serve their highly placed clients—extremely wealthy individuals, influential government officials, industrial titans, almost all operating across borders—by offering them the latest financial products, tailored for them, in large amounts, and in the most discreet or confidential manner that the client preferred. In other words, the breakthrough was less in the substance of the deal than in its manner of execution. The other unique aspect of the Rothschild family was its sheer longevity. The House of Rothschild operated as an extreme powerbroker for nearly a century, during which time it shaped the character of high international finance as no other banking institution had.

For example, the Rothschilds were among the first banks to allow people to deposit money in their home country, which could be withdrawn by a client or representative in a second country, in local currency. Thus, a British company could deposit pounds sterling in London for a customer to withdraw them at a Vienna bank in Austrian schillings. The House of Rothschild constituted a clearinghouse for international finance that began with three money centers—London, Frankfurt, and Paris—and was eventually extended to Vienna and Naples in the 1820s, with one of the five brothers residing in each city.

The Rothschilds used this clearinghouse to pioneer advances in the international bond market, with significant financial, political, and social implications. These advances worked in two ways. A state such as Prussia could issue a bond in London, using that bond to borrow British pounds from British citizens, with the promise to repay them interest in British currency, in London. That possibility vastly enlarged the financial prospects of Prussia, which was small and relatively poor, and which could never have been able to tap its own citizens for all the funds it needed, both because they couldn't afford to lend the government more money and because there were not enough of them to create a large enough loan.

The Rothschilds were the first family bank that had the resources to make reliable interest payments from their own local bank branches across the continent to customers of any nationality. They became a true international bank, one of the first. One reason the

Rothschilds were able to make this system work was the resources that the family had amassed, but another was the level of trust they had built up by cultivating personal ties to public officials across the Continent. That's how finance works. If you are confident your funds are in good hands, you'll leave them there. And, of course, the converse is true.

Moreover, the close links among the brothers and the five branches allowed them to capture more information quicker than their competitors could, and therefore to obtain the best interest rates and the best currency values throughout Europe. The Rothschilds could, for instance, take advantage of pricing discrepancies in Europe in ways few could match. If a Danish bond was selling in, say, Paris at a slightly higher price than in Naples, the family was adept at spotting the opportunity for arbitrage, buying up the lower-priced bonds in Naples and selling them for a profit in Paris—pocketing exceptional profits on large volumes.

Soon the family and its multinational reach allowed the Rothschilds, working on behalf of a government or a wealthy individual, to float a loan that could be placed in *several* national markets with the interest being paid to bondholders in several different cities, each in their local currencies. To be able to do this required a deep understanding of every market, knowledge about the shifting values of currencies, shrewd assessment of the creditworthiness of borrowers and lenders, and a logistical capacity to organize such complex operations. The Rothschilds led the development of a sophisticated international bond market not because no one had thought about doing it before but because no one had the capacity—the internal coordination, the logistical capability, the relationships, and the trust among all partners—to make it work in such volumes and with such reliability.

The family was constantly innovating, adding features to the system. In 1818, they created a special bond fund for Prussia—called a "sinking fund"—with an independent supervisor who would buy Prussian bonds on a regular basis, thus reassuring investors that there was some cash up front to amortize the bonds as they came due. In another transaction, this time for Russia in 1822, an additional feature was added. Investors could collect interest and principal in either British *or* Russian currencies. These innovations may seem minor or technical, but they propelled the international bond market. "The significance of this system for nineteenth-century history cannot be overemphasized," wrote the Rothschilds' preeminent chronicler, Niall Ferguson. "For this growing international bond market brought together Europe's true 'capitalists': that elite of people wealthy enough to be able to tie up money in such assets, and shrewd enough to appreciate the advantages of such assets compared with traditional forms of holding wealth [like land]." Bonds could be sold more quickly than land because of the emergence of a lively pan-European trading market, and could accrue large capital gains in relatively short periods of time.

There was another aspect of this evolution that greatly played to the Rothschilds' strengths. The value of a country's bonds fluctuated with the perception of that nation's financial ability to service the debt. The ingredients of creditworthiness in turn relied on investors' views about everything from a nation's underlying revenue streams, such as taxes and tariffs, to its political stability. Those bankers who advised governments on when and how to issue bonds needed to have their trust because these financiers had to have the ability to frankly discuss with ministers what policies—fiscal, trade, investment, political appointments—would be required to be able to issue bonds at favorable prices.

The Rothschilds would for generations cultivate the official ties that put their fingers on the pulse of all that was happening in Europe—everything relating to economic and political developments that could affect the value of existing bonds or the prospects for raising more money with new bonds. They could provide

English investors unparalleled advice on, for example, Prussian bonds because of their intimate knowledge of the Prussian political situation and of the current price of Prussian bonds in cities such as Vienna and Naples. Gathering news became a Rothschild obsession, and it drew the father and brothers closer to the inner circles of power; the closer they got, the better their insights became. The rarefied network of clients and information continued to grow.

By the early 1800s, the House of Rothschild was constantly upgrading its internal courier service to make it faster and more secure. They built redundancy into the system, using pigeons and stagecoaches to deliver copies of the same message on different routes. They launched a fleet of private ships that carried nothing but Rothschild family correspondence and sailed even in violent storms. The family codes grew more elaborate, starting with the Hebrew script and advancing from there. In 1814, for example, there were frequent lags in the Frankfurt post office between the time a letter was received at the post office and the time when it would leave for delivery to its intended recipient. Amschel asked his brothers to send exchange-rate reports in colored envelopes: blue if the rate rose, red if it fell, so he could send a messenger to the post office to peer into his incoming mail bins and get a read on the news a day before the letter was actually brought to him. Because the Rothschild couriers developed the most advanced communication system in Europe, it came to be used by top political leaders, including French prime minister François-René de Chateaubriand, Austrian foreign minister Klemens Wenzel von Metternich, and England's Queen Victoria.

Mayer Lays Down the Family Law

Sometime around 1810, Mayer built a new four-story house in the Judengasse. He would run his business from there, and he would die there. His roots were in the Judengasse, and leaving would probably have been unthinkable, especially for Gutle. Even as her sons became the leading bankers in Europe, with mansions all over the Continent, she would stay in this last home, where she outlived Mayer by thirty-seven years.

The same year, 1810, Mayer created a partnership agreement that gave each son an equal share in the family business and was to bind them after his death. Each could commit the firm to new business, but no partner was to engage in business independently of the other. Only the sons and their male offspring could become partners. Other family members should not be brought into the business, even the husbands of the Rothschild daughters. If any of the partners died, each would have to renounce the right of their widows to contest the amount to be given to their estate. Mayer probably realized that other great banks had lost their lofty positions when the founding family began to break up, and he was determined to avoid that fate. The agreement would run for ten years, as a formal matter, but Mayer's influence was such that it would hold his sons together for many decades.

When Mayer Amschel Rothschild died on September 19, 1812, at the age of sixty-eight, he was buried in the ancient cemetery at the bottom of the Judengasse. On his deathbed, he said to his eldest son, Amschel, "Keep your brothers together and you will become the richest men in Germany." The sons would in fact outdo their father's forecast, becoming so wealthy that aristocracies across Europe had to grudgingly accept a Jewish family into their ranks. It was a long way from the Frankfurt ghetto where they were all born and raised.

Powerful Beyond Imagination

The Rothschild sons remained strikingly true to their father's vision, no more so than with regard to the rigid rules that required the reinvestment of profits back into the firm. This gave the family a capital cushion, allowing them to ride out bad times that killed off rivals and to bail out troubled family branches. In the 1840s, three decades after Mayer's death, for example, the London operation rescued the brothers in Paris and Vienna, after economic conditions and financial misjudgments put those branches in jeopardy.

The Rothschilds became an increasingly dominant force in international finance after Mayer's death, owing to the sheer scale and sophistication of their operations. In 1815, just three years after Mayer died, the House of Rothschild and Baring Brothers, its chief rival, had more or less the same capital—between 375,000 and 500,000 British pounds. In the next ten years, the resources of the five brothers would grow to more than ten times that of Barings making the Rothschild bank the world's largest, as measured by capital.

By the 1820s, the Rothschild sons were expanding the empire from their bases in Frankfurt, London, Paris, Vienna, and Naples. The most influential bank in Europe was now the most influential bank in the world, and as the Napoleonic wars subsided, the Rothschilds turned from financing war to funding the peace. As the industrial revolution was gaining speed, for example, they became the leading financiers of the European railway system and of mining companies around the world. They helped France and other governments pay off war indemnities and rebuild their dilapidated infrastructures. They acted as unofficial diplomats in relaying confidential information between foreign and finance ministers. They extended personal loans to the most important officials of the day, such as Prince Metternich and his aides. The Austrian prince did not always agree with the family, but he came to them time and again for advice and financial help. In 1823, for example, Metternich asked the Rothschilds to help him loosen the repayment terms on Austria's debts to Britain, which they successfully did.

The Rothschilds built mansions, while collecting noble titles and awards from governments all over Europe. They extended the geographical reach of their financial domain to Russia and Latin America, and continued to build on their extraordinary financial clout. In 1825, they bailed out the Bank of England, and in 1855 they did the same for the Bank of France. They lent directly to states such as Prussia and helped arrange loans from capital-rich nations such as England to those that needed funding. They helped the Austrian Empire to preserve its established borders by financing the dispatch of troops against rebellious provinces. With the agreement of major powers, they also underwrote the independence of countries like Belgium, which was carved out of the Netherlands in 1830.

Today, it is sometimes said that a nation's elite can be divided between men of power—usually top political figures—and men of money. It is often a fuzzy distinction, but for the Rothschilds, there was no difference: they were clearly both. They had political influence and they had outsized fortunes, and the two were intertwined. The Rothschild mythology took on such dimensions that they were often seen as more powerful than royalty. When Nathan died in 1836 he may have been the richest person in Britain, and today he remains arguably the most influential man in British financial history. His funeral was attended by not just the ruling financial elite in England but by ambassadors from throughout Europe. When his brother James—the last of the five sons—died in 1868, an observer said that the funeral was "more like that of an emperor than of a private individual," with some forty thousand people passing through the drawing room and six thousand watching from the courtyard.

In the end, the world of finance influences more than investors' financial returns or the profitability of an industry. The titans of banking shape the political and social structure because they decide who gets money and at what price. They often determine who goes bankrupt and who survives. They apportion funds among governments, and among big and small businesses, because they decide who is creditworthy and who is not. They make the rich richer, but they also discover the up-and-comers with new ideas. In fact, by laying the foundations of the international bond market, which today is among the largest of all financial markets, the Rothschilds also helped undo the nineteenth-century status quo. They created a class of investors who could amass great wealth, even if they did not possess hereditary claims to land. This spelled the end of the long centuries in which power and wealth were concentrated in the hands of the landholding aristocracy. "The emergence of global capitalism," wrote historian Fritz Stern, spread power "from men of power and birth to men of wealth and aspiration."

Finance is the circulatory system of any country, as well as of the global economy. A smoothly functioning system is central to national growth and prosperity, to international trade, international economic growth and development, and to fewer global crashes than otherwise would take place. But few financial systems can operate in a vacuum; given porous borders, they are always linked to, and influenced by, other national systems. Thus, they are truly borderless. They are therefore the very definition of globalization. The *global* financial system allows the world to grow faster because it channels savings—excess money—into places where the money is needed, to facilitate trade, for example, or to build roads, ports, bridges, or new companies. Global finance is also precarious because problems in one part of the world can spread like a contagious disease. Mayer Amschel Rothschild was one of the great pioneers of figuring out how to expand the possibilities of global finance and also how to deal with the consequences of the periodic crises that it spawns.

The House of Rothschild helped governments with all manner of their needs. It enlarged their options, expanded their geographical reach, and created an international bond market that vastly enhanced new financing possibilities. Without the Rothschilds and others like them, the British Empire would have had a difficult time expanding as it did in the days of Robert Clive and the days of his successors—financing its conquests, enlarging its territories, constructing its elaborate infrastructure of highways, ports, and trading operations. Indeed, even today no single financial institution could match the power Mayer and his sons had, for they rolled up into one family operation what the International Monetary Fund, the World Bank, Goldman Sachs, and Citigroup *together* would have a hard time doing today.

Of course, the topography of global finance has changed dramatically since the heyday of the House of Rothschild. Today there are no family-owned global institutions of any significance. Huge publicly listed banks, asset managers, private equity firms, hedge funds, and insurance companies dominate and operate around the globe. Regulators are powerful and ubiquitous.

But some things remain constant. Global finance is not for the fainthearted; it is too complex, too volatile, too dependent on uncontrollable political events within and among countries. Global finance also relies heavily on trust between the suppliers and consumers of money. Mayer Amschel Rothschild and his sons were the essence of trustworthiness. Garnering trust has many dimensions, one of which is accountability. If a banker is held responsible for his mistakes, then his customer has more confidence in him. The Rothschilds could not hide behind public corporations that today essentially shield top individuals from the legal liability of big mistakes, as we have seen in the failure of prosecutors to charge and convict senior financial officials in the global crisis of 2008–9. Unfortunately, few institutions today can command the confidence that the Rothschilds engendered and that is essential to a healthy global economy.

Other common elements between then and now: Rothschild was a master of secrecy; today, despite regulatory and investor pressure for more transparency, many financial institutions' activities are still considerably and deliberately opaque to governments and the public. Rothschild was not above representing all sides to a transaction—what we would now call "conflict of interest"—and neither are some of today's most illustrious banks. Rothschild thrived on influencing governments with his insider status; that game is being played today in spades by the top financial institutions.

A healthy global financial system is essential to meeting the aspirations of billions of people entering the middle class; to fund the growth of cities; to build the global networks of roads, airports, ports, and telecommunications the world needs; to underwrite the expansion of health care, education, and other vital social services; and to raise the trillions of dollars that will be necessary to deal with climate change. The centrality of finance and the search for more and different kinds of funding and credit—the kind of innovation and expansion epitomized by the House of Rothschild—remain pivotal to our interconnected and expanding world economy. And if we look for the most important historical models for effective, enlightened global bankers, Mayer Amschel Rothschild is at the top of the list.

Chapter V

CYRUS FIELD

The Tycoon Who Wired the Atlantic

1819–1892

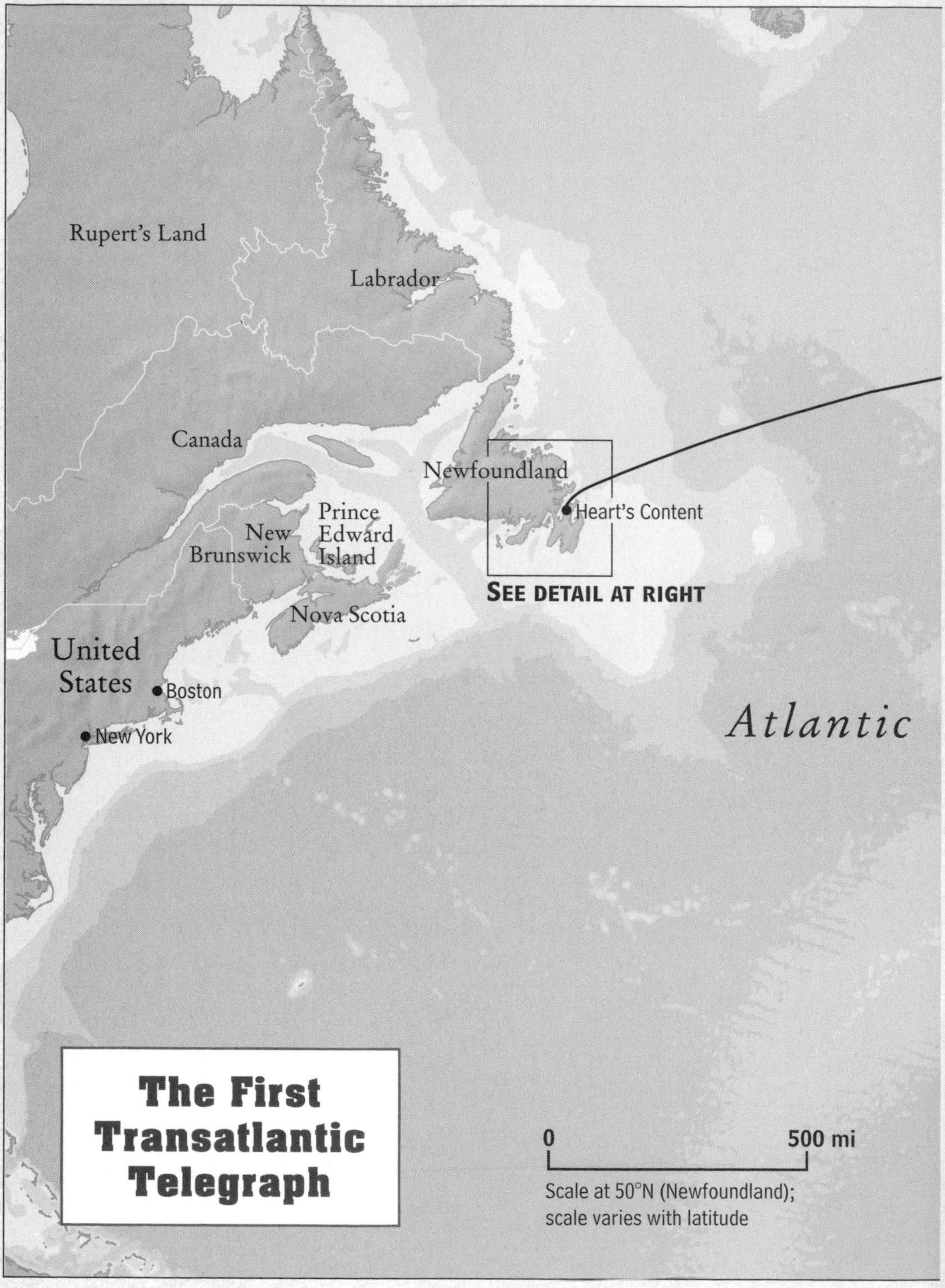

Rupert's Land
Labrador
Canada
Newfoundland
Heart's Content
New Brunswick
Prince Edward Island
Nova Scotia
SEE DETAIL AT RIGHT
United States
Boston
New York
Atlantic

The First Transatlantic Telegraph

0 500 mi
Scale at 50°N (Newfoundland); scale varies with latitude

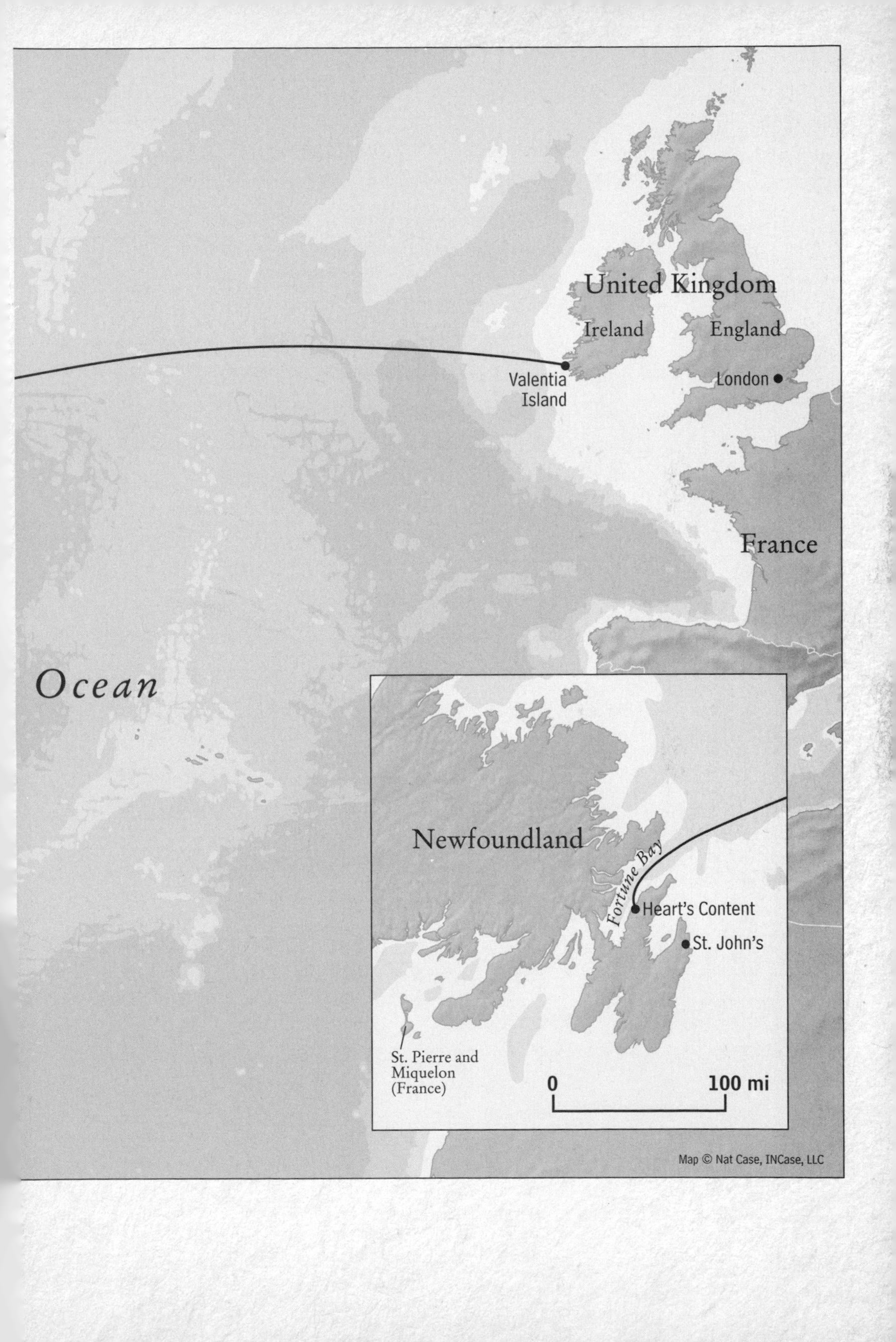

United Kingdom
Ireland
England
Valentia Island
London
France
Ocean
Newfoundland
Fortune Bay
Heart's Content
St. John's
St. Pierre and Miquelon (France)
0
100 mi
Map © Nat Case, INCase, LLC

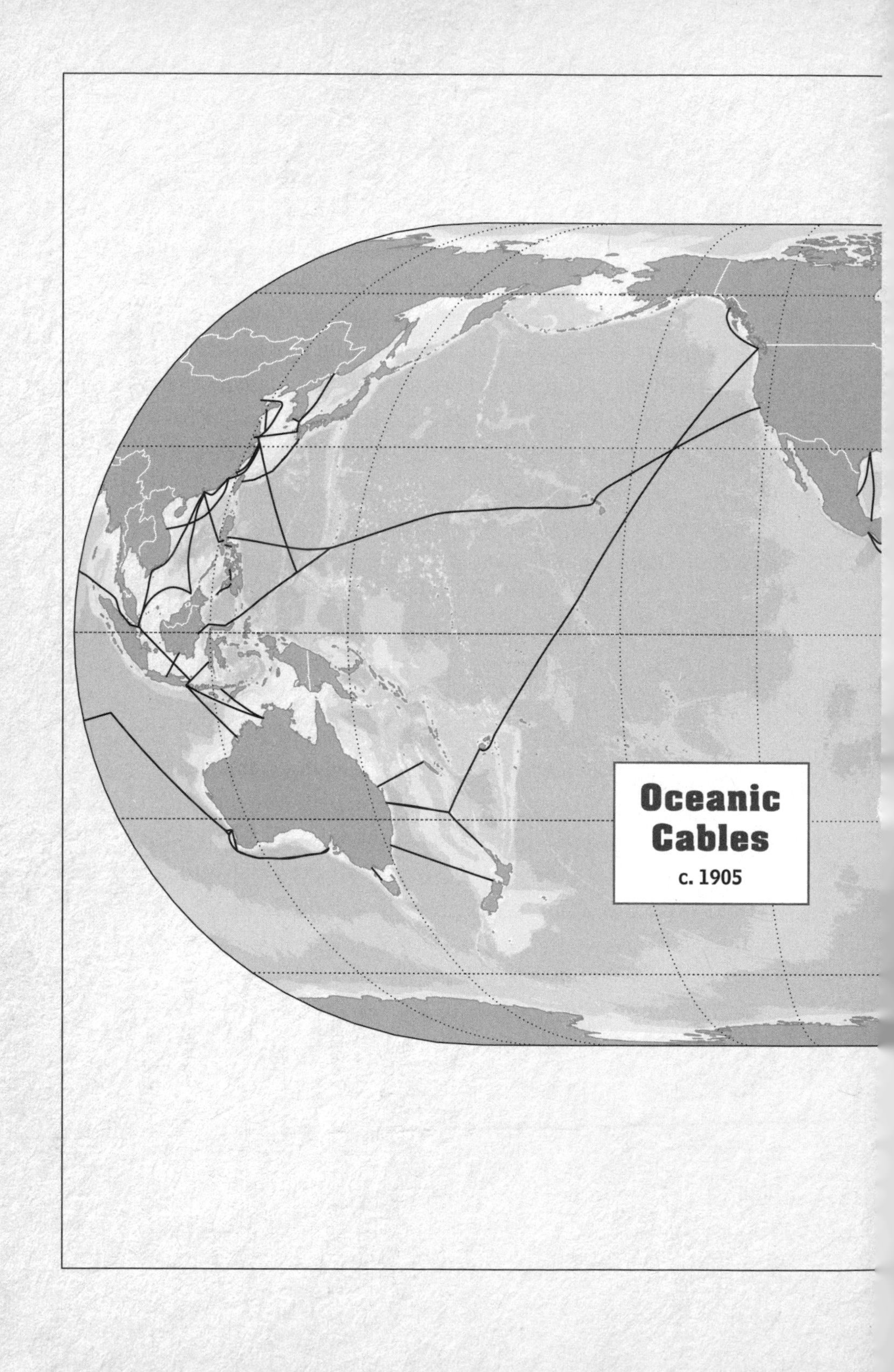

Oceanic
Cables
c. 1905

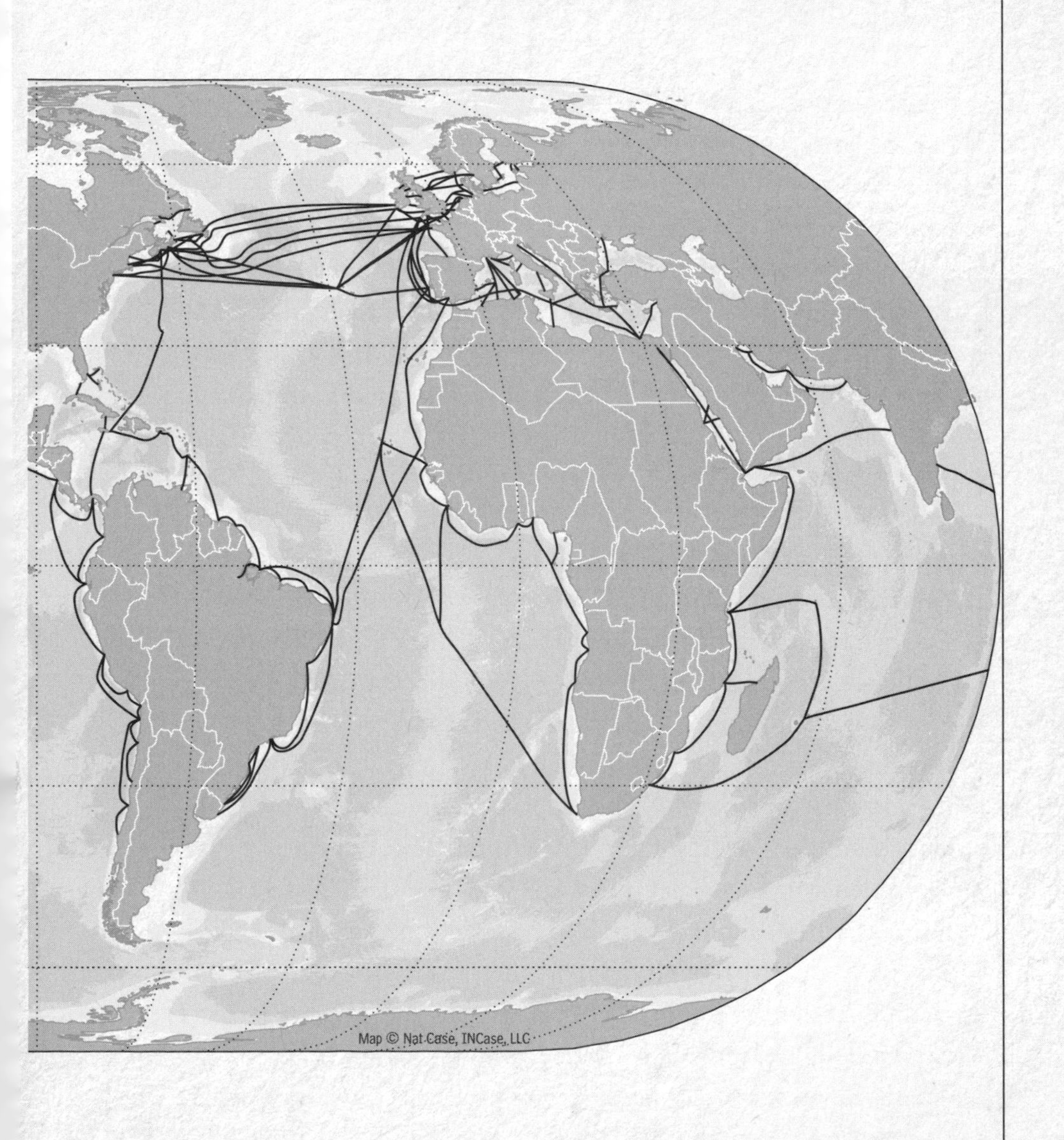
Map © Nat Case, INCase, LLC

In the winter of 1854, Cyrus Field, who had recently retired as a paper industry tycoon at the age of thirty-four, was looking for something new to do. His brother Matthew, an engineer, had introduced him to a young Canadian named Frederick Newton Gisborne, also an engineer, who was in Manhattan trying desperately to raise funds for a grand plan to link Newfoundland to New York by telegraph cable. It was not immediately clear to Field why he—a titan of America's commercial capital—should spend a fortune to establish telegraph contact with an island known mainly for cod fishermen.

In a meeting at Field's posh mansion in the Gramercy Park neighborhood of Manhattan, Gisborne laid out the rationale. He explained that ships from Europe reached Newfoundland considerably faster than they could reach New York City. So if news could be telegraphed from the port of St. John, on the east coast of Newfoundland, then across the island and on to New York, it would cut the two-week passage of letters across the Atlantic by about three or four days. Gisborne had secured land rights and funding from the government of Newfoundland, but he had run into problems trying to lay cable through the densely forested and deeply inhospitable terrain, and now he was out of money.

Cyrus could not see what his brother and Gisborne were so excited about. He knew little about eastern Canada other than that it was still under British control, and he saw no profit to be gained in the huge trouble and expense of running a cable to this British provincial outpost, particularly not just to knock a few days off a transatlantic route. Field ushered Gisborne out the door, then wandered into his high-ceilinged study and started looking for Newfoundland on his map of the world. What caught his eye was not the distance between New York and Newfoundland but the fact that the latter was so much closer to Europe than was New York. He wondered whether it would be possible to connect not only the United States to Newfoundland, but also Newfoundland to Ireland, thereby completing a transatlantic link that Gisborne had barely mentioned. A transatlantic cable could conceivably cut the communication lag between Europe and America not by just a few days, but from a matter of weeks to a matter of seconds. Now *that*, thought Field, was worth getting excited about.

Field could not have known that his epiphany would become an overwhelming personal obsession, turning this retired businessman with a background in fancy stationery into the man who orchestrated what still stands as arguably the single greatest leap forward in the history of global communications, with everlasting consequences for war and peace, diplomacy and commerce, and the spread of ideas and culture across borders. It is inconceivable that he could have imagined that what he would accomplish would become the forerunner of global telephone service, global radio, global TV, and, of course, the global Internet.

Field would of course have been aware that a telegraph network had been up and running within the United States since the 1840s and that it already linked the two American coasts with many of the cities and towns in between. He might even have read about the few underwater cables that had been laid along the European coast, from England to France, from England to Ireland, and from Corsica to Sardinia. However, these cables crossed shallow waterways, no more than 110 miles in width, and most provided unreliable connections.

A transatlantic cable was a vastly different undertaking. It would have to be about two hundred times as long as the most extensive underwater cable in operation. It would require ships of a size that did not yet exist, laying cables of a strength that had never been attained, at ocean depths that had never been measured.

Even thinking about an undersea cable between Newfoundland and Ireland was a step beyond the bounds of existing knowledge. "It was a bit as if someone in the 1950s, reading of the success of the Russian Sputnik, had decided to organize a manned expedition to Mars," wrote historian John Steele Gordon.

A Team of American Dreamers

Field decided to ask his brother Matthew, the engineer, whether a transatlantic cable was, in fact, feasible. Matthew posed two key questions that he said had to be answered before he could render an opinion: Was enough known about the depth and topography of the Atlantic Ocean floor to make it possible to lay a cable on it? And could a wire two thousand miles long carry an electrical impulse that would be strong enough to transmit messages from one end to another? Cyrus and Matthew wrote to two distinguished experts: Lieutenant Matthew Fontaine Maury, the head of the Naval Observatory in Washington and one of the leading oceanographers of the era, and Samuel Morse, the inventor of the telegraph in the United States. Both replied that the cable was theoretically feasible.

Field then started to think about acquiring the land and financial concessions Gisborne had won from the Newfoundland legislature. For that he would have to pay off Gisborne's debts, totaling $50,000, and offer him additional compensation. He met with Peter Cooper, a neighbor who had built a fortune in the glue, steel, locomotive, and telegraph cable businesses. Cooper was enthusiastic but said he would invest only if Cyrus could find several other partners to spread the risk. Field then recruited Moses Taylor, one of the country's largest importers; Marshall Roberts, a major shipowner; and Chandler White, another paper industry tycoon. These five men then decided to put up all the money themselves. While all saw the chance to expand their fortunes, several, particularly Cooper, were also excited about the social vision of linking the world through telegraph networks, improving international understanding, accelerating economic and social progress, and reducing the threat of war.

Without Field's contagious enthusiasm for the project, it is doubtful the other investors would have joined. None was an engineer or a scientist, and none had a clear grip on the challenges ahead. They trusted Field because they shared a background that reflected the optimism of mid-nineteenth-century America. They were self-made men who had already succeeded by thinking big and by seizing opportunities that came their way.

After meeting with Gisborne to get more details, the partners met alone for five successive evenings in Field's home, spreading maps on a big table and discussing what specifically needed to be done, as well as the costs that would be incurred. On March 10, 1854, less than three months after Cyrus Field first met Gisborne, they agreed to form the New York, Newfoundland and London Telegraph Company. It would acquire Gisborne's company, but only if the partners could first persuade the government of Newfoundland to let them take over the project. Cyrus Field, Matthew Field, and Chandler White then sailed for Newfoundland, where the government—probably figuring it was the best offer they would get, and also knowing that a successful venture would put their remote province on the world map—gave the Americans monopoly rights for cable connections to the United States and to Great Britain for fifty years. The province asked Field and his associates to guarantee interest payments on bonds it wanted to issue worth 50,000 British pounds.

On Sunday, May 6, 1854, the investors reconvened in Field's mansion at 6:00 a.m., the only time they could all gather on short notice. It took fifteen minutes to approve the charter of the new company and to collect commitments that totaled $1.5 million, the equivalent of about $40 million in today's dollars. Peter Cooper was made company president and Field was made chief minister without portfolio, since it was agreed that as the key deal maker and operator, he would be moving around too much to have any specific corporate responsibility or title. Gisborne was made an adviser and Samuel Morse was given a small interest in the company (in part because Field knew they would eventually need Morse's patents).

A Scheme Born of Faith and Ignorance

Cyrus Field was born in Stockbridge, Massachusetts, and descended from early English settlers in America. His father was an ordained minister, his mother a hardworking woman who had grown up in New England's small towns. He was the seventh of nine siblings raised in a puritanical household of modest means, for whom evening entertainment often involved theological discussion. Among his favorite passages of the books he was forced to read was this one from *Pilgrim's Progress*: "To *know* is a thing which pleaseth talkers and boasters; but to *do* is that which pleaseth God." Field's younger brother Henry said that Cyrus was conspicuous by his restlessness and his impatience. He was also orderly and meticulous, habitually making detailed lists of nearly everything—things to do, key historical dates, money owed. When Cyrus was twelve his father asked him to keep track of the family's finances.

At the age of sixteen, Field left home to become an errand boy for A. T. Stewart, the largest dry-goods dealer in New York City. Over the next three years he became a successful salesclerk, then left to become bookkeeper and general manager of his brother's paper mill in Lee, Massachusetts. He soon left the office to become a traveling company representative, energetically courting wholesale customers and demonstrating a gift for salesmanship. In 1840, he married Mary Byron Stone, a childhood friend, with whom he would eventually have seven children. That same year he left his brother's mill to become a partner at Root & Co., a New York City paper wholesaler, totally unaware that the firm was in dire financial difficulty.

Shortly after Field arrived, Root & Co. declared bankruptcy. The management gave Field the authority to settle with all the creditors for 30 cents on the dollar. It was a painful experience, but Field's straightforward and courteous manner earned the goodwill of the creditors, who, after all, knew he had not caused the problem. Armed now with experience at three different companies and a wealth of business relationships, he decided to open Cyrus W. Field & Co. in 1843. His plan was to fill a gap in the wholesale paper business—selling fine paper for upscale books, stationery, and financial transactions; high-quality printing of bonds, prospectuses, and contracts; and customizing stationery for important clients. America was booming, its appetite for luxuries was growing, and Field & Co. prospered. Between 1846 and 1849, company sales totaled around $1 million, an extraordinary sum for the time. Cyrus W. Field & Co. would become the largest wholesale paper company in New York City and Field himself eventually became one of the fifty wealthiest men in the city.

He proceeded to live in accordance with his substantial means. He moved to the Gramercy Park mansion and hired a French interior decorator to furnish his home with Louis XIV furniture, Italian drapes, and Persian rugs. He built an extensive library and a solarium, as well as a stable for his carriage and horses. On orders from his doctor, Field had to slow down his torrid work schedule. Hoping to mix relaxation and business, he began to travel the world for the first time, opening his eyes to developments outside the United States. In 1849, he and his wife booked passage to Europe, where he maintained a frenetic schedule. In five months their itinerary took them to many of the leading cities of England, Scotland, Ireland, Italy, Austria, Germany, and

Belgium, each an intense inspection tour for the business mogul looking for new ideas.

Impressed by European technology, Field shipped home goods that would help his paper products compete with European exports, including printing equipment and raw materials such as bleaching powder and soda ash. He was particularly in awe of industrial England, where railroads, telegraph machines, mills, and foundries were all more advanced than in the United States, and he was inspired by the British spirit of innovation and optimism, all reflecting an empire on the move. Britain was the world's leading naval power, in control of the seas and of global finance and trade. London, the heart of it all, was a construction zone, with men building roads, ports, telegraph lines, even sewers and prisons befitting the leading city in the world. In 1853, Field took another voyage, this time to Central and South America, including Panama, Colombia, Peru, Bolivia, and Ecuador. Struck by the isolation of towns and cities in countries with few roads, telegraph lines, or newspapers, and the lack of connections between these countries and the rest of the world, he thought about the potential for expanding trade and investment if these links could be established.

Field's paper business flourished in his absence. In 1853, his extraordinary wealth assured, he decided to repay creditors of Root & Co. the 70 percent of the debt they had not received, despite having no legal obligation to do so. The creditors were astonished and grateful, and Field's reputation in the business community soared. Now he was well positioned to raise money for a new venture when his brother first met Gisborne.

The Child of Boom Times

It is perhaps not surprising that Cyrus Field could envision something as monumental as the transatlantic cable. America was following rapidly on the heels of the industrial revolution in Europe and big ideas were transforming the country. Beginning in the 1820s with the opening of the Erie Canal and other inland waterways, followed by the steam engine, the wave of railroad construction in the 1830s, and the spread of the US telegraph in the 1840s, the nation was expanding in all directions. Interest in education and science was soaring, industry was starting to grow from coast to coast, international trade was exploding, subsistence farming was giving way to commercial agriculture, and millions of people were moving from rural areas to cities. A new doctrine of "Manifest Destiny"—the theory that America was destined to expand not just across the North American Continent but to Asia and to Latin America—was catching hold. The last two-thirds of the nineteenth century was an era of self-made industrial tycoons, men such as Jay Gould, Cornelius Vanderbilt, Andrew Carnegie, and John D. Rockefeller. The big difference between Britain and America was that British fortunes were rising with the empire and particularly on the global commerce it spawned, while American fortunes were much more likely to be made exploiting the enormous home market.

This fundamental distinction between cosmopolitan Britain and provincial America would play a defining role in Field's campaign to link the two continents by telegraph. While trains were a huge leap forward in transportation, they provided the same basic service as the horse—hauling goods and people over land, albeit at unprecedented speed and volume. The telegraph was by contrast almost magical in its impact, converting physical words into invisible code, reducing the long journey of the printed letter to a mere flash across the wires. The telegraph outpaced the speed of the horse, the pigeon, or the train by a margin unimaginable before its invention.

The first patents for a telegraph machine were handed out in 1837: one in England to William Fothergill Cooke and Sir Charles

Wheatstone, and one in the United States to Samuel F. B. Morse, who two years earlier had invented the code that would prove crucial to transmitting numbers and letters by cable. In 1843, Morse obtained government funding to demonstrate his system of dots and dashes to skeptical lawmakers, and the next year he put on a show, sending a now-famous message from a Baltimore railroad depot to the Supreme Court building in Washington: "What hath God wrought." Around the same time, Cooke was demonstrating in England the viability of a longer cable, running thirteen miles from Paddington in Central London to West Drayton on the western edge of the city.

The electric telegraph was a breakthrough not only in speed but also in cost, and it took off quickly. In the United States, there was one working telegraph line in 1846, and from that standing start the network exploded to twelve thousand miles of lines run by twenty different companies in 1850, and twenty-four thousand miles of lines in 1852. Within a few short years sending a telegram had become a part of everyday life in the United States and in continental Europe, where cables were expanding from Germany and Austria into Russia, France, Spain, Switzerland, and Italy.

England, however, had a major problem. An island nation with extensive commercial interests in Europe and across a far-flung empire, it could not connect to the Continent or to its colonies by the telegraph because it faced the seemingly impossible task of laying wire under the English Channel. And the United States faced an even tougher challenge if it was to string a cable to its largest and most important trading partners in Europe. In the late 1840s, Morse and other experts had begun to predict that a transatlantic cable was feasible, but when Field and his partners came on the scene offering to get the job done, it was still a leap into the unknown. "No one who knew anything about telegraphy would be foolish enough to risk building a transatlantic telegraph;

besides, it would cost a fortune," wrote historian Tom Standage. "So it's hardly surprising that Cyrus W. Field, the man who eventually tried to do it, was both ignorant of telegraphy and extremely wealthy."

Bringing Britain and America on Board

Field started work on the transatlantic connection with what looked like the easy part, running cable approximately 500 miles overland across Newfoundland, from the capital of St. John on the easternmost shore to Channel-Port aux Basques at the southern tip. Operating in woods that had grown no more hospitable since Gisborne first tried it, the job was a nightmare. Delays dogged the cable through the forest, and continued as the company struggled to find ships to lay cable under the 150-mile waterway between Channel-Port aux Basques and the coast of mainland Canada. The cable finally made its way down to New York and opened to telegram traffic in 1856, but the delays had cost the company at least a third of its capital, and Field was forced to raise more money. He was now convinced that a project as big as the transatlantic cable would need ships and money that could only come from both the British and American governments. In America, he shuttled between New York City and Washington seeking support from President Franklin Pierce and the Departments of State and Navy. Once he had tapped all his American contacts, Field began his search for more funding in England, where he used every trick in his salesman's arsenal to pitch the transatlantic cable.

During these years, Field would travel back and forth across the Atlantic every few months. In England he built close ties with everyone who could be remotely helpful, including government ministers, rich entrepreneurs, and top engineers and scientists. He was a formidable fund-raiser, with a gift for tailoring colorful imagery to his audience. In making a presentation to a wealthy breeder of large show dogs, Field asked him to imagine a large dog spanning the Atlantic Ocean. If you pinch his tail in Liverpool, Field explained, the dog will bark in New York.

The British were receptive. Under their law, however, Field had to create a company that was majority-owned by Englishmen in order to solicit them as investors, so he created the Atlantic Telegraph Company (in which he was also a major shareholder). It was this British-owned company that would carry the financial burdens of the venture for the next decade and see the project through to its end.

In England, Field also began to explore the challenges of laying submarine cables. He was determined to employ the best available technology, even at a time when there was no consensus as to what was the most effective way to solve even basic issues. For example, no one knew how thick a transatlantic cable needed to be, or how its internal wiring should be configured. Heated arguments arose among the scientists over how to propel and monitor a telegraph signal across some two thousand miles of cable.

There was also no agreement on how to lay a cable this long across the ocean. It would be too heavy to carry on one ship, so it probably had to be built in two pieces that would be carried on two ships and spliced together. Opinions differed on whether the splice should be made in the middle of the Atlantic, with both ships then spooling out cable toward opposite shores, or at one shore, with both ships working straight across, stopping to splice when the first ship ran out of cable. Field was no engineer, but he grew to understand that as the chief organizer, eventually he would have to call an end to these debates and move the project ahead.

He persuaded the British government to supply a ship, the HMS *Agamemnon*, to lay the transatlantic cable. It would also conduct soundings to verify the existing surveys of the eastern Atlantic seabed. London also agreed to pay substantial fees to use the cable, once it was complete. In Washington, members of Congress were generally skeptical of any joint ventures with Britain. But Congress managed to pass a bill of support by a narrow margin in early 1857, on the last day of the Franklin Pierce presidency. Then the new administration of President James Buchanan agreed to provide financial subsidies, a partial survey of the seabed, and a second ship to lay cable. That ship would be the USS *Niagara*, powered by both steam and sail, the largest such vessel in the world at the time.

With backing by the British and American governments in hand, Field sailed again for England to put pressure on the two cable manufacturers to deliver the wire, which needed to be strong, pliable, and five times longer than any wire that had ever been manufactured. The wire also needed to withstand a wide range of temperatures and perform at extreme depth. Field secured the wire, and he also convinced the cable companies to assume the responsibility and liability of spooling or "paying out" the cable, an operation that would prove very risky.

A Pioneer in a Hurry

In the summer of 1857, on the eve of the first attempt to lay the cable, the *New York Herald* proclaimed that the project was the "grandest work which has ever been attempted by the genius and enterprise of man." Energized by public enthusiasm, Field was eager to get started. As the largest stakeholder in both the British and American companies backing this project, he had a huge financial incentive to move fast. He was concerned with potential competition from other cable companies that were also eyeing transatlantic routes and were now approaching Washington for help. The risk with Field's extreme haste, however, was that it would push the project to launch before the engineers had settled key issues, like testing whether a cable designed in two pieces by two companies using different designs would splice together smoothly, or whether the newly built cable payout machines would work at sea.

The two giant cable-laying vessels—the *Agamemnon* and the *Niagara*—came together with their escort ships in London. Field had decided that the project would begin by anchoring the cable at the telegraph station in Valentia Island, Ireland. The two ships would then set out for Newfoundland, with the vessel *Niagara* laying the first stretch of cable, through which both vessels would be able to stay in touch with Valentia Island. On August 4, small boats filled Valentia Island's harbor and dignitaries from Ireland and England packed the town for a festive send-off, with parties, a military band, a cricket match, and rousing oratory. Filled with emotion, Field used his speech to invite the eight hundred assembled guests to visit him in Gramercy Park. "I cannot bind myself to more," he said, "and shall merely say, 'what God has joined together, let no man put asunder.' "

The first few days of the mission proceeded well. The weather was clear and warm and the cable was paying out smoothly. Field's youngest brother, Henry, served as chronicler of the voyage and captured the mood on board the *Niagara* this way: "There was a strange unnatural silence on the ship. Men paced the deck with soft and muffled tread, speaking only in whispers, as if a loud voice or a heavy footfall might snap the vital cord. So much had they grown to feel the enterprise that the cable seemed to them like a human creature, on whose fate they hung, as if it were to decide their own destiny."

On August 11, 1857, a week into the work, the signal from Valentia suddenly went dead. The crew was in a state of shock, with no idea what to do. After two and a half hours, the signal revived, but it brought with it little joy, only angst and confusion over why it had stopped in the first place, why it had returned, and what these unknowns meant for the grand venture. At 3:45

the next morning Field was awakened by shouts from the upper deck. The cable had snapped, plunging to the bottom of the sea. More than half a million dollars of investments had vanished. Field immediately gathered all hands on deck and talked about trying again within a few months. He asked that the ships stay in place and that his team try to analyze what was to blame: the weight of the cable, the strength of the payout machinery, or something else entirely? He remained calm and alert, trying to gather information that he could apply to his next attempt. On his way back to London to meet with his directors, Field wrote to his family, "Do not think I am discouraged, or am in low spirits, for I am not."

The British directors would remain steadfast but decided to wait another year in order to better prepare for the next voyage. Field then returned to New York and found an America seized by one of its recurring national economic panics. The stock market was collapsing, and investors were not in the mood to put money into a wildly risky scheme with a track record comprised of one catastrophic setback. Through skills of persuasion that even Prince Henry might have admired, Field managed to keep the support of both the British and American governments, and he prepared to try again the following year. In the interim, engineers refined the payout machine, making it lighter, simpler, and more sensitive to the pressure of the cable. They also improved the insulation materials that protected the wire and developed a new instrument that could detect when the electrical current changed intensity, an early warning signal of trouble to come.

Field decided that this time the two ships would make the splice in the middle of the Atlantic, and then each go in a different direction paying out the cable toward opposite shores. That way the ships could wait at the rendezvous point for good weather and then complete their mission in half the time, lessening the chance that a storm would disrupt their work.

In June 1858, the two ships set out with their escorts from Ireland. While the weather was perfect at the start, within days the barometer fell, the wind picked up, and the small fleet was caught in heavy winds and rain. From the *Niagara*, Field and the others could see the smaller *Agamemnon* pitching violently, in danger of sinking as coal spilled onto the deck and the precious cable came perilously close to breaking loose. Watching the storm lash the other ship, a reporter on the *Niagara* wrote, "The masts were rapidly getting worse, the deck coil [threatened to unravel] with each tremendous plunge, and, even if both held, it was evident that the ship itself would soon strain to pieces if the weather continued so."

The *Agamemnon* survived, barely, and days later moved into position alongside the *Niagara* to splice the wire. The ships then went their respective ways, the American ship toward Newfoundland, the British ship toward Ireland, maintaining communication via the cable that now connected them. When they were about three miles apart, the cable snapped. The ships came back together, made a new splice, and set off once more. Within twenty-four hours, the cable snapped again, and the ships returned to try a new splice and to start out for a third time in just four days. This time Field decided to abort the mission, suspecting some unseen obstacle deep below the waves. "Of all the many mishaps connected with the Atlantic telegraph," wrote Cyrus's brother Henry Field from aboard the *Agamemnon*, "this is the worst and most disheartening . . . since it proves that there may be some fatal obstacles to success at the bottom of the ocean, which can never be guarded against."

Field rushed back to London to see his directors. On July 13, 1858, he found himself in the same boardroom where, six weeks before, the group had given him an enthusiastic send-off. The mood had changed dramatically. The company's chairman, Sir William Brown, said, "I think there is nothing to be done but dispose of what is left on the best terms we can." Vice Chairman T. H. Brooking agreed, calling the undertaking "hopeless." But Field begged for another try. The ships, crews, and cable were ready to go, he pleaded, and all he needed were some supplies and repairs to the *Agamemnon*. The board reluctantly consented, swayed by the argument that valuable experience had been gained by the past missions. Nevertheless, they warned Field that this had to be the last attempt.

"The Christopher Columbus of America"

On July 17, 1858, the fleet set sail for the third time, now without any hoopla. Even some of the crew were skeptical, believing "that the Company was possessed by a kind of insanity, of which they would soon be cured by another bitter experience," according to Henry Field. The ships met in the mid-Atlantic on July 29, made the splice, and parted ways. The cable held, but soon a series of new crises arose. The *Niagara* drifted so far off course that Field feared it would not have enough cable to reach shore by this unplanned circuitous route. Later it appeared that the *Agamemnon* would run out of fuel. One day the signal between the two ships on the cable that joined them stopped, just as in the very first mission, and, just as before, it mysteriously resumed. Nevertheless, on August 4, the *Agamemnon* reached Ireland, one day after the *Niagara* had reached Newfoundland shortly past midnight, with Field on board.

Unable to wait till sunrise, Cyrus set out for shore in a rowboat at 1:45 a.m. and made his way to the telegraph shack in the small fishing village of Heart's Content, Newfoundland, a small port town with a lone church and perhaps sixty houses eighty miles northwest of St. John. There he woke the two men who were sleeping. "The cable has been laid!" he exclaimed. But the two were maintenance men with no knowledge of what had been going on in the Atlantic and, in any event, an operator was for some reason not available to send a telegram. The nearest point for transmission was fifteen miles away, forcing Field to make the journey on foot and corral two volunteers to send messages for him. One message went to his wife, Mary. "Arrived here yesterday. All well. The Atlantic cable successfully laid," he wrote. Other cables with a lot more detail went to Peter Cooper, to the Associated Press, to President Buchanan in Washington, and to the directors of the Atlantic Telegraph Company in London.

The world had come to assume the project was dead, so the public reaction was first one of shock, then of frenzied jubilation. Businesses on both sides of the Atlantic suspended work to celebrate, and prayer vigils were held in churches and parks. "Since the discovery of Columbus, nothing has been done in any degree comparable to the vast enlargement which has been given to the sphere of human activity," said the *Times* of London. "The Atlantic is dried up and we became in reality as in wish one country," George Templeton Strong, a New York diarist, wrote on August 5. "All Wall Street stirred up into excitement this morning . . . by the screeching newsboys with their extras. . . . The transmission of a single message from shore to shore will be memorable in the world's history for though I dare say this cable will give out before long, it will be the first successful experiment in binding the two continents together, and the communication will soon be permanently established."

President Buchanan telegraphed congratulations to Field. Queen Victoria sent a cable congratulating the American president. Ever the businessman concerned with his company, board of directors, and shareholders, Field was overjoyed by a cable from George Saward, the secretary of the Atlantic Telegraph Company in London. "I most heartedly rejoice with you," Saward wrote, "for the name Cyrus W. Field will now go onward to immortality, as long as that of the Atlantic Telegraph should be known to mankind."

Field traveled to New York City, where he was honored with a one-hundred-gun salute, parades, fireworks, banquets, and a host of formal honors. His picture appeared on ribbons, buttons, headbands, and banners. Tiffany's made souvenirs out of the same material used for the cables. Newspapers were full of praise for "Cyrus the Great" and "the Christopher Columbus of America." The pandemonium continued for two weeks, spreading to cities throughout the country.

The cable's impact became clear immediately. On August 27, 1858, authorities in London learned that a mutiny against British rule in India had been put down, and they were able to cable an order that same day to halt troops who were at that moment boarding ships in Canada, bound for India to help suppress the insurrection. The British figured they saved a small fortune by being able to cancel that unnecessary redeployment from Canada to India. The changes were transformative in every walk of life, from New Yorkers who learned in real time, also on August 27, that the Chinese emperor had agreed to pay war indemnities to England and France and open the Middle Kingdom to foreign trade, to the traders in London and New York who received nearly instant word on what the market across the ocean was doing. Americans and Englishmen "were used to having the Atlantic Ocean be two weeks, three weeks, six weeks wide, and suddenly here it was ten minutes wide, and it was just a miracle," said one historian.

It was just a tantalizing glimpse of the information age to come, for all was still not well. The cable was operating only sporadically; some messages were getting through but many more were not. Queen Victoria's ninety-nine-word message to President Buchanan took sixteen and a half hours to transmit, and by far the greatest amount of traffic between Europe and the United States consisted of technicians testing the line. At a dinner on September 1, 1858, less than a month after the cable had been connected, Field was passed a telegram that contained just two words, "Please inform . . ." The next morning he discovered that these had been the last transmissions on the transatlantic telegraph and that the cable had fallen silent. On September 6, the company placed an announcement to that effect in the London *Times*.

Suspicions arose that the entire enterprise was a hoax to pump up the price of stock in the companies of which Fields was the chief shareholder. The American public, so giddy weeks before, was now outraged. The reaction in England was more sparing because the British were accustomed to the vagaries of new telegraph lines, given earlier failures they had experienced with other undersea cables. Still, Field was humiliated and went into hiding at home, away from the glare of the media and the public. He refused to give up, however, and in May 1859, eight months after the cable fell silent, he left again for England.

The British to the Rescue, Again

Although Field had failed three times, he had also made considerable gains. His company had strung two thousand miles of cable and transmitted 366 readable messages across this wire in just the month of August 1858. In the process Field's team had advanced the technology of electromagnetic transmission far enough to make the dispatch of messages across such a distance possible, his company had collected extensive data on conductivity through the cable, and it had vastly improved the machinery to lay cable. In 1859, British authorities established a committee of government and company experts along with leading scientists and engineers to explain what went wrong and how to create functional submarine cables. Meanwhile, Britain would suspend support for undersea telegraph projects until the committee came up with an answer.

The timing was tragically apt, for the United States had turned completely inward when the Civil War erupted in 1860. America would have no time or interest in visionary schemes, and Field turned to helping the North any way he could, including using his London ties to court support for the Union cause in Britain, an often uphill assignment since sympathies ran deeply in favor of the American South.

The British commission of experts would deliberate for much of the war, generating and studying thousands of pages of highly technical analysis. On July 13, 1863, it delivered its report, suggesting improvements in the technologies for building, insulating, testing, and laying cable, and new ways to measure conductivity. It took a swipe at Field by insinuating that he had been too hasty in his preparations, but its conclusion rang with optimism, stating that a properly designed transatlantic cable could not only work successfully "in the first instance, but may reasonably be relied upon to continue for many years."

This endorsement of the technology was just what Field needed. By the end of 1863 he had left the Civil War behind and was back in England, imploring his directors for one more try. He knew that he would have to tap into new sources of money and that the lead would once again have to come from England, where, unlike in America, his credibility was not so suspect. He approached the British government, which agreed to make him chief ambassador and organizer for the project on land, while command of the work at sea would shift to more technically experienced hands.

Field then approached the cable suppliers, who had merged into a new giant—the Telegraph Construction and Maintenance Company—that offered to work for the cost of labor and materials plus 20 percent of the shares. Field eagerly accepted and went on to pitch the scheme to British industrialist Thomas Brassey. The largest railway contractor in the United Kingdom, Brassey had no experience in the telegraph industry, but he did have money, ambition, a cutting intellect, and experience in international deals. He put Field through a "cross-examination" that made the American tycoon feel like he "was in a witness box," then agreed to be one of the underwriters of the next expedition. With Brassey on board, Field was in a much better position to line up other British investors.

A second break came in the form of a gargantuan ship called the SS *Great Eastern*. Constructed by one of nineteenth-century England's most accomplished engineers, Isambard Kingdom Brunel, it was 692 feet long with a gross displacement of 18,915 tons—more than five times the tonnage of any vessel built up to that time. Its size dwarfed that of the *Niagara* and *Agamemnon* combined; indeed, it would be forty years before a bigger ship would be built. It contained the most powerful engines ever used in a seagoing vessel and could just about circumnavigate the globe without refueling. Designed for tourism, it turned out to be much too big for the market, so its owners were eager to deploy the ship to new ends. The *Great Eastern*, properly outfitted, could carry the entire load of cable on its own.

Field went back to work, helping to bring together the technicians who would keep pushing to strengthen the weak links—the cable and its insulation—and to make the payout machines even more sensitive to tugs on the cable, reducing the chances of a break. By June 1865, he was ready to load the *Great Eastern*, which took on a crew of five hundred along with 7,000 tons of cable, 2,000 tons of water tanks, 1,500 tons of coal, not to mention a floating farmyard with a cow, ten bulls, twenty pigs, scores of sheep, geese, turkeys, hens, and eighteen thousand eggs.

The ship sailed on June 24, 1865, and ran into trouble almost immediately, each snafu now announced by a giant gong that issued a noise resembling a thunderbolt and sent all hands scrambling on deck. Early the first morning the cable signal began to flicker, and the great ship had to retrace its course for ten miles, reel in the cable, and replace a defective part of it. That took thirty-six tense, frustrating hours. On Saturday, July 29, the same thing happened again, this time in mid-ocean, at the cost of twenty-four more hours. With no specific role on this voyage, Field served just as an observer but nevertheless raced out of his cabin to the control room at each peal of the gong.

On August 2, major trouble struck. As shifting winds forced a sudden change in course six hundred miles off the Newfoundland coast, the cable got stuck on an iron protrusion on the deck and snapped, disappearing into the ocean. The chief engineer walked into the room where the officers had gathered. "It is all over," he said. "It is gone!" Momentarily stunned, Field thought for a moment and replied, "Well, it's so. I must go down [to my cabin] and prepare a new prospectus. This thing is to be done."

The ship's captain was determined to reclaim the lost cable, and the *Great Eastern* carried an anchor-shaped grapnel designed for the recovery job. Over the next nine days, the crew made four attempts, each time locating the cable and beginning to lift it to the surface, only to see the grapnel break. The mission was aborted, and Field returned to London, fearful that this fourth failure would make his board close down the project for good.

He was wrong. Sensing they were close to success, the British directors and investors were still enthusiastic. In a few weeks Field had raised enough new money for another voyage. The owners of the Telegraph Construction and Maintenance Company and the *Great Eastern* renewed the terms under which they would participate, the same ones as before. The next voyage was set for the following summer. Meanwhile, engineers set to work on the *Great Eastern*, making sure that the cable could not again get hung up on deck.

"Field Did What No Other Man Could Do"

The following summer, in July 1866, the *Great Eastern* returned to Valentia Island, with Field again on board as an observer. The ship set sail for Newfoundland, paying out cable at a steady clip that might have been considered monotonous had the star-crossed expedition not grown so accustomed to disasters around every corner. With everyone on high alert, the ship hit the half-way point, where Field cabled Valentia for the news of the day. The crew was riveted by word that Austria, Prussia, and Italy were massing for war, not to mention news of the latest prices on the London stock exchange. From then on, the crew posted a roundup of global news on a bulletin board twice a day, gathering a small crowd.

After fourteen oddly uneventful days at sea, the ship picked its way through fog toward Heart's Content, where many of the houses were now flying the Union Jack alongside the Stars and Stripes. Two British ships came out to escort the *Great Eastern* into port, where the crew connected the cable, and Field rowed to shore to send a cable to his wife. "Heart's Content, July 27 [1866]. We arrived here at nine o'clock this morning," it read. "All well. Thank God the cable is laid, and is in perfect working order."

From that moment, the excitement rose around the world—again. Cable messages poured into Heart's Content. Queen Victoria and the new American president, Andrew Johnson, exchanged greetings. Field received congratulatory messages from President Johnson, Secretary of State William Henry Seward, and Britain's Chancellor of the Exchequer, William Gladstone. "On this day," wrote historian Chester Hearn, "the world read closing quotations on Wall Street, prices on the Brussels grain market, and the fact that Congress had readmitted Tennessee into the Union."

By late August, Field and the crew of the *Great Eastern* were also able to retrace the steps of the previous year's ill-fated voyage, fishing the lost cable off the bottom of the ocean and completing the task of laying it all the way to Nova Scotia. Within one month, Field had managed to complete the dream of a lifetime twice over, and it was the second success—recovering the lost cable—that proved the most emotional, perhaps because it represented resurrection, or maybe because it proved that the new age was indeed on solid ground. "I went to my cabin; I locked the door," he recalled in his diary. "I could no longer restrain my tears—crying like a child, and full of gratitude to God." A year later, during a speech in New York, Field described his emotions at the knowledge that there were now well-established cable connections between Europe and the United States:

> It has been a long hard struggle. Nearly thirteen years of anxious watching and ceaseless toil. Often my heart has been ready to sink. Many times, when wandering in the forests of Newfoundland, in the pelting rain, or on the deck of ships, on dark stormy nights—alone, far from home—I have almost accused myself of madness and folly to sacrifice the peace of my family, and all the hopes of life, for what might prove, after all but a dream. I have seen my companions one after another falling by my side, and feared that I, too, might not live to see the end. And yet one hope has led me on, and I have prayed that I might not taste of death till this work was accomplished. That prayer is answered, and now, beyond all acknowledgments to men, is the feeling of gratitude to Almighty God.

With two cables up and running, communication across the Atlantic expanded very quickly. In 1858, it had taken sixteen hours for a message from Queen Victoria to reach the US president, and by 1866 it took just seven minutes. After 1866, the cable was transmitting 7.36 words per minute, and one thousand messages per month. Those numbers would climb exponentially in subsequent years.

Spanning the once fearsome barrier of the oceans would put globalization in overdrive. The fast flow of the telegraph would open new markets for suppliers, new choices around the world for purchasers of machine tools and equipment, and new ways for managers to exercise control of their far-flung operations. Shipowners could now communicate with their captains as they moved from port to port, greatly increasing efficiency. Governments could manage their diplomats and generals in real time. The instant transmission of prices for stocks, bonds, and commodities would tie the markets of New York, Chicago, London, Brussels, and Amsterdam together as never before, reducing the blind risk of investing and leading to a huge increase in the size of these markets. The *New York Evening Post* wrote on July 30, 1866, "The Atlantic Cable will tend to equalize prices and will eliminate from the transactions in bonds, in merchandise and in commodities, an element of uncertainty." The next day the *Post* began to publish stock prices from the London market.

Even as the world was celebrating Field's accomplishments, rival entrepreneurs were following in his tracks. Media entrepreneur Paul Julius Reuter, the founder of today's global news agency, was the next to lay an Atlantic cable, connecting Germany to England later in 1866, and France to the United States in 1869. In the late 1860s telegraph companies would run new cables from Europe to India, Hong Kong, Japan, and Australia, and in the next decade they would connect South America to this expanding worldwide grid. Between 1870 and 1900, in fact, the length of global communication cables increased tenfold, and it doubled again in the next decade. Soon every major nation in the world would be in nearly constant communication with every other.

Field was able to sell some stock at considerable profit, multiplying his substantial wealth. While again honored in New York and London, this time Congress gave him a unanimous vote of thanks and a specially minted medal. At forty-seven, he was considered one of the giants of his era. Peter Cooper, the great tycoon and philanthropist and an early partner in the cable venture, said in a magazine interview, "If any of the foremost men of our time had died, their places would have been filled, but Field did what no other man could do; he saved the scheme and brought victory out of despair."

The Sad and Reckless Final Days

Field went on to become one of the titans of New York, buying up newspapers, a skyscraper, four prime blocks of Manhattan real estate, numerous companies, and a major share in the city's elevated railway. All the while, he was veering from investor to speculator, borrowing heavily, and buying stocks on credit. We can only guess what pressures motivated him to take this reckless course, a sharp departure from the habits of careful accounting and unambiguous integrity that had defined his early career. We know from the extravagant way he lived in Gramercy Park that he had a taste for luxury; perhaps now he was trying to keep up with the Vanderbilts and the Rockefellers. After his heroic accomplishment with the transatlantic cable, maybe he developed an insatiable appetite to do monumental things that were publicly applauded. Could it be that in his later years he found satisfaction only when his adrenaline level was sky high, when his schemes sailed like his cable ships always had, on the edge of crisis?

In the summer of 1881, New York's boom turned to bust. The stock and property markets tumbled. Some of Field's business partners nefariously squeezed him out of ventures. Everything went wrong for Cyrus Field. As he was in the throes of financial ruin, his daughter was institutionalized with a mental illness and his son was convicted of financial fraud. The 1880s would be a painful decade, culminating in 1891 with the death of his much-loved wife, Mary, at the age of seventy-four. Cyrus Field died in his sleep on July 12 the following year. He was seventy-two. His tombstone read:

CYRUS WEST FIELD
TO WHOSE COURAGE, ENERGY AND
PERSEVERANCE THE WORLD OWES
THE ATLANTIC TELEGRAPH

Legacy

Field went into business in the 1830s. At that time, information moved around the world no faster than it had for centuries, dependent as it was on sail and wind conditions. In his lifetime the world of communications was about to change more dramatically than at any time since Johannes Gutenberg invented the printing press. Wrote political scientist Debora Spar, "In 1830 a message from London to New York or Bombay took almost as long to reach its destination as it had in the days of Vasco da Gama or Magellan." (That is, early to mid-sixteenth century.) In the early 1600s, for example, it took between fourteen days and four months—depending on the weather—for a ship to cross the Atlantic. When steamships arrived, the transit time was reduced to around ten days. Thanks to the oceanic cable, by the time of Field's death in 1892, communication between the United States and Europe was just about instantaneous, and the same was true of communications around the world.

The speed of advance was mind-boggling. In 1838, Samuel Morse presented Congress with the prototype for the telegraph. By the 1850s, twenty-three thousand miles of telegraph wires crisscrossed the United States, and by 1880 almost one hundred thousand miles of subterranean cable had been laid around the world, reshaping international trade, politics, and diplomacy. In linking the United States to Europe in real time, Field made it much more difficult for America to maintain the isolation that had always been its preferred position in global affairs. The transatlantic cable was the communications portal of the American Century, a conduit that accelerated the expansion of US business, culture, and political influence to every corner of the globe.

The telegraph not only linked the world in real time but became a bridge to subsequent international communications breakthroughs—the radio in the 1920s, the telephone in the 1950s, the Internet in the 1990s. Even the magic of the wireless Internet rests on a solid foundation of wire cables, just like the magic of the telegraph. In retrospect, the transatlantic telegraph was a far more dramatic breakthrough than the Internet. There is no doubt that the Internet is a disruptive technology, but by the time it came along, real-time communication around the world already existed. We could already talk by phone to anyone, anywhere, and we could watch events on the far side of the globe—a man landing on the moon, a war in Vietnam, a violent protest in China—in our living rooms. But when that telegraph line was hooked up to Heart's Content in 1866, the world changed immediately in a truly discontinuous way.

Tomorrow's Master Builders

Field was the quintessential master builder, starting with nothing and creating the environment in which a highly complex project could flourish. An enthusiastic promoter with a solid grip on the practical challenges, he was the glue holding together men with vastly different roles and expertise, pushing all parties to reexamine their thinking after each failure and to remain optimistic that success was around the corner. He spent much of his time talking to the contractors and scientists, who came to appreciate his incessant curiosity, his contagious sense of urgency, and his exuberant optimism. The transatlantic cable would eventually have been built without Field, but it is hard to imagine how it could have been done as early as it was, when the technology was so underdeveloped, without someone like him to lead, inspire, and organize the financing, the technology, the wide array of talent, and, of course, the design and execution of the entire project.

Cyrus Field embodied many features that resonate with us today, particularly our obsession with new technology that is developed by innovative firms, small and large. He was an entrepreneur of the first order, envisioning a grand idea and overcoming one obstacle after another to achieve it. He was a great team leader, assembling experts of all kinds—many of whom disagreed with one another—and molding an effective team that relentlessly pursued a single goal. In that regard he seemed able to sublimate his own ego in order to let the talent around him come to the fore. Today there is a big focus on public-private partnerships—intensive cooperation among business, finance, and government to pool resources and share risks. We saw evidence of this in Prince Henry's efforts to partner with commercial merchants, and the pattern is equally clear in the way Cyrus worked so closely with governments on both sides of the Atlantic.

But most of all, Cyrus represents the category of leaders—like the architects of the Egyptian pyramids and the engineers who built the Roman aqueducts—who complete projects so grand that they change the human understanding of what is possible.

The transatlantic telegraph, sometimes referred to as the "Victorian Internet," replaced ships powered by sail and steam when it came to the transmission of information around the world. Since communication is perhaps the most powerful force accelerating globalization, Field's accomplishment was critical to furthering all that Genghis Khan, Prince Henry, and Mayer Amschel Rothschild began. Field's transatlantic cable, and the cables that quickly followed it in other parts of the world, shrank time and space, making the world smaller and the spread of international influences of all kinds infinitely faster.

The transoceanic cables became the linear ancestors of international telephone service; of radio and television that could be beamed from one country to another; of the communications satellites that circle above us or the high-speed broadband networks that join us all in cyberspace. It is a little-known fact that nearly 95 percent of communications traffic between continents—including e-mail, phone calls, videos, and financial transfers—travels not by air or through space but via underwater fiber-optic cable—close to one million miles of it. And the demand is growing. In late August 2014, for example, a consortium of six global companies committed to build and operate a new transpacific cable system to connect the United States and Japan in order to accommodate intense demand for broadband, mobile, and other traffic that includes data, voice, messaging, conferencing, and all manner of digital entertainment. Thanks to Cyrus Field, we've seen this movie before.

THE ART ARCHIVE AT ART RESOURCE, NY

Chapter VI

JOHN D. ROCKEFELLER

The Titan Who Built the Energy Industry and Also Launched Global Philanthropy

1839–1937

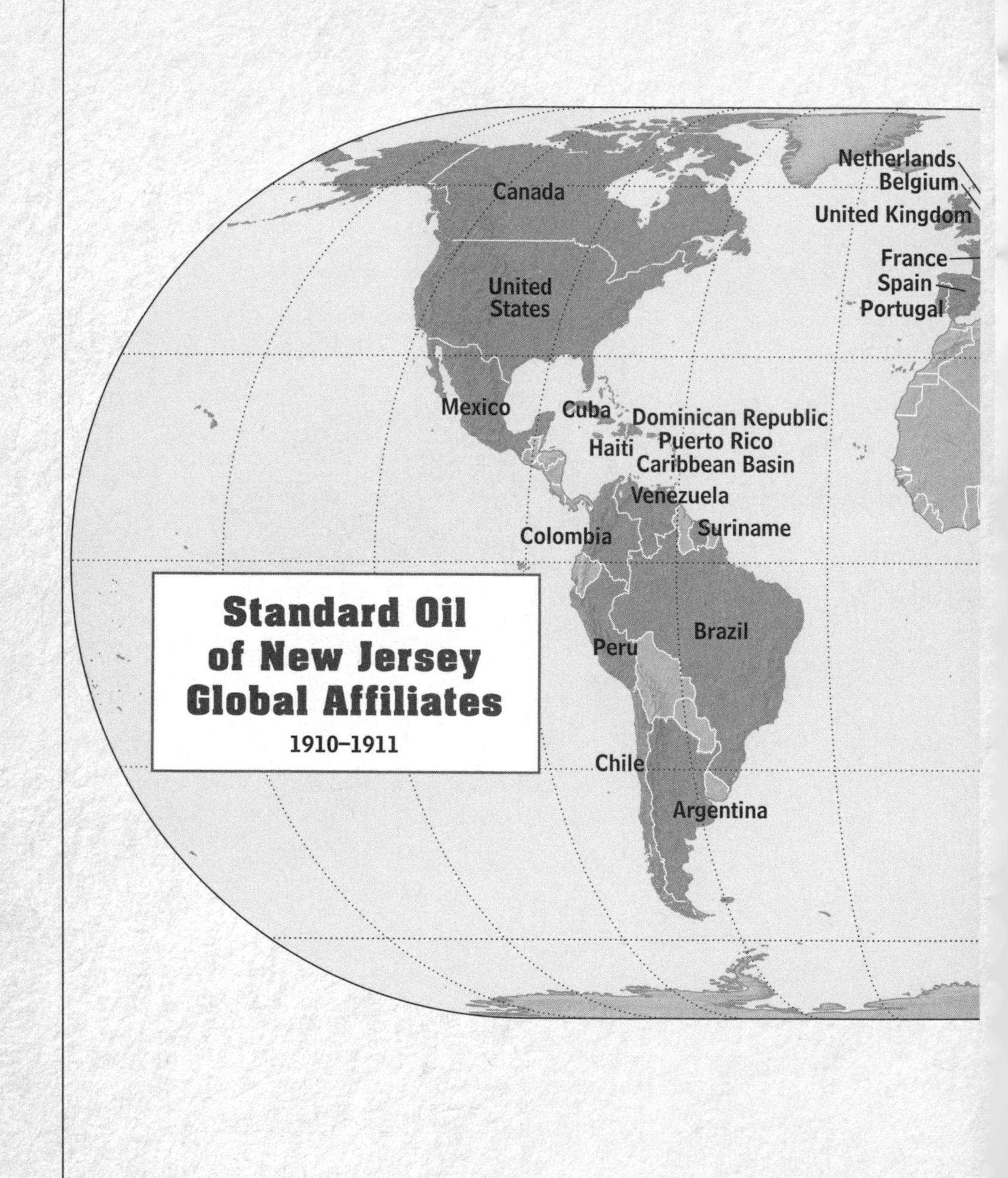
Canada
United States
Mexico
Cuba
Dominican Republic
Haiti
Puerto Rico
Caribbean Basin
Venezuela
Suriname
Colombia
Brazil
Peru
Chile
Argentina
Netherlands
Belgium
United Kingdom
France
Spain
Portugal
Standard Oil of New Jersey Global Affiliates
1910–1911

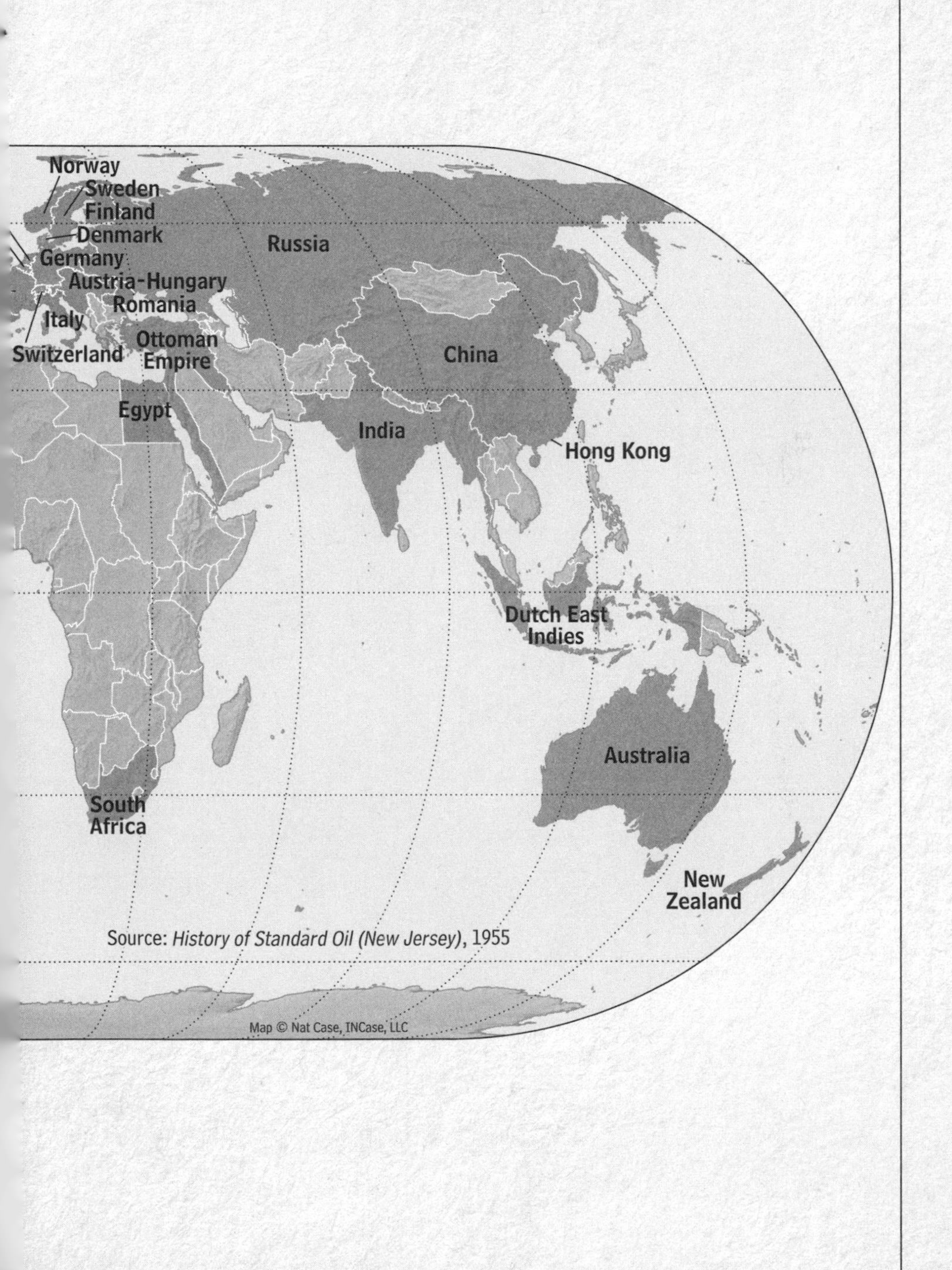
Norway
Sweden
Finland
Denmark
Germany
Austria-Hungary
Romania
Italy
Switzerland
Ottoman Empire
Russia
China
Egypt
India
Hong Kong
Dutch East Indies
Australia
South Africa
New Zealand
Source: *History of Standard Oil (New Jersey)*, 1955
Map © Nat Case, INCase, LLC

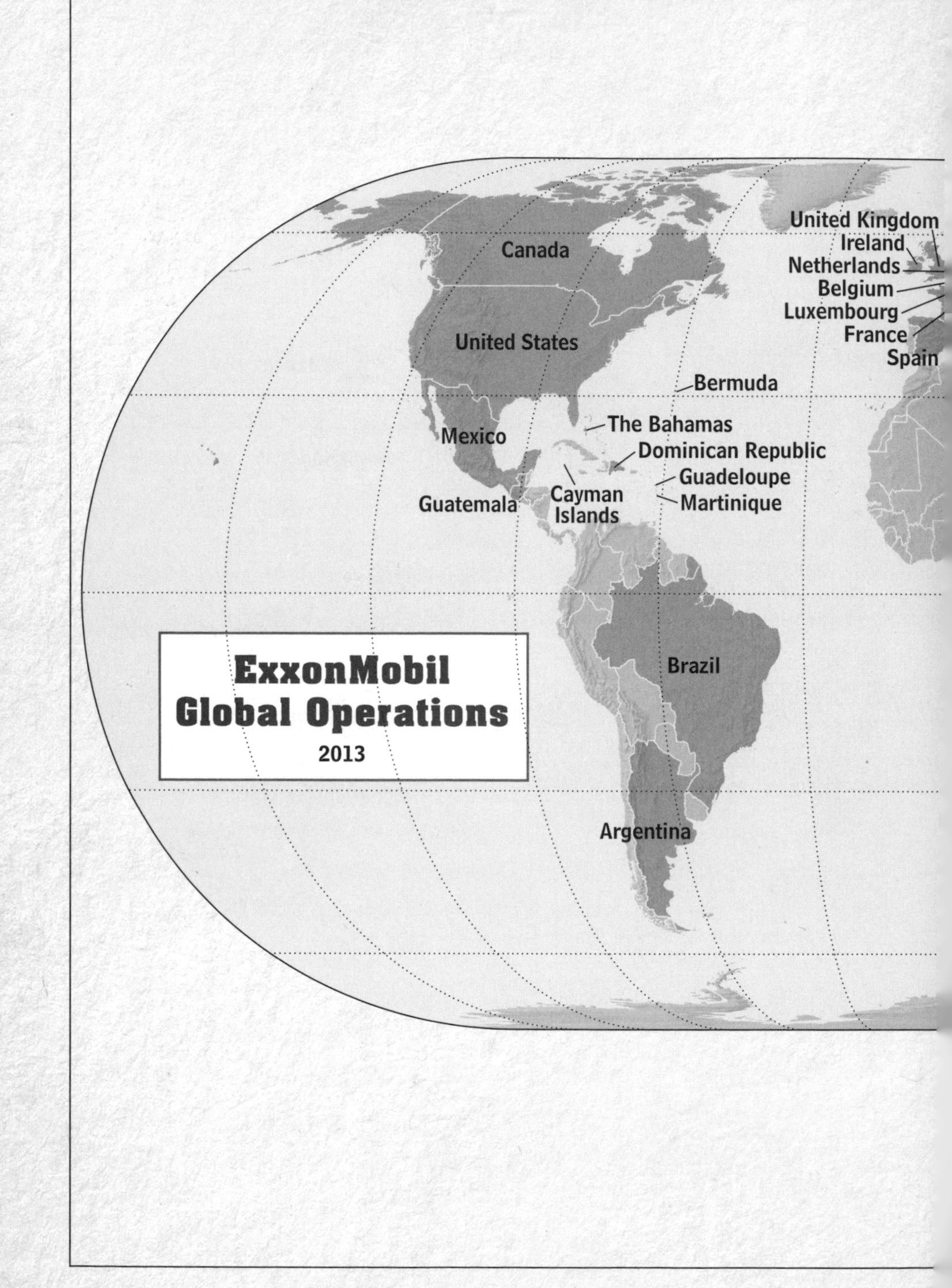

ExxonMobil Global Operations
2013
Canada
United States
Mexico
Guatemala
Cayman Islands
Bermuda
The Bahamas
Dominican Republic
Guadeloupe
Martinique
Brazil
Argentina
United Kingdom
Ireland
Netherlands
Belgium
Luxembourg
France
Spain

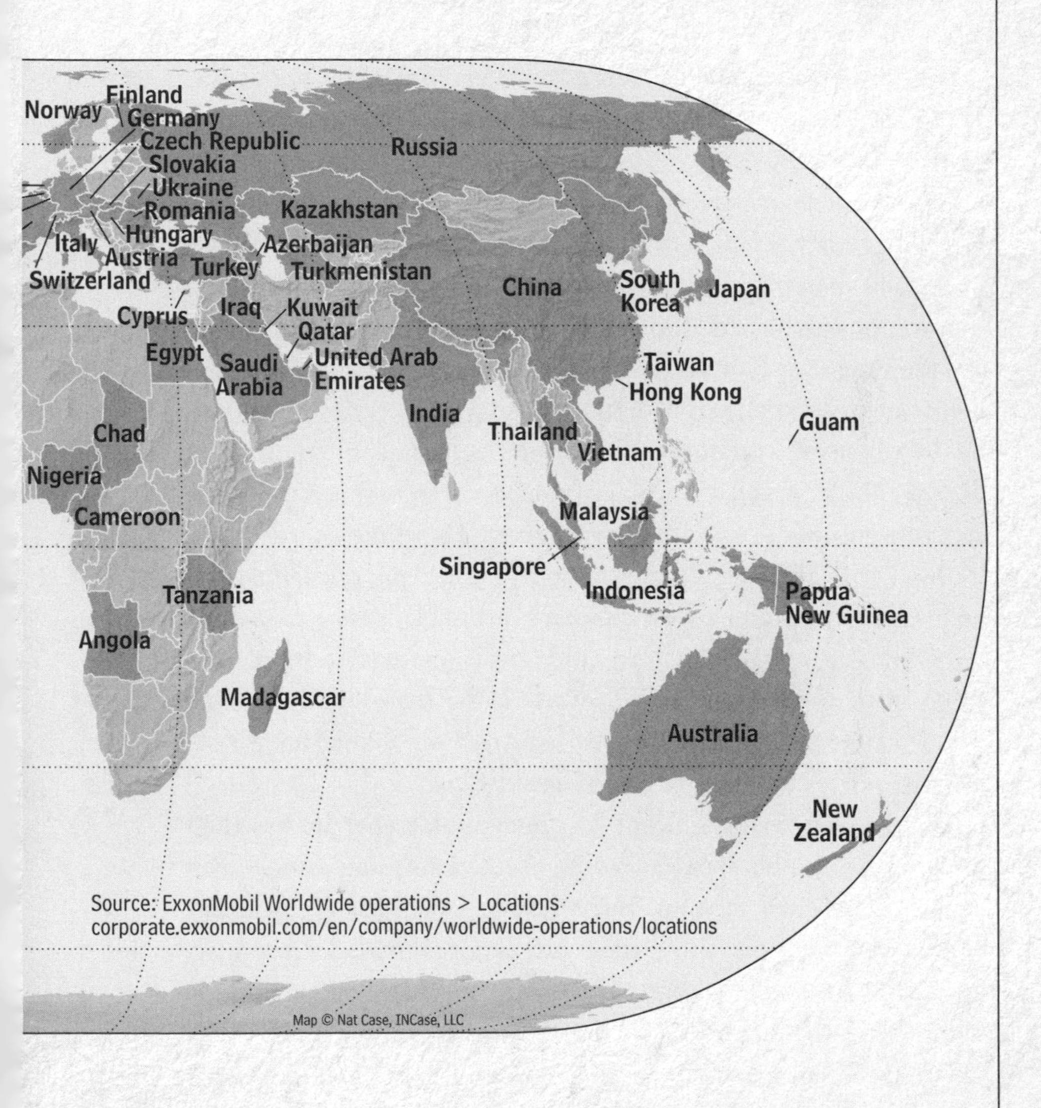
Norway
Finland
Germany
Czech Republic
Slovakia
Ukraine
Romania
Hungary
Italy
Austria
Switzerland
Russia
Kazakhstan
Azerbaijan
Turkey
Turkmenistan
Cyprus
Iraq
Kuwait
Qatar
Egypt
Saudi Arabia
United Arab Emirates
China
South Korea
Japan
Taiwan
Hong Kong
India
Thailand
Vietnam
Guam
Chad
Nigeria
Cameroon
Malaysia
Singapore
Indonesia
Papua New Guinea
Tanzania
Angola
Madagascar
Australia
New Zealand
Source: ExxonMobil Worldwide operations > Locations
corporate.exxonmobil.com/en/company/worldwide-operations/locations
Map © Nat Case, INCase, LLC

John D. Rockefeller secured his first job at the age of sixteen as an assistant bookkeeper, a job that seemed perfectly suited to a young man who had a way with—even a love for—numbers. It was 1855, and Rockefeller signed on to keep the ledgers and pay the bills for Hewitt and Tuttle, Cleveland merchants who invested in real estate and also traded in iron ore, marble, and foodstuffs. Rockefeller learned to handle a wide range of commercial transactions, from arranging shipments by rail, canal, and lake to managing property holdings, including houses, warehouses, and office buildings. He didn't just do the job, he reveled in it, verifying every bookkeeping entry, spotting minute errors, collecting every last dime of rent, and looking critically at colleagues who did not bring the same doggedness and passion to these chores. For him, ledgers were sacred books that not only helped to guide decision, expose fraud, and gauge performance but also protected one from "fallible emotion." Ledgers rooted an imprecise world in "a solid empirical reality," writes biographer Ron Chernow.

Rockefeller was promoted to chief bookkeeper in less than two years, and he started trading flour, ham, and other farm goods on his own account. But when he asked for a significant raise, much to his disappointment it was not granted. In response, on April 1, 1858, not quite twenty, he decided to go out on his own. In less than three years on the job he had saved a full year's salary, $800, and to that he added $1,000, borrowed from his flamboyant father, William. He partnered with an older friend named Maurice B. Clark to establish Clark and Rockefeller, and began trading in grain, fish, meat, water, salt, lime plaster, and other commodities. As the new business got rolling Rockefeller proved himself an adept fund-raiser, enthusiastically and relentlessly pitching his new enterprise to Cleveland banks. Within one year, Rockefeller tripled his former income and had his firm poised to profit spectacularly during the coming Civil War, when food prices would soar on rising demand from the military and other industries that would feed the war efforts.

At this stage, John D. Rockefeller was just getting started. He would eventually be labeled the world's wealthiest man. He became the driving force of the global petroleum industry, with its extensive exploration, drilling, and distribution tentacles—an industry that never stopped driving global economic growth, trade, and investment across borders, not to mention having a decisive impact on geopolitics. And beyond that, he used his astounding wealth to establish another industry—global philanthropy, a business that filled the vacuum between the international activities of governments and those of multinational corporations, and focused on challenges such as health and education around the world. On one hand, Rockefeller is alone among the characters in this book in being identified with two spectacular achievements, for the rest spent their entire professional lives pursuing one big goal. On the other hand, though, both thrusts—oil and philanthropy—were two integrated sides of the same man, two sides of someone whose upbringing caused him to donate part of his income to charity from a very early age. Without oil wealth he would not have had the wherewithal to transform global health and education. Without his far-reaching philanthropic achievements he would not have been a role model for many millionaires and billionaires after him who made an impact on the world well beyond the commercial enterprises they ran.

Raised with a Love for God and Money

One of six children, John Rockefeller had been born in Richford, a small town in central New York, to a hell-raising father and a

God-fearing mother. As odd a couple as there ever was, William and Eliza Rockefeller would move their family several times during John's early adolescence. The wandering eventually led them in 1853 to a home on the outskirts of Cleveland—a city that the railroads were about to discover and, like Chicago, St. Louis, and others, was on the cusp of rapid industrialization.

William was a boisterous, itinerant merchant who peddled snake-oil medicines in the classic fashion. A natural showman, he dressed colorfully, flashed wads of cash, and regaled his audiences with stories of his adventures and conquests, some partly true. Always coming and going, William was an occasional presence at his own home, which was shadowed by gossip about his philandering, and worse. He was accused of raping the family's maid and was known to be a bigamist, with a second family initially not far away from the first, until the Rockefellers moved to Cleveland. There were tales of his thievery and gambling, which brought repeated humiliations to his wife and children.

John grew up in a state of constant insecurity, never knowing where his father was, when the next rumors would strike, or if there would be money to pay the bills. He attended good secondary schools and excelled in math, but he dropped out of high school two months before graduation—possibly because his father was strapped for money. He enrolled in a three-month course at E. G. Folsom's Commercial College, where he studied double-entry bookkeeping, essentials of banking, commercial law, business history, and clear penmanship. Then he went straight to work.

John learned much from his father's mistakes, but there was more to the relationship than that. For all his extravagances, William was scrupulous about maintaining accounts and paying his debts. He lent money to his son at commercial rates of interest and often demanded the loan be repaid ahead of time, thereby compelling John to develop the habit of maintaining protective reserves. He taught his son how to bargain for goods and how to appraise products and people. As a child, John would purchase candy by the pound and sell single pieces at a large profit, sometimes lending the proceeds to his father—with interest.

Eliza tolerated her marriage to a loudmouthed scoundrel in stoic silence. Frugal, highly disciplined, and somber, she was totally devoted to the Church, possessing a moral righteousness that was visible to all, if painful at times to those close to her. During William's long absences, Eliza put John in charge of cutting wood, milking the cow, tending the garden, shopping for food, watching his siblings, and handling the tight family budget. John never appeared in school pictures because he was ashamed of his worn clothes. He often wore his sisters' hand-me-downs until he was eight. His mother would whip him with a cane for disobeying rules, always calmly and with an explanation.

Eliza's discipline could be harsh, but she was also a force of order in a home often stretched to the breaking point by her husband's irresponsible behavior. She would ask for John's advice, confide in him about her anxieties, and, of central importance to his future, raise him inside the Church. Eliza took her children to Baptist services every Sunday and led them in prayer every night. John sang in the church chorus and taught Sunday school as a teenager. The Church seems to have strengthened in John a belief that he acted on for the rest of his days: that while life is deeply insecure, his own moral foundation was rock solid.

It is not surprising that John grew up to find the sacred in a disciplined reading of ledgers, given how powerfully commerce and religion came together in his early life. When John started out in his boyhood business as a candy trader, Eliza always insisted that he contribute a portion of his proceeds to the Church—which he did for the rest of his life.

A Calculating Man among Gamblers

When John was partnering with Clark in their new agricultural trading venture, the dawn of a new global trade was unfolding about 135 miles to the east, with the discovery of oil in Pennsylvania. The oil rush had been triggered in part by the work of Yale University chemistry professor Benjamin Silliman, which showed that petroleum could be refined into a fuel to light lamps and a grease to lubricate machines—products with vast commercial potential in a young nation bursting with new houses, offices, shops, factories, and railroads.

One of the early entrants was a company called Seneca Oil Corp., which dispatched Edwin L. Drake to drill wells in the remote western Pennsylvania village of Titusville. Drake's trials and tribulations are reminiscent of those Cyrus Field suffered as he struggled to build the transatlantic cable. Successes and failures, accompanied by wild mood swings, were the order of the day. Improvisation—building their own tools, scrambling for workers—was necessary. After five years of agonizing effort, on August 27, 1859, Drake's drills reached sixty-nine and a half feet into the earth and on Sunday, August 28, liquid gold gushed out of the ground. "The drillers started shouting 'oil, oil,' " wrote historian Allan Nevins. "With good reason might Drake feel triumphant. He had written his name on an imperishable page of history . . ."

Fortune seekers streamed into the small village, building wells everywhere, even on the main streets—a reflection of the growing, restless entrepreneurial class in America during and after the Civil War. In the next few years, other gushers were struck in western Pennsylvania. At first the oil was processed in small, sometimes ramshackle refineries, tossed up around the drilling sites. But as word of Drake's successes spread throughout the country, larger refineries began to appear in Cleveland, mostly because of its position as a major railroad hub. The industry had come to Rockefeller's doorstep.

From the beginning, the oil business was a wild one, perfectly suited to a man with a lifetime of practice in finding order amid chaos. The market was young, and supply and demand often fell dramatically out of sync, leading to wide price fluctuations. These swings were aggravated by recurring predictions—which persist to this day but have yet to prove true—that the supply of oil was on the verge of depletion. In 1861, for example, prices of crude went from 10 cents to $10 per barrel. In 1864, the range was between $4 and $12. Producers found themselves rich one day, bankrupt soon after. The industry attracted a motley assortment of investors, many reckless speculators among them, and they would prove easy prey for a methodical man like Rockefeller, one with a highly strategic bent, too.

John Rockefeller entered the energy business gradually. In retrospect, it was a thoughtful, incremental strategy, an exquisite combination of caution, experimentation, vision, and execution. He operated like a master chess player, taking one thoughtful step at a time, but decisively and rapidly. In 1863, Rockefeller and Clark purchased half the working capital of a refinery owned by Clark's friend, Samuel Andrews. Because he was already working as a middleman between producers and consumers, John found this to be a comfortable point of entry. Shortly afterward he married Celestia Spelman, who, like his mother, had put the Church at the center of her life. At this time he had a large income, considerable savings, and steadily growing wealth. He maintained significant involvement in his church and expanded his financial donations to the church. By 1865, however, the partnership with Clark had turned sour, and John decided to buy him out. Being able to act on his own insights and instincts was liberating. The deal was done with a handshake. "I ever point to the day when I separated myself from [Clark] as the beginning of the success I have made in my life," John said years later.

Later that year Rockefeller proceeded to buy the largest refinery in Cleveland, with other purchases following. The acquisitions reflected a long-term confidence in the industry. He did not hesitate to borrow to fund his new facilities. His hands-on operational role was also notable for attention to its every detail. Unlike many of his rivals, he managed by hard numbers, constantly paring expenses. He decided to save money by building as many of the components of the business as possible in-house—his own barrels, his own pipes, his own joints, his own hauling wagons. Oftentimes he could be seen at the refinery at daybreak, observing workers rolling out barrels and stacking hoops. A colleague once said, "The only time I saw John Rockefeller enthusiastic was when a report came in from the creek that his buyer had secured a cargo of oil at a figure much below the market price. He bounded from his chair with a shout of joy, danced up and down, hugged me, threw up his hat, acted like such a madman that I have never forgotten it." In a culture of buccaneers who lived big and fell hard, John Rockefeller was a quiet, deliberate strategist with an obsession for long-term fundamentals.

Rockefeller's empire grew quickly. From the start, oil was in heavy demand around the world. The United States began exporting oil to Europe soon after Drake's discovery, and Cleveland emerged as a major transport and refining hub for the US industry. By 1863, the Cleveland area had twenty refineries, producing 103,691 gallons of refined products, one-fourth of which was exported. And by 1868 Rockefeller owned the largest set of refineries not only in Cleveland but in the world.

Almost alone among his contemporaries, Rockefeller saw the oil industry as one interrelated system—drilling in the backwoods of Pennsylvania, refining, transporting, distributing the refined product all over the world, and selling it at gas pumps all over the country. He saw all these businesses as different sides of the same box. He also realized early on that building this box would take a lot of capital, and that this investment would be under constant threat, due to the wild price swings and hyper-competition that infected the oil trade. It took an unusual mind to imagine that one could impose some stability on this global trade, and Rockefeller saw how to do it. He would have to build a global company that dominated and united the fragmented industry, one with enough influence on all the railroads, pipelines, and shipping companies to dampen competitive pressures and keep its costs under strict control.

The Pioneer and the Predator

Rockefeller reached his prime years during a golden age for big business in America. It was the same Gilded Age in which Cyrus Field excelled, and in the same way the man and the times came together. Globalization was gathering force as capital markets became linked across borders, trade expanded, and immigration soared. Industrial transformation was in full swing, propelled by breakthroughs in transportation and telecommunications, advances in education, the broad move of farmers to the cities, and the huge influx of labor and talent from abroad. National businesses were springing up everywhere—iron and steel, glass, meatpacking, flour mills, and leather tanneries. All of this created wide-open opportunities for aggressive industrialists. In 1860, for instance, the United States had 140,433 manufacturing establishments, with capital of $1,010,000,000. Ten years later, it had nearly twice as many establishments (252,148), with total capital of $2,118,208,769.

The government was large enough to provide law and order, and too weak to threaten the rise of great tycoons. Businessmen enjoyed the benefit of bankruptcy laws and legal protection for private property and the sanctity of contracts, but they faced no income taxes and no effective antimonopoly laws. "Institutions were still sufficiently plastic to be molded by men of strength," wrote biographer Allan Nevins. "The United States more than any other nation produced genuine captains of industry, men who thought and operated on a grand scale . . . [and] the pioneer in the American movement toward industrial concentration was Rockefeller."

In this unique era, the oil industry was rapidly emerging as a key to prosperity not only in the United States and Britain, the latter being the world's industrial and financial leader, but also in continental Europe and in Asia. In 1859, the United States produced 2,000 barrels of oil for the year. By 1870, the annual total was 5.5 million. "A commodity that had been a curiosity when President Abraham Lincoln was nominated had become a necessity before he was murdered," wrote Nevins.

With lightning speed, Rockefeller consolidated his control over refining, storing, shipping, sales, and eventually the extraction of oil in the United States and abroad. He combined ruthlessly aggressive tactics with innovative organizational moves to enhance and streamline the management of Standard Oil, the corporate name under which all his oil-related interests operated.

By 1872, he had bought up nearly all the refineries in Cleveland, Philadelphia, West Virginia, New York, New Jersey, and New England, using a variety of cutthroat tactics to drive competitors out of business. He hoarded chemicals, barrels, and other supplies critical to the refining industry, in order to create artificial shortages. He drove rivals to a point of desperation at which they had to sell, then bought them out at low prices, in an atmosphere of airtight secrecy. Often the names of the companies he bought were not changed, and to outsiders and the public they appeared to remain independent, even though they were part of Rockefeller's juggernaut.

It was Rockefeller's manipulation of the railroads for which he would become most notorious. As his business grew and he became a prime customer of the railroads, Rockefeller began to demand that they pay him an exclusive rebate for every barrel he shipped. In secret, he even forced the train companies to give him a rebate on oil shipments of his competitors—so-called drawbacks—in effect forcing his rivals to subsidize Standard Oil (although they did not know that at the time). Rockefeller negotiated increasingly better deals for himself by promising the railroads even larger cargoes, thereby making the railroads conspiratorial allies of his budding monopoly. Standard Oil was not the only company to use these tricks, as others soon followed. As biographer Ron Chernow wrote, "The proliferation of rebates hastened the shift toward an integrated national economy, top heavy with great companies enjoying preferential freight rates."

The bigger Standard Oil became, the more clearly smaller companies saw the hopelessness of competing with it. The larger it was, the more leverage Rockefeller had, and the more savings it could extract from all its suppliers, not just the railroads. With a stronger balance sheet also came more leverage over banks, and ever lower costs for financing.

In later years railroad rebates and drawbacks would become the centerpiece in the government and public indictment of Rockefeller, not because they were novel but because of the sheer scale, the secrecy, and, in the case of the "drawbacks," the brutality with which they were used to crush and take over the competition. Journalists stitched together human-interest stories about Rockefeller's victims to paint a picture of the oil tycoon as a predatory monster. But most of his tactics were not illegal or even that unusual in late-nineteenth-century America, where cutthroat competition thrived in the absence of laws to restrain it.

Rationality from the Top

The rise of Standard Oil is attributable not only to Rockefeller's aggressive tactics but also to his management acumen. His company was better managed than the others. He practiced extreme cost control. He pushed his company to innovate, expanding the range of products that could be made from refined petroleum to include all kinds of grease for machinery, paraffin, Vaseline, paint remover: over three hundred products altogether. (At this point, the gas-powered automobile had yet to be invented.)

For his entire career, Rockefeller understood the value of surrounding himself with exceptional talent, which he did by identifying and acquiring experienced managers with every new takeover. He figured out from the start not just how to survive the price cycles in the industry but how to prosper from them. Thus, he kept deep financial reserves not only as a cushion for his own operations but in order to have the ability to buy assets at their lowest prices, all of it tracked in his meticulously kept ledgers.

It is also hard to exaggerate how hands-on he was. He studied and grasped every aspect of the business and achieved an encyclopedic knowledge of how things worked. Since the early days when he could be seen at the railroad tracks urging freight handlers to move more quickly, he constantly visited plants, interrogated employees for information, noted places where greater efficiency could be achieved—all the more commendable because he kept the broader picture, the long-term vision and strategy, in clear view at the same time.

Rockefeller was also a great delegator. He would often implore his colleagues that neither he nor they should do anything that someone else could do. "You are responsible," he once told his staff, "but as soon as you can, get one whom you can rely on, train him in the work, sit down, cock up your heels, and think out some way for the Standard Oil to make money."

He deeply understood internal communications, coordination, and motivation. This was evident from the very beginning when he created a system that allowed his refineries to exchange information on how to save costs, or how to use certain materials, or how to train workers. By the 1890s, Standard Oil was a global and smoothly functioning machine. It had a streamlined executive board and operating committees overseeing each function from production to marketing. Every committee adhered to standard operating procedures. Processes were established for exchanging intelligence and other information among committees. The entire setup was akin to what we would see in today's best multinational firms.

In his leadership style, Rockefeller was far from the autocrat we might have expected to see. His language was "we," not "I." He ruled, yes, but as the center of a circle of powerful colleagues. He strove to make decisions by consensus, and he would postpone most decisions until there was near-unanimous agreement. "[He] was not a dictator in the organization, but simply *primus inter pares* in the executive committee that ruled it, the principal planner and coordinator," wrote one biographer.

He believed in treating workers exceptionally well. He paid high wages, encouraged shorter hours, and showed great consideration for his staff. He was in fact a cheerleader for the company and its employees, and often used the specter of competition to ratchet up enthusiasm for the firm, using attacks on the company to bring it closer together. Rockefeller's colleagues saw him as unfailingly courteous, patient, quiet, modest, and never angry or emotional. In the end the railroad rebates and all of the issues associated with them were only one ingredient in his success, even though they dominated public perceptions.

Stability at All Costs

For Rockefeller, there was a more philosophical consideration, too. He simply did not believe in the Jeffersonian vision of a country characterized by unfettered competition among many small independent entities, nor was he a proponent of Adam Smith's "invisible hand." He felt that in a world bursting with possibilities, capitalism should be highly organized and managed. He favored a nation run by big industrial monopolies. He looked favorably on new ways for companies in distinct industries to "cooperate"—pools, consortiums, monopolies, and other arrangements that created a more stable and predictable environment. Industrial scale, he felt, was required to mobilize the large amounts of capital needed to invest in new factories, roads, and telegraph lines—the basic hardware of a rising industrial power.

It was also critical in his view to cut costs by bringing under one roof the entire chain of operations, from extracting oil to delivering it. Rockefeller saw how violent boom-bust cycles of the oil industry would, time and again, push small refineries and independent producers into bankruptcy in droves. In Rockefeller's mind, only a giant could survive the unavoidable roller-coaster ride of an economy prone to bouts of deflation and inflation, recessions and booms. The United States did not have a central bank at the time and Rockefeller was trying to manufacture stability. Years later in his memoirs, he explained how he felt throughout his career. "The day of individual competition in large affairs is past and gone," he wrote. "You might just as well argue that we should go back to hand labor and throw away our efficient machines. . . . It is too late to argue about the advantages of industrial combinations. They are a necessity. And if Americans are to have a privilege of extending their business in all the states of the union, and into foreign countries as well, they are a necessity on a large scale."

Rockefeller aimed to build a monopoly, and by 1887 Standard Oil controlled 90 percent of US refining capacity. It had a near monopoly on distribution and sales, too. It had the railroads in its pocket and it had bought up most of the pipelines. In the late 1880s, Rockefeller began buying leases on newly tapped oil fields in Ohio, and by 1891 he had become a major producer, responsible for over 25 percent of US oil output. By the turn of the century, Rockefeller's oil empire encompassed one hundred thousand people, four thousand miles of pipelines, and five thousand oil tank cars.

In the late nineteenth century, the legal means to create a national holding company did not exist, so Standard Oil used every organizational and legal maneuver, secret and overt, to be able to run his company as a national entity. By 1882, for example, Standard Oil comprised at least forty separate companies across many states and countries. There were attempts to create the equivalent of a holding company, first the Standard Oil Trust, then Standard Oil of Ohio (1890), then Standard Oil of New Jersey. It may not have been one company legally, but it was managed that way and it acted that way. Whatever Standard Oil was called, however, until the late 1880s, John Rockefeller, surrounded by a highly talented managerial team, tightly controlled a far-flung empire that included fast-growing operations outside the United States.

Dominating the Global Arena

Rockefeller's success in the global oil business sprang from his dominant position in the American market, soaring demand abroad, and his ability to seize the moment. Only two years after the discovery of oil in Pennsylvania, the United States shipped its first oil-based export, kerosene.

Rockefeller was quick to exploit the global markets. In 1866, he sent his brother William to New York City to launch a new company, Rockefeller and Company, to oversee its growing foreign sales. By then he was the preeminent refiner in Cleveland, and two-thirds of the city's kerosene output was sent abroad. The new global office was originally established at 181 Pearl Street, but two decades later it moved to 26 Broadway, a location that became one of the world's most famous business addresses, "Shorthand for the trust itself, evoking its mystery, power and efficiency."

By the 1870s, kerosene was America's most valuable nonfarm export, and the fourth-largest overall. And 90 percent of all kerosene exports from the United States were passing through companies controlled by Standard Oil.

From the very start, overseas demand grew dramatically, and had a profound impact on the boom-bust price cycle. If French and German buyers heard news of a new gusher in Ohio—news that traveled fast over the transatlantic telegraph—they would hold off before purchasing more oil, knowing that the new supply was likely to push down prices. In this way, from the inception of the industry, oil prices were set globally.

At first, Rockefeller's strategy was to market refined products through foreign sales agents, combined with some of his own salesmen. As a preeminent American company, Standard Oil received extensive assistance from US embassies and consulates abroad. Foreign service officers, who often received side payments from the company, supplied intelligence on commercial conditions, compliance with contracts, and also made regular business pitches on behalf of Rockefeller's operations.

By the late 1880s, the company took more direct control of its overseas operations. In addition to establishing joint ventures with foreign companies, Standard Oil created its own importing and distribution companies, such as the Anglo American Oil Company in England and the Deutsche-Amerikanische Petroleum Gesellschaft in Bremen. Not surprisingly, Rockefeller's operation became increasingly sophisticated as it studied ways to surmount all kinds of tax, tariff, and other trade barriers.

Rockefeller moved to New York in 1883, and soon the family was buying or building all manner of facilities in the harbors of New York, Philadelphia, and Baltimore—warehouses, docks, transfer machinery—to accommodate ever-larger shipments from their refineries. The move to New York was important also because it brought Rockefeller closer to the big Wall Street banks. For most of the period between 1870 and the late 1890s, Standard Oil dominated global markets, and it had outgrown the capabilities of Cleveland financiers.

By the turn of the century, the company was building its own fleet of ships and exporting 50,000 barrels of oil a day to Europe alone. Standard Oil was selling not only in the major European cities but in Shanghai, Calcutta, Yokohama, Singapore, and other commercial centers. The company spared no effort to develop and serve foreign customers. In England, for example, the lamps were not designed to burn American kerosene, so Standard Oil built a factory that manufactured and sold the right kind of lamps. In Mexico it used donkeys to deliver candles to remote villages.

Although Standard Oil had virtually no competition in the United States, it did have rivals abroad. The first major competitors emerged from Russia in the 1880s, spearheaded by two European family dynasties, the Nobels from Sweden and the Paris branch of the House of Rothschild. These family operations had discovered a rich supply of oil in the region of Baku on the Caspian Sea, and they moved quickly to supply the Near East and Europe. In the 1890s, the Dutch presented a new challenge in Asia in the form of Royal Dutch Shell.

Rockefeller and Standard Oil were in full battle mode to combat foreign rivals, sending emissaries to assess precisely what was happening abroad, enlisting the help of US embassies, and setting up new marketing arrangements throughout Europe and Asia. Standard Oil saw itself as engaged in a war to control the global market, which was growing as rapidly as the US market. Rockefeller was the hands-on commanding general, running the war room from its headquarters in New York.

Standard Oil responded to its foreign rivals in ways reminiscent of how it met competition in the United States, including more aggressive marketing, more attention to customer preferences, and tighter quality control. Most of all, it used the outsize profits that it was earning on its monopoly position in the United States to cut foreign prices to well below what its competitors charged, triggering vicious price wars. Standard Oil also became adept at spreading malicious rumors about the quality of its competitors' products. These tactics were widespread in the industry, but Rockefeller's uniquely competitive style was a critical force in shaping the nature of the international oil business in its early days.

Retirement, Vilification, and Breakup

By the late 1890s, roughly three decades after Drake's discovery of oil and Rockefeller's entry into the business, dark clouds were gathering for Standard Oil in the political arena. The emerging progressive political movement was demanding that the US government step in to break up large trusts, to regulate the working hours and conditions of blue-collar labor, and to protect consumers from monopoly pricing. In 1890, the Progressives had won their first big victory with passage of the Sherman Antitrust Act. The newspapers, which heavily favored the Progressives, were calling for authorities to use the act as a tool to punish the trusts that controlled sugar, steel, tobacco, and other industries. But Standard Oil soon became the most prominent target for the groundswell of political and public resentment against big business, and Rockefeller was its personification.

By 1896, however, Rockefeller had stopped going to the office. He was tired, had been gaining weight, and was diagnosed with alopecia, a failure of the immune system that caused hair loss. He did not publicly announce his retirement, however, and continued to hold the title of president of Standard Oil of New Jersey, then the holding company for all of Rockefeller's oil interests.

In retrospect, the lack of a formal break with active service and the absence of any public announcement to that effect was a grave mistake. As the company grew more powerful, those who took over from Rockefeller appeared more arrogant and more prone to manipulate the political system. They were more aggressive in raising prices than Rockefeller, and they employed slash-and-burn competitive tactics with even less discretion. Although Rockefeller no longer shouldered the responsibility of his former title and position, it seems no one outside the company knew that, and the government in Washington and the states, as well as the public and the media, attributed everything that his successors did to John Rockefeller himself.

Decades earlier, Rockefeller had been the subject of a scathing attack by journalist Henry Demarest Lloyd in the *Chicago Tribune* and the *Atlantic Monthly*, and in 1894 Lloyd came out swinging again in a new book, *Wealth Against Commonwealth*. It was read by hundreds of thousands of citizens and inspired countless sermons and editorials, ultimately becoming what Nevins has called "one of the most famous polemics in American history." The most searing and well-known indictment of Rockefeller, however, came from investigative reporter Ida Tarbell in nineteen installments in *McClure's* magazine in 1902 and in her 1904 book, *The History of the Standard Oil Company*. This diatribe ran 550 pages with another 240 pages of source material, describing Standard Oil as an industrial combine of awesome power and organization, all built by fraud, coercion, discrimination, and deceit.

Rockefeller did not fight back. He refused to rebut public criticism, forfeiting the opportunity to present his side of major public controversies. He believed in dignified silence, not understanding it could be read as an admission of guilt. In 1906, a colleague said to him, "It is your own fault, Mr. Rockefeller, you refuse to see reporters or to make known your side of the case." Rockefeller replied, "So it is my fault. I cannot break a rule of fifty years standing. A man is judged by his acts more than by his views and opinions, and so do I wish to be viewed." Some ten years later, Rockefeller revealed to an interviewer how he really felt. "So much of the clamor against rebates . . . came from people who knew nothing about business. . . . As a matter of fact the Standard Oil Company has been one of the greatest, if not the greatest, of uplifters we have in this country—or in any country. All of which has inured to the benefit of the towns and cities the country over; not only in our country but the world over. And that is a very pleasant reflection now as I look back."

All the media attention was accompanied by a rising tide of government investigations, lawsuits by states and the federal government, and endless public hearings. When President Theodore Roosevelt was elected in 1904, trust busting was at the center of his agenda, and the die was cast for Standard Oil. A federal suit against the company and sixty-five corporations it was alleged to have controlled was filed on November 18, 1906. Standard Oil was charged with monopolizing the oil industry and conspiring to restrain trade through railroad rebates, abuses of pipeline monopoly, predatory pricing, industrial espionage, and secret ownership of competitors. The case was not in much doubt. On May 15, 1911, in *Standard Oil Co. of New Jersey v. United States*, the Supreme Court found Rockefeller's company guilty of gross violations of the Sherman Antitrust Act and ordered it to divest itself of all subsidiaries. Said Chief Justice Edward White, "No disinterested person can survey the period in question [since 1870] without being irresistibly drawn to the conclusion that the very genius for commercial development and organization . . . soon begat an intent and purpose to exclude others . . . from the right to trade." Rockefeller received the news on the golf course. He turned calmly to his partner, Father J. P. Lennon from the local church, and said, "Father Lennon, have you some money?" The priest said no and asked why. Rockefeller replied, "Buy Standard Oil." He couldn't have been more prescient. In the next few years, the value of the divested entities soared.

At the time of the breakup, Standard Oil was a vast interconnected empire. It transported more than 80 percent of all oil produced in Pennsylvania, Ohio, and Indiana—the major petroleum-producing regions in the United States at the time. It refined more than 75 percent of all US crude. It controlled 75 percent of all retail sales of kerosene in America and was responsible for 80 percent of all US kerosene sold abroad. It supplied 90 percent of all the lubricating oil bought by the railroads. It even deployed its own shipping fleet, possessing seventy-eight steamers and nineteen sailing vessels.

The Supreme Court's decision broke up Rockefeller's empire into thirty-eight companies, with various components controlling certain territories or specific functions, such as drilling or transportation. But for many years afterward, tacit cooperation among the former Rockefeller enterprises was the order of the day, driven by material dependence, cultural similarities, and personal ties among executives. A de facto monopoly stayed intact for many years.

Still, the breakup eventually did create a number of powerful independent companies, which would become the leading players in the global oil industry. Standard Oil's offspring would include

Standard Oil of New Jersey (Exxon), Standard Oil of New York (Mobil), Standard Oil of Indiana (Amoco), Standard Oil of California (Chevron), ARCO, Conoco, and many more. Some of them have merged in recent years, creating new giants such as ExxonMobil, but none rival the original.

These new companies became the embodiment of modern multinational corporations, straddling continents with their global transportation and communications systems. They negotiated with, and often dominated, governments. They created highly sophisticated management systems to rebuild what Rockefeller had envisioned: vertical integration from exploration, to refining, to sales at the corner retail station; systems to mobilize finance on a massive scale; systems to oversee a multinational and multiethnic workforce. Like many contemporary multinationals, it was often unclear to whom these global goliaths were accountable.

Within months of Standard Oil's breakup, the public was clamoring for a share of the pieces, and the price of stock in the successor companies soared, doubling or tripling in many instances. Some had reached 500 percent their original value by the 1920s. No one came out better than John Rockefeller, who owned major stakes in each of the successor companies, amounting to a quarter of the total value. He became what many thought to be the world's richest man. (In 2014, *Forbes* estimated that in 1937, when Rockefeller died, his assets would have equaled $340 billion, or more than four times Bill Gates's net worth.)

Giving It Away

One can be forgiven for thinking that John Rockefeller was two men, one obsessed with making money, the other with giving it away; one the master at rapacious corporate tactics to slay his competitors, the other dedicated to improving conditions for humanity. Critics argued that he lived his life in two stages, the second one devoted to philanthropy as an image-building exercise to soften the image he had gained as a greedy and unscrupulous businessman during the first stage. But that explanation cannot account for the aspects of Rockefeller's life that usually get much less attention than his business career. While he was destroying industry rivals, he was leading a loving, caring life at home and in his own offices, where he was always respectful of others. He was a family man with many outside interests.

Foremost among those interests was the Church, the moral foundation he had inherited from his mother. Rockefeller's charity began early and it increased steadily as his income and wealth grew. To him, business and God were both callings, for which money was his source of power and leverage. "It may be urged that the daily vocation of life is one thing, and the work of philanthropy quite another. I have no sympathy with this notion," he wrote later in life. "The man who plans to do all his giving on Sunday is a poor prop for the institutions of the country." Charity could have given his business success a larger purpose. We should not rule out, either, a general sense of obligation or noblesse oblige that came with massive wealth accumulated in the Gilded Age. Rockefeller was conscious of the social problems spawned by an age of industrialization and urbanization. Later in life he wrote, "In this country we have come to the period when we can well afford to ask the ablest men to devote more of their time, thought and money to the public well-being."

Rockefeller wasn't alone in this kind of endeavor; in fact, Andrew Carnegie's investment in libraries around the world preceded Rockefeller's first major gifts. But looking at the entire picture—the scope, the philosophy, the strategy, the organization—the oil baron became the father of modern global philanthropy.

Rockefeller gave away hundreds of millions of dollars for a multitude of projects. Four of the biggest and most transformative were the establishment of the University of Chicago (1890);

the underwriting of educational improvement across the nation under the auspices of the General Education Board (1902); the Rockefeller Institute for Medical Research (1901), later renamed Rockefeller University; and the Rockefeller Foundation (1913). The common aim of these latter two institutions was to push advances in science and health by improving the tools and discipline of research efforts, and breaking down barriers between different fields of knowledge. In an unprecedented way, both had a global focus from their inception.

The Rockefeller Institute for Medical Research, for example, was the first biomedical research institute to be established in the United States with the addition of a hospital in 1907. Its early focus was on pneumonia, influenza, diphtheria, typhoid fever, and other infectious diseases that were among the leading causes of death around the world at the beginning of the twentieth century. The institute adopted the best aspects of others in the field—the Koch Institute in Berlin, the Pasteur Institute in Paris, the Lister Institute in London—but it also pioneered a new model, by abolishing formal departments and emphasizing cross-disciplinary work, focusing only on research as opposed to teaching and clinical work, and providing generous grants for scientists to follow their own interests, as opposed to the agenda of the director. (Years later, teaching and clinical activities were added.) The global mind-set was reflected not just in the research agendas—diseases that were killing people around the world—but also in the early recruitment of scientists from France, Austria, Japan, Russia, and Germany.

Later the Rockefeller Foundation was established with a charter that was even more self-consciously global. It was, in fact, the world's first global foundation, with an explicit mission of helping humanity "throughout the world." From its start, the foundation supported work to help poor and vulnerable populations in Asia, Africa, and Latin America, providing advice, giving financial support, and building institutions.

Within a decade it became the largest grant-making institution in the world. In its first several decades, the foundation mounted a program to eradicate hookworm in Mexico, supported critical research on yellow fever in Africa, and bankrolled great biomedical scientists such as Hideyo Noguchi from Japan. It built medical schools in China, Ghana, Uganda, and elsewhere. It trained health officials and built health care systems all over the developing world. It has since established and expanded programs in nutrition and agriculture, building centers of research such as the International Rice Research Institute in the Philippines, the International Maize and Wheat Improvement Center in Mexico, and the International Center for Tropical Agriculture in Colombia, all of which would play central roles in the green revolution that vastly increased farm output in the 1960s and '70s. Its International Health Division, established at the outset, became involved in eighty different countries.

What is so critical about *global* philanthropy in the overall picture of globalization itself? Rockefeller very deliberately envisioned and established institutions that could develop and execute ideas of universal applicability. Perhaps it was because he inherently realized the limitations of a company like Standard Oil in terms of its mission and role in the world, or perhaps his worldwide business gave him more than the usual understanding of the social challenges in such regions as Asia, Africa, and Latin America. At some level he also might have known that these philanthropic institutions—and others for which they could serve as a model—would operate in the space that fell between the work of governments and companies, work that should be depoliticized, that should embody the latest scientific and technological knowledge that focused on the well-being of human beings, like health and education. His ideas were fundamental to the evolution of some of the most important institutional pillars of globalization today—from organizations that bear the Rockefeller name to others like the Ford Foundation, the Bill and Melinda Gates Foundation, or the Gordon and Betty Moore Foundation. In the decade that ended in 2008, international grant making by American foundations alone had been increasing faster than domestic grant making, reaching over $6 billion. That was about a quarter of all US foreign aid for economic development.

What also made these endeavors so significant—as significant to science and medicine as Standard Oil was to the development of energy and industry—was the characteristic Rockefeller combination of bold vision and disciplined execution. Rockefeller hired Reverend Frederick Gates, a Baptist minister, to help him administer his philanthropic endeavors. Dispensing Rockefeller's escalating fortune in an organized way required some clear principles. Among them: Surround yourself with the most impressive experts in the field and carefully identify the avenues of research that would add most to existing knowledge. Focus on the roots of the problem, aiming for prevention rather than just mitigating the symptoms.

Rockefeller was obsessed with encouraging self-reliance, as opposed to fostering a culture of dependency. He was among the early proponents of matching grants, seeking to ensure that his donation was never the sole gift and that the receiving institution was not dependent on him alone. He was determined to hire excellent managers and an exceptionally knowledgeable board, but otherwise he did not interfere with the operations of the organizations he established. Rockefeller wanted to professionalize big philanthropy, to run it like a highly focused and competitive business, dedicated to producing cutting-edge research. "An individual institution of learning can have only a narrow sphere. It can only reach a limited number of people. But every new fact discovered, every widening of the boundaries of human knowledge by research, becomes universally known to all institutions of learning, and becomes a benefaction to the whole race," he said.

In his lifetime, Rockefeller's generosity could not overcome his brutal business reputation, but today that image has softened, and the name is more closely associated with giving than taking.

In his retirement, which would last almost four decades, he evidenced a great and uncharacteristic capacity to relax. Among other activities, he enjoyed biking, horse riding, swimming, and skating. He had a particular love of landscape gardening, which he once described as "the art of laying out roads and paths and work of that kind," perhaps a metaphor for his innate capacity to think strategically. He continued his intense involvement in the church, attending services, teaching classes, contributing funds, maintaining his calm. He built an estate at Pocantico Hills, outside of New York City. In this de facto fortress, he lived his final years almost as a fugitive and recluse. On May 23, 1937, at the age of ninety-eight, he died of heart failure in his sleep. Two days later, a five-minute silence was observed in all of Standard Oil's successor firms, now active in every corner of the world. After a simple family funeral, he was buried in Cleveland, where he had started as a lowly clerk. His obituary in the *New York Times* called him not only the richest man in the world but "the pioneer of efficient business organization and of the modern corporation, the most powerful capitalist of his age, and the greatest philanthropist and patron of higher education, scientific research and public health in the history of the world."

No one did more to shape the DNA of the global oil industry than John D. Rockefeller. Although his most intense activity took place more than a century ago, and although the global picture for energy is exceedingly dynamic, petroleum is still the lifeblood of global economic activity. It is still the world's most important commodity, still a key ingredient in fueling all kinds of industry including automotive, aviation, shipping, mining, construction, and many others that span the globe. It is still a central component of geopolitics. The offspring of Rockefeller's Standard Oil—such as ExxonMobil and Chevron—still constitute some of the largest multinational companies on earth, linking nations via trade, investment, technology research facilities, and jobs. At the same time, they also contribute to one of the biggest threats to the world: destructive climate change. Rockefeller's business was both a prime element of globalization and a manifestation of it.

Standard Oil and the companies that grew out of it epitomize the influence that multinational companies have had on globalization. Harkening back to Robert Clive and the East India Company and looking ahead to Andy Grove and Intel (in Chapter IX), these organizations have been at least as important as governments in providing the glue that links together disparate societies around the world. At times companies like the East India Company and Standard Oil had the unambiguous support of their home countries, including the full backing of their diplomatic and military assets. By the mid-twentieth century, multinationals had independent clout by virtue of the essential goods and services they provided in a world so hungry for them; indeed, corporations such as General Electric (GE), International Business Machines Corporation (IBM), and Intel became essential components for modernization in Europe, Asia, Africa, the Middle East, and Latin America. With the power they accrued, they became deeply enmeshed in the internal affairs of countries—in labor policies, in the shape of financial regulation, even in the determination of who controlled the government. For better or worse, the East India Company, Standard Oil, and those who followed helped define what globalization is, how it works, and who benefits and who loses from it.

Rockefeller also did more than anyone else to establish the evolving model for running global philanthropy on a major scale. Rockefeller University, whose roots go back a century, is still at the cutting edge of biotechnological innovation. The Rockefeller Foundation is still making breakthroughs in health and nutrition, preparing to launch another green revolution, and focusing on improving the management of the world's multiplying cities. The Bill and Melinda Gates Foundation, perhaps the most visible and best financed of nongovernmental organizations devoted to global health and associated issues, seems to be a direct descendant of what Rockefeller built. This much is certain: the more complex globalization becomes, the more such organizations will be essential to work alongside governments and corporations to manage some of the biggest challenges that mankind faces.

John D. Rockefeller was lauded for his business acumen and for his generosity, and vilified for his rapacious business practices, and both the praise and the criticism were well earned. But few men in history have done more to power the world's growth and development, to address some of its most pressing social problems, and to make the world more economically integrated.

MARTINE FRANCK, MAGNUM PHOTOS

Chapter VII

JEAN MONNET

The Diplomat Who Reinvented Europe

1888–1979

Norway
Denmark
Ireland
United Kingdom
Netherlands
Belgium
West Germany
Luxembourg
France
Switzerland
Portugal
Spain
Morocco
Algeria
(part of France)
Tunisia
Map © Nat Case, INCase, LLC

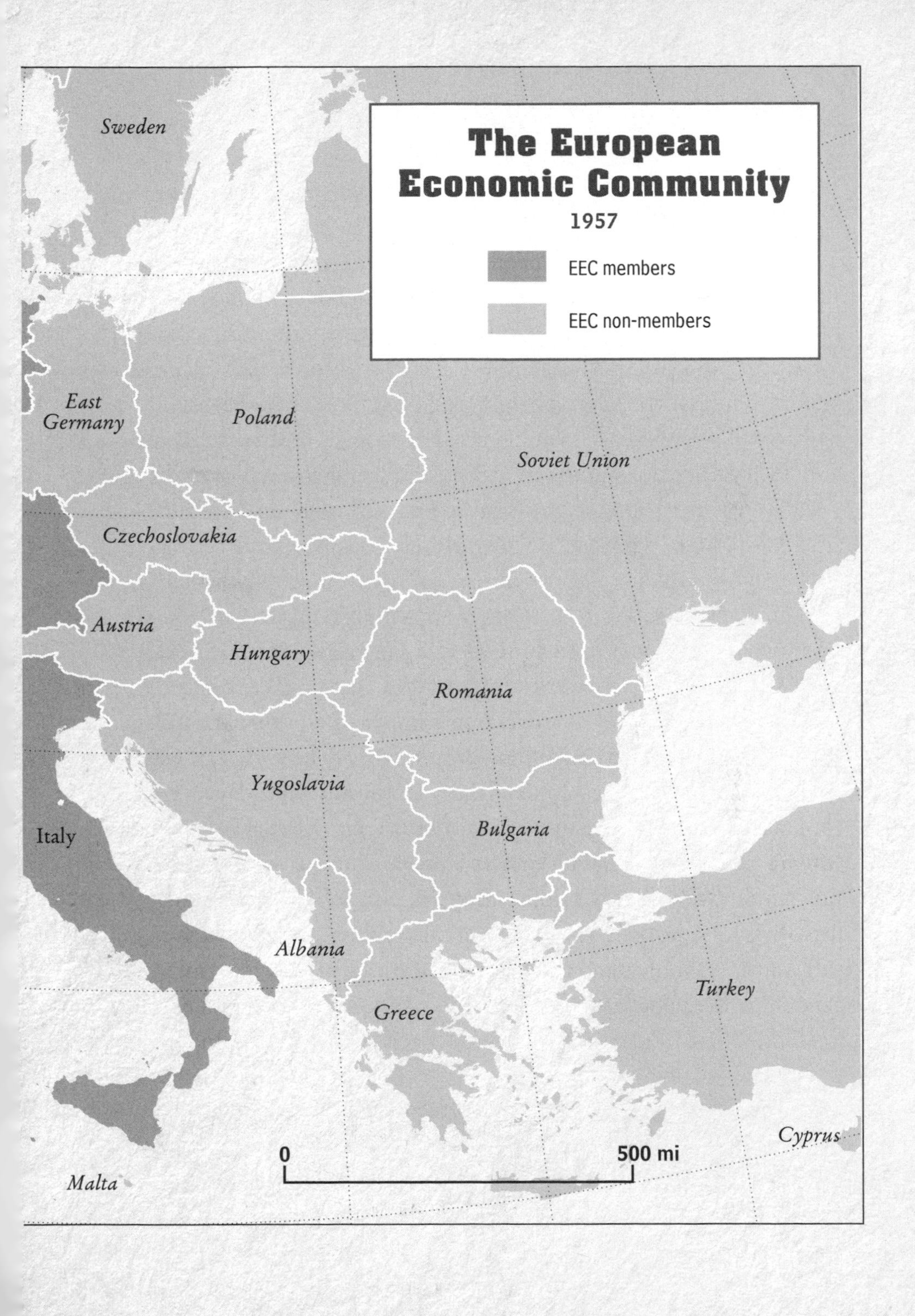

The European Economic Community
1957
EEC members
EEC non-members
Sweden
East Germany
Poland
Soviet Union
Czechoslovakia
Austria
Hungary
Romania
Yugoslavia
Italy
Bulgaria
Albania
Greece
Turkey
Cyprus
Malta
0
500 mi

Norway
Denmark
Ireland
United Kingdom
Netherlands
Belgium
Germany
Luxembourg
Czech
France
Switzerland
Austria
Slovenia
Italy
Portugal
Spain
Morocco
Algeria
Tunisia
Malta*
Map © Nat Case, INCase, LLC

The European Union
2015
Eurozone members
Other European Union members
European Union non-members
Finland
Sweden
Estonia
Latvia
Lithuania
Belarus
Russia
Poland
Ukraine
Republic
Slovakia
Moldova
Hungary
Romania
Croatia
Bosnia and Herzegovina
Serbia
Bulgaria
Kosovo
Montenegro
Macedonia
Albania
Turkey
Greece
Syria
Cyprus*
Lebanon
*Cyprus and Malta are EU and Eurozone members

It was August 1914, the eve of World War I. Although the twenty-five-year-old Jean Monnet had never had a job in government and was physically unfit for the military, he wanted to speak to the prime minister of France about how to prepare the allies for the coming conflagration. His father, a cognac producer, thought the idea was presumptuous, to say the least. Nevertheless, the elder Monnet's lawyer was close to Prime Minister René Viviani, and he set up an introduction for the second week of September. Having traveled the Continent selling cognac for his father, young Monnet knew that the British and French ways of doing business were substantially different and, he told the prime minister, he felt strongly that if the two nations were to have any chance to stand up to the Axis powers, they would have to create a new organization to align and execute a common strategy. When Viviani responded that the government already had means in place to oversee cooperation with the allies, Monnet explained his view of the difference between allied communication—the act of meeting and discussing issues of common concern—and the actual implementation of one common policy.

Monnet's presentation evidently proved persuasive enough that the prime minister steered Monnet into a job as a liaison between Paris and London, coordinating the delivery of war supplies. Monnet later reflected on why he had instinctually bypassed everyone below the top official. "It was not some conceited need to make myself important, but a simple idea of what would be most effective," he wrote. "First, have an idea, then look for the man who can put it to work." This singular perspective—a nuanced parsing of the distinction between ideas and actions, and how to connect the two—would animate Monnet's life for the next twenty-five years. It would give him the influence in government circles—despite the fact that he held no official office—to become the architect of new supranational institutions that would bring the ideal of a united Europe closer to fruition. He thus led one of the most far-reaching experiments in the history of globalization, for it constituted no less than the abolishment of borders that separated sovereign states when it came to trade and regulatory barriers, and ultimately to substituting a common currency managed by one central bank in place of nineteen separate national monetary systems. "The European Union is an unparalleled historical experiment in governance," wrote Professor Kathleen R. McNamara of Georgetown University. "There is no other example in modern times of such an intensive effort to establish a peaceful, prosperous political community beyond the nation-state."

In the Cognac Trade "One Did One Thing, Slowly"

In grammar school Monnet was so restless that he was given a special dispensation to get up and move around class during lessons. When he left school in 1904 at the age of sixteen to join the family cognac business, his father sent him to Great Britain to learn English and to work as an apprentice. For two years he lived with a family of wine merchants and reveled in the challenge of trying to understand a new culture. He was fascinated by the City of London, the financial district where relationships were so cozy that they created what was effectively a closed club.

After two years, Monnet set out on his first voyage to find markets and clients for J. G. Monnet & Co. His father's advice: "Don't take any books. No one can do your thinking for you. Look out the window, talk to people. Pay attention to your neighbor." As a salesman Monnet convinced the important Hudson's

Bay Co. to carry his father's brandy in Canada. He helped his firm enter the American, British, and Egyptian markets. Other trips took him to Russia and China, where he demonstrated a fine eye for cultural nuances. "In China you have to know how to wait," he once said. "In the United States you have to know how to come back. Two forms of patience to which cognac, itself the fruit of time, is a good preparation."

We can only guess how Monnet had achieved so sophisticated a view on international relations at such a young age. He grew up in a large house in the Cognac region, about halfway up the Atlantic coast of France, where his family entertained liquor distributors and salesmen from Britain, Germany, Scandinavia, America, and beyond. He spent hours listening to tales of adventure and news of faraway places, and saw that his father's business depended not only on personal relationships but also on an ability to judge who among these contacts had the most reliable information on distant markets. Working in a cognac cooperative, the problems of one member became the problems of the group, and the challenge of growing cognac grapes would test the patience of them all. "One did one thing, slowly and with concentration," Monnet once said. Monnet himself would later trace his perspective to his father, a man of big ideas, but especially to his mother, a woman who knew how to get things done. "Unlike my father, who could daydream, or lose his temper, or if necessary escape on some journey," Monnet said, "my mother was tied to reality, and she always brought the family back to it. I may have my father's imagination, but my mother taught me that nothing can be achieved unless it is built on reality. She distrusted ideas as such. She wanted to know what was to be done with them." Monnet was an exquisite blend of both parents. Thus, he could not only envision big things but execute them, too.

World War I: Battle Tests for His Big Idea

After World War I broke out and Prime Minister Viviani paved the way for Monnet to work in the London office of the French Civil Supplies Service, the young Frenchman was in constant motion between Britain and France. His frustration mounted as he saw Paris and London giving mere lip service to collaboration. Their national purchasing bodies were competing with each other, bidding up the prices of essential raw materials from countries such as Argentina and Australia. Britain, which had a longer global reach, saw itself as too powerful and independent to compromise with France or comply with requests to supply French forces.

Witnessing firsthand the absence of coherent planning, he drew at least two conclusions. First, leaving the civilian war effort in the hands of naturally competitive private companies from the two countries was undercutting the efforts of both nations. Second, the allies could not pursue their national interests separately and then hope to coordinate their actions successfully. They had to pool interests at the start, creating a totally different mind-set and strategy on both sides. They had to act as one entity.

Monnet pushed for closer allied coordination and achieved his first breakthrough by persuading the British and French to establish the "Wheat Executive," with equal representation from both nations. The team placed joint orders for wheat, which was in short supply worldwide. In intense bargaining, representatives of the two nations, coordinated by Monnet, managed to figure out how this staple food would be parceled out between the two sides. This buyers' cooperative helped keep the price of wheat low and the supply flowing.

Monnet's influence continued to grow because he knew how to push for deeper collaboration in a manner that never seemed to put the interests of France above those of the alliance. By 1917 he was head of the French mission in London for all food and shipping. His efforts led to the establishment of the Allied Maritime Transport Council, which oversaw the procurement of war matériel, including pig iron and steel for guns, ammunition, and ships, as well as coal, wood, and other fuels. British diplomatic historian Harold Nicolson later wrote that the Allied Maritime Transport Council was, in its time, the most advanced experiment in international cooperation ever attempted.

By the time the war ended on November 11, 1918, Monnet had developed a network of relationships at the highest government levels in France and Britain, and he had established himself as a man of one big idea—the need for deep allied collaboration based on the delegation of sovereign powers to a person or institution with the authority to make common decisions. He had seen how high officials made decisions under extreme stress, what they needed by way of information, how they did or did not rely on their staffs. He also had his first glimpse at the workings of an awesome power that was to influence virtually everything he accomplished in his professional life—the United States. As *New York Times* journalist James Reston wrote in his obituary of Monnet in 1979, "He realized over the Allied arguments during the First World War that France could not defend Britain nor Britain defend Britain, unless, with the vital assistance of the United States, they all agreed on common policies to defend a common civilization."

At the time of the armistice, Monnet was just thirty. When Britain's Sir Eric Drummond was made the secretary-general of the new League of Nations, he picked Monnet as his deputy. From its inception the league was hamstrung by the need to get the unanimous agreement of its member nations before it could act, by its lack of enforcement powers, and by America's refusal to join. Monnet was deeply disappointed, having seen that by preserving the principle of full sovereignty, the league guaranteed that the self-interest of member states would always triumph over the common interest. But he still held out hope for gradual progress. "Cooperation between nations will grow from their getting to know one another better, and from interpenetration between their constituent elements and those of their neighbors," he wrote—too optimistically, as it turned out—in a memorandum in May 1919.

This viewpoint—that the habit of cooperation required small incremental steps that would build on each other—was to become an article of faith for him. He would later describe his view on how human perspectives evolve through the parable of a man who scales a mountain, seeing many different views that improve as he climbs. The top offers the best view, so it is important to give the man incentive to keep climbing, however slowly. As he ascends, he will see things differently. The key is to keep him moving up, Monnet believed.

A Diversion into Chinese Railways

Monnet led several major initiatives at the league, but in 1922 his father called saying that the family business was in trouble. He returned home and, with astute judgment about focus and organization, revived the company. Four years later, he was hired on as vice president and managing partner of the Paris office of Blair and Company, an American investment bank.

At the time, American investment banks were major actors in the post–World War I recovery and were underwriting extensive loans for European countries. Using his high-level connections, Monnet negotiated loans for Poland, Romania, Yugoslavia, and Bulgaria, coming into close contact with central bankers and finance ministers from Europe and the United States. At a time when international finance was dominated by Americans, Monnet developed ties with top Wall Street bankers and lawyers who were well connected in Washington and often moved in and out of high US government jobs. He thus came to know what would become the inner circles of the Roosevelt, Truman, Eisenhower, and Kennedy administrations, including future secretaries of state such as Dean Acheson and John Foster Dulles, as well as other men on whom presidents would rely to aid European recovery after World War II, including John J. McCloy, George Ball, and Averell Harriman.

When Blair and Company merged with Bank of America in 1929, Monnet moved to San Francisco to be vice chairman. The merger soon soured, however, and Monnet returned to Paris and became an independent financial consultant and then a partner in a small firm, Monnet, Murnane & Co. A highlight of his career in this period was his work in China. Hired by Chinese finance minister T. V. Soong to organize the country's chaotic economic development program, Monnet saw that foreign lenders were pouring money into China without getting much done on the ground or earning much of a financial return. Always conscious of the problems of fragmented interests, he managed to reorganize the scattered Chinese banks into a consortium called the China Development Finance Corporation, which became the central conduit both for channeling foreign capital and for raising matching investments from Chinese citizens. The operation was considered a great success and helped fund the construction of an extensive railway system for the Middle Kingdom.

Uniting Europe against Hitler

By the fall of 1938, Nazi aggression was fanning fears throughout Europe. France and England foresaw another major war and Monnet was eager to help his country defend itself. Because securing help from the United States was critical and because Monnet had extensive American connections, the French government made him an official emissary and dispatched him that October to meet with President Franklin D. Roosevelt in his country home at Hyde Park, New York. Roosevelt, who was not accompanied by any aides, said isolationist sentiment in the United States was too strong for America to get directly involved in the war that he knew was coming. Monnet talked about the need for France to acquire warplanes from America on strictly commercial terms, so the deals could not be criticized in the United States as American intervention in a European war. He also asked whether French warplanes could be repositioned in the United States if the German air force began bombing French airports. Monnet would later recall how Roosevelt, working from memory, had taken out a slip of paper to tally the combined weapons production potential of the United States, England, and France compared to that of Germany. Monnet asked if he could keep the scrap of paper, which was potentially explosive if it became public, and FDR casually turned it over to him, deeply impressing Monnet with both the president's grip on the facts and his trusting nature. Roosevelt did not respond definitively to Monnet's requests, however, and instead sent him to see Secretary of the Treasury Henry Morgenthau Jr.

From that point on Monnet became increasingly tied to the Allied war effort. He was the force behind a joint Anglo-French campaign to pressure Washington to sell the Allies more weapons and to help the United States understand that even if it remained officially neutral, it would need to vastly enlarge its defense production to meet not just its own needs but those of its allies.

On June 25, 1940, French resistance to Germany collapsed and Paris surrendered to Berlin. While France was no longer in a position to buy military hardware, Britain still was. Prime Minister Winston Churchill gave Monnet a British passport to help England secure arms from the Roosevelt administration. The

Frenchman became involved in American decisions on military production, staying in direct contact with Secretary of War John McCloy, Roosevelt adviser Harry Hopkins, and Supreme Court Justice Felix Frankfurter, another Roosevelt intimate. He later recalled the ceaseless discussions with Roosevelt's inner circle over dinner, via telephone, and in written notes at all hours of the night. Behind the scenes, he helped forge the relationship between Roosevelt and Churchill that became a cornerstone of the Allied war effort, even authoring cables from each of these leaders to the other. Without Monnet's constant prodding, it is unlikely that the United States would have ramped up its production of war matériel so early. John Maynard Keynes, the great British economist, said Monnet's efforts shortened the war by a year, saving countless lives and avoiding even more widespread destruction.

Consolidating the French Forces in Exile

When Japan bombed Pearl Harbor on December 7, 1941, the United States officially entered the war. It began producing armaments at maximum speed. It also ramped up cooperation with Great Britain, so Monnet was no longer needed in Washington as a liaison to London. He was looking for another role when the Allies landed in North Africa on November 8, 1942, a first step toward the liberation of France. In February 1943, Roosevelt's top aide, Harry Hopkins, asked Monnet to assess the equipment needs of the exiled French forces who were massing in Algiers. Monnet found the French leadership badly fragmented, with General Charles de Gaulle in London claiming to be the true voice of France in exile, General Henri Honoré Giraud in Algeria claiming to speak for the "free French" in North Africa, and two other would-be leaders in occupied France, one heading the pro-Nazi government, the other overseeing the armed resistance to that government. Monnet had to quiet the rivalries among the exile and resistance forces before he could secure Washington's help for the free French, and he proved himself a formidable mediator in this role. By June 1943 he had organized the Comité Français de Libération Nationale (CFLN), comprising all of the French resistance and exile groups. The work introduced Monnet to General Charles de Gaulle, who eventually took over the CFLN. There were great tensions between de Gaulle, an ardent nationalist who believed he was the embodiment of France, and Monnet, the internationalist who put faith in collective action. But for now the two men needed each other.

With the French resistance forces united, and liberation approaching, Monnet returned to Washington as the official representative of de Gaulle's government in exile, anxious to participate in the postwar reconstruction of Europe. He knew that many of the big decisions would be made by Washington, and, far more than any other Frenchman, he had earned the trust of top American officials during the war. The problem was that FDR did not want to recognize de Gaulle as the leader of France until the French general was chosen in a democratic election. Since that would take a while, Monnet took the patient and incremental approach to Washington that would become his signature style: "Rather than make a frontal attack on the problem of diplomatic recognition, which for the moment was insoluble, I believed that by seeking the maximum common interests, both material and psychological, it would eventually be possible to create a *de facto* situation, which of necessity soon became *de jure*," he wrote. That is exactly what happened.

When Paris was liberated on August 25, 1944, France was in terrible condition, having lost more than a quarter of its national wealth during the war. Its national income, adjusted for inflation, was little more than half its peak in 1929. Three million French prisoners and deportees were coming home amid shortages of coal, locomotives, medical supplies, blankets, footwear, baby linen, and many other necessities. The national infrastructure was barely functioning, the average age of its machinery was twenty-five years, and the currency had been destroyed. Monnet had a long talk with de Gaulle about postwar France. "You speak of greatness," said Monnet, "but today the French are small. There will only be greatness when the French are at a stature to warrant it. . . . For this purpose they must modernize—because at this moment they are not modern. Materially the country needs to be transformed." De Gaulle replied, "You are certainly right. Do you want to try?" The nationalist and the internationalist were in agreement on one overriding objective: France must have a plan to regenerate the economy, and Monnet had to lead it.

The Future of France in Seven Pages

In November 1945, Monnet and his family moved back to Paris, where he would remain for the rest of his life. He briefly considered running for office but realized politics was not his calling. His achievements had come from influencing politicians, not from being one himself. He understood that while political leaders had to skim the surface of many problems, he could go deep into an issue and become invaluable in achieving big dreams.

De Gaulle gave Monnet a role tailored for him. He was made the equivalent of a minister of planning, reporting to the prime minister, but with no official tie to government, so as not to slow him down or provoke the jealousy of other ministers. He set up his office in the Bristol Hotel, which had been commandeered by the government. Space was tight and he was given just a few rooms, a board over a bathtub serving as one desk. He assembled a small staff and worked at his usual frenzied pace to produce France's first national economic plan.

Monnet found that every ministry was competing for limited sources of funding, energy, and manpower, and he set out to define common interests and goals that all sides could support. It was the same type of challenge of trying to coalesce disparate interests, antagonistic to one another. He faced it in dealing with Britain and France in World War I, with the League of Nations, with the Chinese railroads, with the United States and its allies in World War II, and with the fractured French resistance forces in Algiers. His approach of uniting the government in Paris was a novel one for France, which had no tradition of bringing civil servants, CEOs, and labor leaders together in an attempt to define the national interest. Because the country was broke and the United States alone had money to spend abroad, Monnet also needed to create a plan that would attract American financial support.

He wrote a seven-page outline for France's modernization and presented it to de Gaulle in December 1945. The paper was notable for its clarity and simplicity, the enduring features of all of Monnet's plans. It called for government-wide collaboration in setting goals, and modernization commissions for critical sectors such as coal, electricity, transport, and farm machinery. The commissions would report to a Central Planning Commission with representatives of labor, management, civil servants, and every political party. The planning commission would be a standalone agency, run by Monnet and reporting only to the prime minister. De Gaulle quickly approved it.

That outline became a two-hundred-page blueprint that went into operation about one year later with detailed guidelines for a uniquely collective national effort, focusing on energy, steel, and related infrastructure as the basis of France's national regeneration. Though lack of funding threatened the plan at the start, within months the United States had come to its aid due to the kind of serendipity that favors the well prepared. In June 1947, worried about the potential for a Communist takeover in war-weakened Europe, Washington announced a financial rescue operation in the form of the Marshall Plan, named for US secretary of state George Marshall. The plan asked Europe to come together to list its needs for recovery and reconstruction. At a conference of European governments the next month it was clear that, thanks to Monnet, France was better organized than other countries to apply aid effectively, and that its internally coordinated modernization plans should serve as a model for Europe. France received the lion's share of American aid, fully funding Monnet's plan for French modernization.

Europe Comes Together

The Marshall Plan changed everything, pushing Europe in exactly the direction Monnet wanted it to go. The war had left tens of millions of people dead on the Continent, farms and small towns devastated, cities turned to rubble. Countless homeless people roamed the streets, dazed and hungry. Law and order had collapsed in many places, armed gangs terrorized neighborhoods, and ethnic cleansing was rampant. Great tension now arose between America and Europe. In the first instance Washington saw the need for Europeans to work together on recovery. However, Europeans saw themselves as sovereign competitors. The United States also wanted a united Europe to share the burdens of global diplomacy and defense and to become a prosperous market for American exports. As a result, Washington was adamant about putting European integration on the broad Allied political agenda. During a trip to the United States, Monnet explained the new environment in a memo he sent to French foreign minister Robert Schuman. "Everything I have seen and reflected on here leads one to a conclusion which is now my profound conviction: that to tackle the present situation, to face the dangers that threaten us, and to match the American effort, the countries of Western Europe must turn their national efforts into a truly European effort. This will be possible only through a Federation of the West." This was not a new idea for Monnet—he had talked about it with various colleagues years earlier, but he felt the time when people would listen was now approaching.

To be sure, many European leaders understood that a return to the prewar status quo ante would not be enough to restore their influence in the newly competitive world. Even before the war, Europe had been in decline. In 1913, it produced 45 percent of the world's industrial goods, but by 1937 that figure had fallen to 34 percent. Now, after the war, Europe faced industrial competition from two giants—the United States to its west and the Union of Soviet Socialist Republics to its east. No European nation was a match for these continent-sized economies, and conditions in Europe were deteriorating with, among other things, inflation and growing shortages of everything from food to transportation equipment.

Europe also faced the challenge of Germany. The Cold War between America and the Soviet Union was in full swing, and Germany was being divided into two countries, west and east. West Germany was recovering very fast, worrying many Europeans that it could once again threaten the peace with its expansionist tendencies. France, in particular, was concerned about resurgent German nationalism and some French leaders even harbored designs on dismantling the German state and taking over its industrial regions. The United States and other allies took a different tack, arguing that to hold down Germany would deeply anger the population, making it ripe for Communist revolution. This debate came to a head in August 1950, when Communist North Korea, backed by China and Russia, invaded South Korea. The United States and much of Europe were convinced that this act of aggression could be the opening salvo in a new world war, this time pitting democracy against communism, and that West Germany would be the next target of the Communist armies.

The need to incorporate a vibrant Germany into a strong new European political framework was increasingly urgent, especially in the minds of America's leaders, but the basis for that arrangement did not exist. In fits and starts Europe had been inching toward a common identity for two millennia, but such cohesion never took hold despite the fact that music, art, theater, literature, and architecture jumped borders with ease. Leaders going back to Augustus Caesar, Charlemagne, Napoleon, and even Hitler had tried to unify the Continent by force. Intellectual movements, including socialism and trade unionism, had promoted a unified Europe, and many organizations had created pan-European religious councils, legislatures, and peace groups. In the years since 1946, leading politicians had called for a European League for Economic Cooperation or a United States of Europe, and others had established the Union of European Federalists in Paris, the Socialist Movement for the United States of Europe in London, the Council of Europe, and the Organisation for European Economic Co-operation.

All of these efforts, however, fell well short of the Europe that Monnet envisioned because they were all based on a traditional idea of national sovereignty, which held that the sole concern of government is to pursue its national self-interest. This view not only encouraged proud nationalism but it also exposed anyone who proposed ceding national authority to a greater European power to charges of betraying their own countries. Monnet was therefore treading on ground that was not just politically sensitive but, in the eyes of many politicians, heretical.

Then came a critical turning point. In a meeting that took place from the fifth to the nineteenth of September, 1949, US secretary of state Dean Acheson told French foreign minister Robert Schuman that at their next meeting, set for some yet-to-be-decided day in May 1950, the United States expected to see a French plan for how to deal with Germany. Schuman was on the spot. He knew America would not agree to a nationalistic French plan for the suppression of Germany, but he also knew that French voters would punish any politician who was not perceived as clearly defending French interests on the German issue. Schuman saw no way out of this impossible dilemma. He had eight months to figure out how to repair a divided Continent, and he had no big ideas himself.

The First Experiment in Supranationalism

Monnet was aware of Schuman's problem and saw the opportunity it presented. In March 1950, he went on vacation to Switzerland, where he spent a lot of time on long walks, for many decades his way of relaxing and thinking. Usually he walked alone but occasionally he took a close friend such as George Ball, an American diplomat, who later recalled the experience: "Monnet thought most effectively when he not only talked but walked." He would walk as much as nine miles a day, arriving at his office "electrically charged with priorities and ideas which fell like hail on subordinates' desks."

When he returned from the Swiss mountains he knew exactly how to help Schuman with the German problem. Acting on his own initiative, he gathered a few aides as he began to put together a proposal that he would label the Schuman Plan, reflecting Monnet's understanding that he would gain enormous leverage by crediting others for a powerful idea. His genius was in honing big ideas into simple ones, hammered out in round-the-clock discussions with just a small staff selected without regard for stature, experience, or nationality but only for intelligence and imagination. In draft after draft, they would shape the idea to near perfection. Often Monnet would remain silent or simply ask questions to keep the discussion going, many times during meals in Monnet's home or in the small dining room he maintained in his office. He made everyone feel that whatever he was working on was of singular and monumental importance. He also generated a sense of permanent crisis, working his staff to the bone seven days a week. A Monnet proposal was always refined to a high sheen of conceptual clarity. It was always embedded in the larger political setting, and it was always a call to action.

The Schuman Plan was a simple answer to the complex problem of West Germany. Like many in France and elsewhere in Europe, Monnet was worried about another European war, but he was less afraid that Germany would be the instigator than that it would be the prize over which the West and the Soviets would fight. Convinced that the German problem could not be solved in a conventional way, he thought that any deal would have to change the fundamental context of governance in Europe. And to do that, Monnet concentrated on the problem that he felt was at the heart of Franco-German tension: who would own and control Europe's coal and steel resources, the key to industrial supremacy and the engine of any national war machine? Throughout the twentieth century, German dominance of these critical industries had given it military superiority on the European continent.

Monnet came to this conclusion: the French would have to agree to German industrial revival, but Germany would have to find a way to put French industry on equal footing to its own. Such a plan could be the small and practical first step toward something far more grandiose: a united Europe. "Experience had taught me that one cannot act in general terms, starting from a vague concept," Monnet later wrote. "But anything becomes possible as soon as one can concentrate on one precise point which leads to everything else."

Monnet's concrete idea was to put coal and steel from France, Germany, and four other European nations under the control of a new European Coal and Steel Community. The ECSC would not coordinate national policies; in a dramatic departure from the way governments had behaved for centuries, it would establish a new "High Authority" to develop and administer a single policy for all the nations of Europe's main coal, iron, and steel belt, including parts of Belgium, Luxembourg, Italy, and the Netherlands as well as the Lorraine Valley in France and the Ruhr and Saar regions of Germany. National governments would not have the right to veto decisions of the High Authority, forcing an unprecedented surrender of national authority to an international organization. The supranational powers of the ECSC went way beyond any powers granted to the failed League of Nations, and beyond those given to new institutions that were being established in the wake of World War II, such as the United Nations, the International Monetary Fund, and the World Bank.

In late April 1950, after overseeing nine drafts, Monnet sent his succinctly written document to Schuman, who received it just as he was boarding a Friday train for his weekend home in the country. When he returned to Paris on Monday morning, Schuman told his chief of staff, "I've read it. I'll use it." The proposal was shown to only nine people before the French cabinet approved it on May 9, 1950. The same day Schuman sent a secret emissary to Bonn to show the plan to German chancellor Konrad Adenauer, who left a cabinet meeting to read it. He immediately gave his full support. Neither Schuman nor Adenauer would have underestimated the coming political resistance from national politicians, but they also understood the dangers to Europe if they did not act now, and they were acutely aware that the moment of opportunity was upon them. They were patriots but also global statesmen in the instant when it really mattered.

"That's Right, a Leap in the Dark"

That evening in Paris, about two hundred reporters crowded into a salon adorned with gilded mirrors and chandeliers at the French foreign ministry, where Schuman announced: "It is no longer the moment for vain words but for a bold act. The French government proposes to place the whole of Franco-German coal and steel under a common High Authority, in an organization open to the participation of other countries of Europe." He called the plan "a first step in the federation of Europe" that would "change the destinies of those regions which have long been devoted to the manufacture of munitions of war . . ." With only Monnet's brief outline to go on, Schuman couldn't answer any detailed questions, and as he was leaving, one reporter yelled out, "In other words, it's a leap in the dark?" Schuman replied, "That's right, a leap in the dark." Although few realized it, at that moment the cornerstone of the European Community was cemented in place. In a joint statement, Adenauer and Schuman emphasized the point. "By the signature of this Treaty, the participating Parties give proof of their determination to create the first supranational institution and the true foundation of an organized Europe."

The next month Schuman opened a Paris conference to hammer out a treaty creating the ECSC, using a draft prepared by Monnet. "Never before have states undertaken or even envisaged the joint delegation of part of their national sovereignty to an independent supranational body," said Schuman. As chair of the drafting committee, Monnet's challenge was to reconcile the often clashing histories, goals, and business systems of the founding nations. When talks heated up, the representatives of every country including France tried to preserve their sovereign authority by pushing coordination as opposed to integration. Monnet conducted a patient give-and-take but never allowed the delegates to forget the larger picture. "Remember," he would say, "we are here to build a European Community. The supranational authority is not merely the best means for solving economic problems: it is also the first move towards federation." Although it was Schuman's burden to get the ECSC treaty ratified, Monnet emerged as the architect and builder of this unprecedented transformation in European government.

The treaty included an ECSC executive body called the High Authority, a Council of Ministers, a Common Assembly, and a Court. It called on member countries to reduce customs duties, subsidies, and restrictive trade practices in the coal and steel industries. In order to modernize these industries, the High Authority would control output to meet demand, guide investment in new facilities, create and enforce antitrust laws, and work to improve the economic security and professional training of workers. The High Authority also had the legal right to tax the industry to cover its expenses. To say that these were unprecedented powers for an international organization would be a gross understatement.

Signed in April 1951, the treaty was enveloped in contentious debate but it was ratified in all six member states by the following summer. Whenever any member was stalling, Monnet would mobilize his powerful friends in Washington to apply pressure to pass the unprecedented legislation. The High Authority's operations began in the summer of 1952. It was located in Luxembourg in what was formerly the headquarters of the Grand Duchy's state railway. Germany was now bound to the West and vice versa. The world's first experiment in supranationalism on a grand scale had begun.

Monnet became the first ECSC president. His reputation, as a man who was committed to transcending national self-interest and who had shown no interest in political office, helped establish the community's own supranational credibility. In one of his early speeches as president, he said of the new ECSC assembly, court, and other bodies, "All these institutions can be modified and improved in light of experience. But there is one point on which there will be no turning back: these institutions are supranational . . ." From the beginning, Monnet expounded on his philosophy of institutions as repositories of human experience and knowledge. "Each man begins the world afresh. Only institutions grow wiser; they store up the collective experience; and from this experience and wisdom, men subject to the same laws will gradually find, not that their nature changes but that their behavior does."

Although Monnet presented his proposals in a simple, highly concrete way, exquisitely relevant to the moment at hand, he did hold a more grandiose and theoretical vision. He understood that within most countries, a kind of federation existed in which power and responsibility were shared between central and local authorities, each exercising certain prerogatives. He wanted to apply this concept on a European level, with a European government dealing with issues and policies that were Continent-wide, and the states handling the rest. The political minefield was of course in the details of where the line was drawn.

Monnet became the personification of the High Authority, a leader who saw the world through *European* eyes, not French eyes, or German eyes, or Italian eyes. He recruited a staff from throughout Europe that became the first cadre of "Eurocrats." He wanted to modify people's behavior so that they didn't think like nationalists. "If we succeeded in proving that men from different countries could follow the same text, work on the same problem with the same data, and eliminate all ulterior motives and mutual suspicions, then we should have helped to change the course of international relations," he said.

After less than three years on the job Monnet submitted his

resignation from the ECSC, to take effect on June 9, 1955. He felt he had done all he could and his interest was less in administering the High Authority than in helping to build a broader federation of European states. Historians have given the operation of the ECSC mixed grades. It did manage to expand trade in coal and steel, although that might have happened anyway given the improving economy of Europe. It helped remove some transportation barriers, such as discriminatory rail rates, and improved the social safety net for workers.

But Monnet's strength was his vision, not managerial detail. He was in constant tension with the representatives of the business associations, and he never did manage to fully eliminate the coal, iron, and steel cartels, or to dismantle many restrictive business practices, or to attract significant new investments for modernization. He also fomented an almost wartime atmosphere of crisis and emergency. He presided over endless investigations, analyses, surveys, and reports that undercut the effectiveness of the High Authority. He was also consumed with external relations, cultivating support in the United States and trying to get Britain to join the ECSC—which it never did. He pushed for the establishment of a second supranational organization, the European Atomic Energy Community (EAEC or Euratom), which succeeded, and for a third, the European Defence Community, which did not.

A more competent manager might have done a better job running the ECSC day to day, but in the long run Monnet did what he set out to do—establish a supranational institution that would be the springboard for further integration. Europe's first experiment in supranationalism would become to the European Union what the Articles of Confederation were to the United States—the initial legal and political framework that laid the groundwork for a stronger union.

Europe Grows in Monnet's Image

Although Monnet left the ECSC in June 1955, the momentum he had created continued to build. He remained the most vocal, most persistent, and most influential advocate for new supranational structures. He became a powerful lobbyist representing large European companies pushing for the establishment of a Common Market, a free trade zone for member states that came into existence on January 1, 1958, along with the European Atomic Energy Community. Together with the ECSC, the new entity was called the European Economic Community (EEC). As soon as that was up and running, Monnet advocated for even more integration, including a common agricultural policy and regular summit meetings of finance ministers, defense ministers, foreign ministers, and heads of state. He and his colleagues also lobbied for the integration of European financial markets, the establishment of a European currency, and the creation of a European Central Bank. He raised the subject of inviting new countries into European institutions. He wanted to see a single capital city for all of Europe's major political institutions, including its executive, parliamentary, and judicial branches. Above all, he wanted to see a true political union—a European government.

Much of what Monnet was pushing came to pass in the following decades. The membership of the EEC expanded, as did the extent of integration. Ancient national prerogatives were eroded every step of the way. In the early 1960s, the member states created a common agricultural policy that basically draws funds from strong economies like Germany to subsidize farmers and the pastoral self-image of nations like France and Spain. In the 1970s, they created a system that transfers funds from rich regions to develop poorer ones, with the aim of bringing the poor regions up to the income level of the rich. In 1993, the EEC gave way to the European Union, which established freedom of movement for goods, people, services, and money across the borders of member states.

As Monnet wished, the EU became a powerhouse in international commerce, able to hold its own with the United States in global trade negotiations. The EU would gain the power to regulate politically flammable issues, such as working conditions, food safety, environmental protection, and the breakup of monopolies. It has also united competing companies in new European multinationals that build planes, helicopters, and spacecraft—industrial organizations that countries on every other continent still jealously guard behind the mantle of national security. Elsewhere a European Investment Bank would provide long-term financing for projects that Brussels assigns a high strategic priority, such as high technology and modern infrastructure.

As restrictions on trade and capital movements fell, trade and investment among the six European nations that were the core of the European Community grew phenomenally quickly. European companies such as Renault and Deutsche Bank expanded across borders. Industrial groups, from food processors to soap manufacturers and trade unions, began banding together in continental associations to better tap the Europe-wide market. British and American corporations also began to see Europe as one market. The companies' entrance on the continental stage in full force increased competition, lowered costs, and widened the choices available to European consumers.

Perhaps the most remarkable step forward came on January 1, 2002, when twelve of the twenty-eight European Union members retired their national currencies from active circulation, and overnight replaced German marks, French francs, Italian lira, and the rest with the new European currency, the euro. The euro would be managed by a new European Central Bank, which would replace twelve national central banks. The operation went surprisingly smoothly. One day Europeans used the currency that had been circulating in their country, sometimes for hundreds of years; the next day they used a new one, decorated with one of the Continent's historic bridges, symbols of the collaboration Monnet had engineered decades before.

"In Crises, Most People Don't Know What to Do. I Do Know."

In 1976, the nine heads of government of the European Community proclaimed Jean Monnet the one and only Honorary Citizen of Europe. Three years later, on March 16, 1979, at the age of ninety, he was laid to rest under gray skies and snow-covered ground in a small cemetery near his country house. Helmut Schmidt, the German chancellor, arrived by helicopter to attend the ceremony in a small fifteenth-century church. He was joined by French president Valéry Giscard d'Estaing plus numerous other dignitaries and friends from throughout Europe, as well as prominent Americans such as senior diplomats George Ball and John McCloy. In 1988, French president François Mitterrand celebrated the centenary anniversary of Monnet's birth by moving his ashes to the Pantheon in Paris, where France's most revered statesmen and saints are buried.

It's never easy to know the full extent of someone's motives. Some historians have said that Monnet was effective because he sought no personal power, never campaigned for a political position, and often let others take political credit for his work. But it could be that Monnet wanted a different kind of power, one that is exercised out of the limelight, where big ideas are translated into action. Perhaps he delighted in being a puppeteer, pulling the strings of major actors on the world stage. He wrote, "Men in power are short of ideas; they lack the time and information; and they want to do good as long as they get the credit for it." He understood that it is when politicians are desperate for ideas that they can take steps no one has contemplated before. "In crises, most people don't know what to do," Monnet once said. "I do know."

Monnet was above all a political entrepreneur, someone who could spot and sell a big, timely idea. He had an extraordinary ability to package complex proposals with attention to minute detail, crystal clarity, and a keen sense of timing—a point demonstrated most dramatically in the quick acceptance by the leaders of France and Germany of his plan for a supranational European Coal and Steel Community.

A close colleague would recall that Monnet wasn't particularly well read or charismatic, he wasn't a good writer or public speaker, and yet he had a great gift of persuasion. Another friend said that he knew how to link an idealistic objective to a pragmatic way of achieving it. A third described him as a reducer of the previously unworkable to workability, his proposals having "the crushing simplicity and elemental clarity of the solutions presented in dreams."

Monnet was so preoccupied with the workability of an idea that he believed that any proposal could succeed only if it were specific to time and place and could not have been effective before or after. He relished a quote from American diplomat Dwight Morrow: "There are two kinds of people—those who want to *be someone* and those who want to *do something*."

He understood just when to capitalize on his wide range of high-level personal relationships and exert some influence. He was a master at building and keeping alliances with influential people.

He believed in the efficacy of public policy and strong governmental direction. He was a planner, and he was not shy about directing outcomes in the market; in today's parlance, he might be called a believer in state capitalism. What made him different from others of that strain—say, General Charles de Gaulle or

British socialists like Prime Minister Clement Attlee—was that he favored international institutions and transnational policies over national governments, which he preferred to keep small and focused.

He absorbed lessons from every new experience and applied them to his next challenge. He used the positions he held in World War I, the League of Nations, the world of private finance, and World War II to refine certain of his skills to the level of fine art. He balanced the craft of extracting information from multiple sources with the discipline of extreme discretion, thus becoming a trusted confidant of the world's most powerful leaders.

Above all, Monnet was driven his whole life by one big idea—the inadequacy of the sovereign state to handle a wide range of contemporary challenges. Officials would one day say that he had a "circumspect manner of inspecting a problem from every angle . . . like a peasant buying a cow." "I have never met a man before or since who had Monnet's single-mindedness," said Jelle Zijlstra, the Netherlands' prime minister.

To be sure, Monnet had his critics. To men like de Gaulle, he was an idealist who refused to recognize the reality that all politics was rooted in nationalism. Many French politicians thought he was too close to the British and Americans, more attuned to Anglo-Saxon than to French thinking. The way he used personal contacts to bypass normal political channels branded him as an elitist, disconnected from the everyday problems of ordinary people. Many critics saw Monnet as a Socialist dedicated to complex government control of market forces; some considered him a wily politician masquerading as a technocrat in order to reshape the political order. Today, it is not unusual to hear that Monnet and all he stood for brought about a "democratic deficit" in Europe, where technocratic elites built institutions that operated with the consent of governments but not of the people. The result, as the argument goes, is a supranational structure that has little democratic legitimacy, which in the critics' view explains why so many of Brussels's edicts are today so unpopular.

It is doubtful that Monnet saw this lack of popular consent as a big problem. He *was* an elitist, after all, and most likely felt that the creeping powers of the European bureaucracies would be matched by the increasing prominence of the elected branches—the parliament and eventually a president or prime minister—such that the EU would come to look much like the United States, with an equivalent level of representative democracy. For Monnet, though, all that would probably seem less important than the creation of more intra-European agencies that actually did good, practical work on the ground. In other words, he would probably have said that if the elites could create a new Europe that offered tangible benefits to its ordinary citizens, then the public would willingly support the result, whatever its other shortcomings.

If ever an organization was the shadow of its human creator, the EU is a clear outline of one man. No leader has ever changed the course of globalization without generating controversy, and whatever critics may say about Monnet, it is hard to disagree with the proposition that the path he created for Europe was a spectacular improvement after many decades of national rivalry and cataclysmic warfare. To an astounding degree, he not only knew what to do in the crisis that followed World War II, but he figured out how to do it.

Monnet's Impact

Whereas the fifty years before the establishment of the Common Market was characterized by two world wars and a depression, the years since 1958 have brought a period of peace and prosperity that few could have imagined in light of European history. In the forty years before 1945, Europe lost at least sixty million people in two great wars, but until the civil war in Yugoslavia some fifty years later it saw no wars, big or small. After the fall of the Soviet Union in 1991, the two halves of the divided continent have been integrating with impressive speed, in large part because the European Union had already brought so much unity to the West. It had created a single home with rules on free trade, immigration, and many other issues such as food safety within its borders, making it a magnet for nations from the East. It had established a common currency managed by a European Central Bank. Despite today's financial and political strains, many caused by the financial crisis of 2008–9 and the subsequent long global recession, plus a crisis in Greece, and despite Russia's invasion of Ukraine, the EU has held together. Its brand of social democracy, combining as it does individual freedom and a strong social safety net, has been a remarkable success of political, economic, and social policy. Aside from becoming a powerful force in international commerce, moreover, the EU has been the essential partner to the United States in shaping a relatively open global trading system based on market forces, binding rules, and adjudication of disputes. Indeed, the steel and coal authority Monnet created was a seed for a model of democratic capitalism and the rule of law throughout the world.

With the addition of many new member countries after the fall of the Iron Curtain in 1989, the EU had by 2014 grown to twenty-eight countries stretching from Ireland to the Black Sea, encompassing over five hundred million citizens, and a GDP of $18 trillion, slightly larger than that of the United States and accounting for about 20 percent of global trade. It is not easy to measure precisely the specific economic gains within Europe from economic integration. After all, no one can calculate what would have happened if the EU had not existed. In addition, any calculations of gains and costs are highly sensitive to the precise time period under consideration. Also, since 2009, the economic trajectory of the EU, like so much of the world, has been deeply and negatively impacted by the global financial crisis and its aftermath. Nevertheless, an analysis by Deutsche Bank concludes that in the period from 2003 to 2013, the EU benefited from significant gains in growth, trade, labor mobility, competition among European firms, and greater consumer choice. A lot of studies exist, each based on somewhat different assumptions, but one of the more credible analyses shows that from 1958 to 2005 economic integration boosted EU GDP by 5 percent; another concluded that between 1992 and 2012 intra-European trade increased from 12 to 22 percent of GDP. Concerning foreign direct investment into the EU, in the early 1980s the amounts were truly negligible, but it has since grown more than twenty times. One analysis showed that with more financial integration and more deregulated labor policies—all under active discussion in Europe—much larger gains could be achieved. Another indicator: In 2004, ten countries joined the EU, led by Poland, Hungary, and the Czech Republic. In the mid-1990s, when their membership negotiations began, most of them had a GDP per capita of about one-quarter or one-third of the existing fifteen EU countries. Ten years after joining, most of the new members' GDP per capita reached two-thirds of the 2014 EU levels of the original fifteen members.

Monnet's work raised all the big issues surrounding European integration, not just for his time but for ours. His work on the ECSC was an early attempt to define the proper division of power between supranational and national authorities. His idea for supranational government ultimately led to a debate about how the views of individuals could be taken into account—a debate about democratic accountability, or lack thereof, that remains front and center in European politics today. His focus on meshing economies together before creating the underlying political integration created a problem that still bedevils officials trying to succeed in making the EU—and especially the eurozone within it—function more effectively. Moreover, he advocated measures that were implemented after he died, such as a common currency, a single European Central Bank, an incipient continental banking union, and the beginnings of a common energy policy. Although he did not identify certain other directions, he created the momentum for some of the new initiatives of recent years such as a common telecommunications policy or a common policy toward digitalization. Top European officials have even begun to call for an intra-European military force. It is no exaggeration to say that in pressing for what is now the European Union, he built the most far-reaching globalized multinational regional grouping in history, and in the process he engineered one of the greatest advances in international relations since the establishment of the nation-state itself in the 1648 Treaty of Westphalia.

Although his focus was on the Continent, the European experiment can be seen as a microcosm of how to organize globalization itself. It should be noted that other parts of the world have looked at the evolution of the EU as a model for their own development. In the dynamic region of Southeast Asia, for example, ten countries are building what they call an ASEAN (Association of Southeast Asian Nations) Economic Community designed to integrate trade, investment, transportation, communication, and regulatory processes. Elsewhere, the African Union has launched negotiations for a continental free-trade area. Similar moves have been afoot in the Caribbean. In this century, it is a good bet that the United States, Canada, and Mexico will constitute a deeply intertwined North American market that looks much more like the EU than we can envision today. To be sure, some fundamental questions exist: Can what works in Europe be emulated elsewhere? In an integrated global system, is more supranational governance a possibility, perhaps beginning as Monnet would have with a narrow, practical application, such as oversight of the resource-rich Arctic Circle? Does Monnet's creation reflect the possibilities for globalization or the limitations? No one can know for sure.

In 1963, President John F. Kennedy sent a letter to Monnet acknowledging his achievements:

> For centuries, emperors, kings and dictators have sought to impose unity on Europe by force. For better or worse, they have failed. But under your inspiration, Europe has moved closer to unity in less than twenty years than it has done before in a thousand. You and your associates have built with the mortar of reason and the brick of economic and political interest. You are transforming Europe by the power of a constructive idea.

Indeed, the essence of globalization is the reduction of borders—precisely what Monnet did. In their own ways, that is what Genghis Khan and Robert Clive did, too, via the creation of empires. That is what Prince Henry did by discovering new lands to open. That is what Mayer Amschel Rothschild did by building a global financial system that allowed citizens in different parts of the globe to participate in it. That is what Cyrus Field did in linking the world by a revolutionary new communications system. And that's what John D. Rockefeller did by organizing the world's fuel supply to propel industrialization, and also by opening new global channels to pump funds into social advancement via philanthropy. But Monnet did something the others didn't. He has been called a statesman, wrote Strobe Talbott, president of the Brookings Institution and former deputy secretary of state. "In fact, he was something far rarer and more consequential—a key figure in the transformation of the concept of statehood itself."

It is of course the case that the EU has been experiencing a number of severe problems. For the better part of the last decade it has been mired in an economic and financial crisis, characterized by, among other things, slow growth, high unemployment particularly among youth, high indebtedness, and stressed banks. The economic problems surrounding Greece have been particularly dire and promise to create severe tensions within the country as well as deep political divisions among the countries of the eurozone and the EU for many years to come. There are other crises, too. Migrants and refugees have been flooding into the EU from North Africa and the Middle East, placing excruciating pressure on long-standing policies in Europe for freedom of movement and creating horrendous humanitarian problems. The United Kingdom is threatening to leave the EU. And Russia has annexed part of Ukraine, threatening EU members such as Poland, Slovakia, Romania, Hungary, Latvia, and Estonia. All of these challenges test the ability of the EU to fashion coherent policies among its members. On top of all that, the European electorate has been deeply disillusioned with current politics and policies out of Brussels. The results include the rise of extreme parties on both the left and particularly the right, leading to the kind of nationalism and xenophobia that Monnet fought against. A day doesn't pass now without serious commentators asking whether the EU can survive the tensions within it. To that, there are at least four forceful counterarguments.

First, Europe's leaders are conscious of the stakes at a most fundamental level. Said Europe's most powerful leader, Angela Merkel, in 2014:

> It took centuries before the peoples and nations of Europe found their way initially to economic and eventually political

> cooperation. One symbol of this process was the signing of the Rome Treaties fifty-seven years ago. These treaties are based on the conviction that European integration was and remains a question of war and peace. Furthermore, it is the key guarantor that we—with our values, our way of life and of doing business—can hold our own, even in the globalized world of the twenty-first century.

Second, a substantial part of the European leadership wants to build a stronger European Economic and Monetary Union (EMU). That kind of vision was on display in the summer of 2015, even as deep Europe-wide tensions surrounded the Greek crisis, in a report signed by the president of the European Commission and supported by the president of the European Council (composed of heads of state), the president of the Eurogroup (finance ministers), the president of the European Central Bank, and the president of the European Parliament. The five presidents praised the benefits of the euro, both economic and political. They set out a detailed agenda to establish policies to put the EMU on firmer footing by building more of a banking union, integrating capital markets, and moving toward much tighter coordination in the fiscal arena. "Europe's Economic and Monetary Union (EMU) today is like a house that was built over decades but only partially finished," said the report. "It is now high time to reinforce its foundations and turn it into what the EMU was meant to be."

Third, the eurozone and the broader EU could well undergo a metamorphosis under the pressures that are building. For example, whatever agreements are struck in the Greek situation, over time Athens may well leave the eurozone. Perhaps others will, too. But the remaining nations may then find it easier to pursue even tighter integration than they previously contemplated. The same could be said about the possibility of the UK leaving the EU; perhaps a tighter European Union would be the result. In any event we should see the crises of the moment in a long historical process that consists of taking two steps forward and one step back.

The fourth counter to skepticism about the survival of the EU is that it has overcome several deep crises before. "I have always believed that Europe would be established through crisis, and that the outcome would be the sum of the outcomes of those crises," said Monnet. And in the last lines of his memoirs, he wrote, "The sovereign nations of the past can no longer solve the problems of the present . . . and the [European] community is only a stage on the way to the organized world of tomorrow." Whether it's terrorism, cybercrime, climate change, financial stability, migration and refugees, equitable growth, or so many of the other big challenges of globalization, it may take several generations to get there, but I am sure he is right.

PETER MARLOW, MAGNUM PHOTOS

Chapter VIII

MARGARET THATCHER

The Iron Lady Who Revived Free Markets

1925–2013

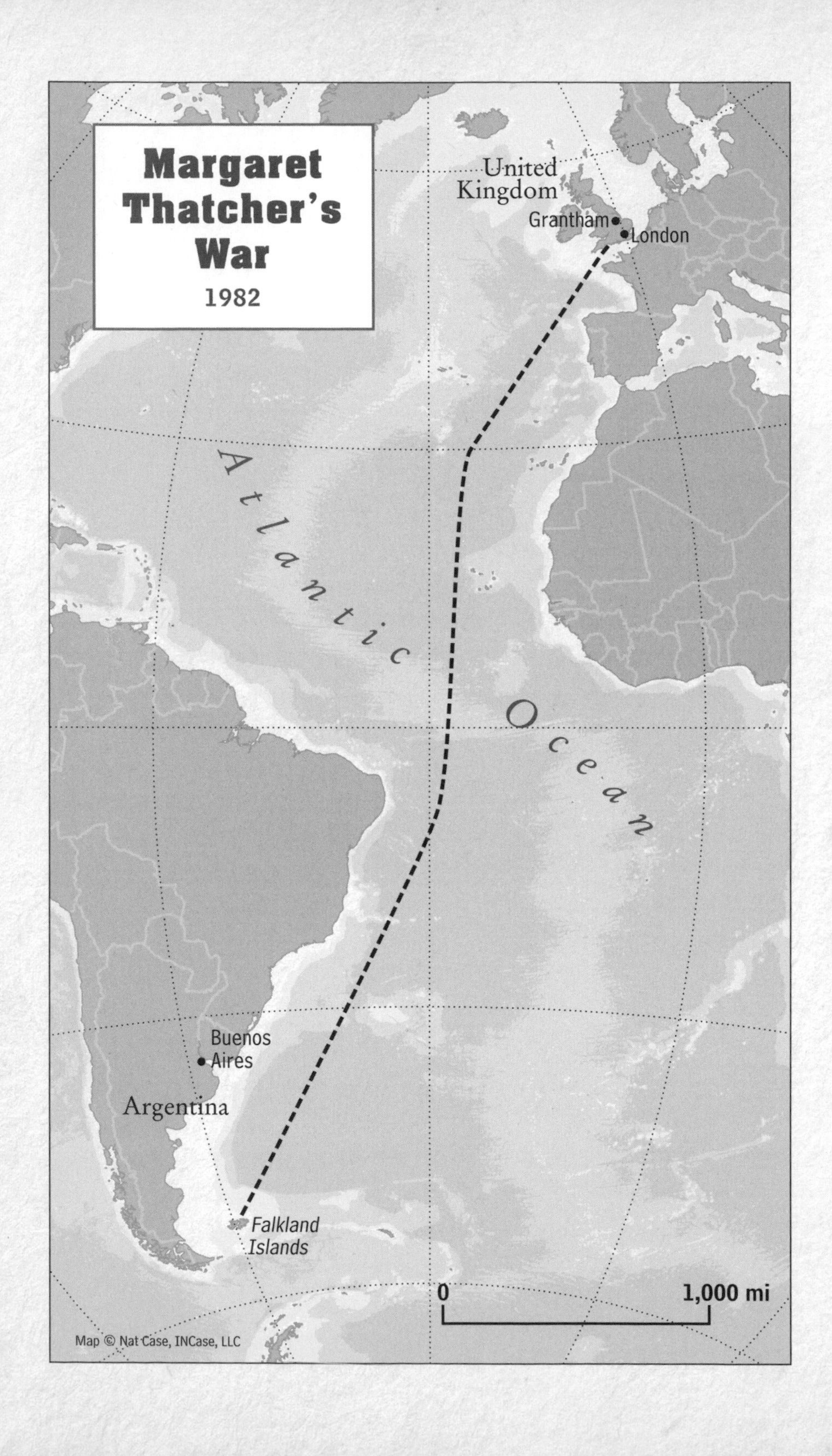
Margaret Thatcher's War
1982
United Kingdom
Grantham
London
Atlantic
Ocean
Buenos Aires
Argentina
Falkland Islands
0
1,000 mi
Map © Nat Case, INCase, LLC

The ten-year-old girl, her short blond hair parted on one side and swept back from her freshly scrubbed face, was in constant motion, folding bright red election leaflets, running back and forth between the Conservative Party headquarters and the voting station, relaying gossip and information. It was November 14, 1935, an election day in England, and Margaret was helping her father, Alfred Roberts, to elect the Conservative candidate from their middle-class hometown of Grantham, about one hundred miles north of London.

Margaret Roberts would one day become Margaret Thatcher, the first woman ever to become prime minister of Great Britain or leader of any major industrial democracy. She would go on to exceed this distinction by becoming the world's most important advocate for freeing trade and investment from government control, selling off state-owned companies to the private sector, weakening labor unions, and, in general, allowing markets to link with one another to cross borders and build a web of connections that became—and remains—the motor force of globalization. Thatcher's contribution to building a more interconnected world can be understood by examining what it means to unlock trade and investment flows from government restrictions. Before she took office, for example, the government owned many companies in such industries as telecommunications, energy, and transportation. Supported by the state, these firms were essentially cloistered monopolies. They had far less incentive to be sleek and modern or to expand around the world in order to compete and survive. But once they were privatized—that is, sold to the public on the stock market—globalizing their operations in order to become more profitable became more urgent. The Thatcher government liberalized regulations constricting the activities of the City of London, commonly referred to as "The City," the United Kingdom's version of Wall Street. This had the effect of creating a financial hub in London that was far more globally competitive, in part because British banks were now free to hook up with foreign firms as never before. A third example is the effect of Thatcher's eviscerating the labor unions. Once that happened, the workforce had to become more competitive to survive, especially because the government embraced freer trade. All this translated into more globalized economic activity. Like a number of others in this book, though, Thatcher left considerable human damage in her wake. Of course, it was nothing like the fallout from Genghis Khan, but what came to be called "Thatcherism" resulted in the decimation of many communities, the impoverishment of many people, sharply rising inequality in an already class-ridden nation, and a society that favored the rich, the educated, and the connected over the rest.

In the mythology that would one day surround Thatcher, her father's grocery store became an iconic symbol of capitalist enterprise, a place where the owners had to make a profit to survive. And to make a profit they had to embody the entrepreneurial ideals: work hard, balance the books, and cultivate good relations with customers. The government—but for wartime rationing and the tariffs it levied on imports—had little impact on the store's success or failure. Indeed, her father's corner store in Grantham would become to her life story what the log cabin was to Abraham Lincoln's.

Thatcher would recall her father as a towering influence on her life, while remembering her mother much less vividly and mostly for her skills as a housewife. "[My father] told me that first you sought out what you believe in," she recounted. "You then apply it. You don't compromise on things that matter." Alfred served in his free time in various civic posts—alderman, mayor, president of the chamber of commerce and of the Rotary Club—and he passed on his interest in public service to his daughter. Watching him direct Rotary Club campaigns to help the needy deeply impressed young Margaret, who developed a lifelong preference for private charity over direct government assistance for the poor.

Alfred was a devout Methodist, and family life revolved around the church. Thatcher would regularly spend almost her entire Sunday in church. Years later, she would talk about her father's approach to faith as sorting out differences between right and wrong. "There were certain things you just didn't do," she recalled. "Duty was very, very strongly ingrained into us. Duties to the Church, duties to your neighborhood and conscientiousness were continually emphasized."

Thatcher grew up in the culture of sacrifice and courage forced on Britain by World War II. From the time she was fourteen, her daily routine was often punctuated by electrical blackouts and rushing to pull on a gas mask when the air raid sirens blew. During the war, bombs dropped on Grantham's factories killed some seventy people and injured another two hundred. While many Grantham residents had urged Britain to stay out of the war, Alfred Roberts believed in rearming to defend the empire, which he saw as a bulwark of law and order and a civilizing influence on foreign lands. He had been a skeptic of the League of Nations and felt he was proven right when the league failed to prevent the rise of Hitler. His daughter would come to agree with him on virtually all counts, and would pursue a muscular foreign policy when she had the chance.

"A Battle Between Two Ways of Life"

Margaret entered Oxford University in October 1944, focusing her studies on chemistry. She joined both the Methodist Study Group and the Oxford University Conservative Association, which gave her a first shot at politics in a circle of like-minded students. When the war ended in 1945, the Labour Party mounted a serious challenge to the Conservative government of Winston

Churchill, warning that in its battle-weary condition Britain could fall back into the depression of the 1930s. Labour argued that the only way to rebuild bomb-blasted cities and to resolve postwar shortages of housing and jobs was to put the state in charge of the economy and in control of key industries. Although the Conservatives opposed this big increase in state power, they nevertheless supported a significant role for government and had a hard time clearly distinguishing themselves from Labour.

Thatcher had no such problem. Even as a university student she gained attention for her ability to explain the distinguishing character of Conservative ideas with conviction and passion. Foreshadowing the contentious issues that would define her career for the next half-century, she banged away at socialism, at government control of industry and social services, at what she felt was the pernicious influence of labor unions in national politics. Like her father, she was a fiery patriot and took an expansive view of the role Britain should play in the world, despite the reality that the empire was in retreat, its finances stretched to the breaking point by war, and its military influence eclipsed by the United States and the Soviet Union.

As an Oxford student Thatcher participated in campus debates and frequently returned to Grantham to warm up audiences who were waiting to hear from professional politicians. She took lessons in public speaking to sharpen her delivery. At school she read voraciously, her ideas shaped by conservative classics like *The Road to Serfdom*, Friedrich Hayek's withering forecast for the future of socialism; *Imperial Commonwealth*, Lord Elton's positive vision for the future of the British Empire; and *Darkness at Noon*, Arthur Koestler's novel about the brutal realities of Soviet communism.

Leaving Oxford in 1947 with a degree in chemistry, Margaret found a job as a research chemist at British Xylonite Plastics. Later she worked in the research department of J. Lyons & Co.

in Hammersmith, testing the quality of cake fillings and ice cream. But politics consumed her nights and weekends. In 1948, she received her first big break when she attended a Conservative Party conference in Llandudno, Wales, and met the chairman of a local party association, John Miller. He was looking for a candidate to run for Parliament in his district, a suburb southeast of London called Dartford. For many years Dartford had voted overwhelmingly for Labour, but Miller thought his forceful young candidate would at least mount a credible campaign. Margaret Roberts went on to lose that election in 1950 and another in 1951, but she impressed everyone with her ideas and her energy. Managing on four hours' sleep, she visited shop workers in their stores and laborers in their factories. She seemed willing to go anywhere, meet anyone, and do anything for publicity, including dressing up as a barmaid to get into men's clubs, pouring beer and being photographed for the newspapers.

The themes in the stump speeches Thatcher delivered in 1950 and 1951 were almost indistinguishable from those she would give three decades later as prime minister: Labour's proposals were a threat to the British way of life; the welfare state robbed citizens of their natural tendency to be self-reliant; managing an economy was like managing a household—expenditures had to match revenues, simple as that; the British Empire had to be maintained; the pound sterling should again be strong; the "Great" in Britain had to be restored. A typical speech would contain language like this:

> We are going into one of the biggest battles this country has ever known—a battle between two ways of life, one which leads inevitably to slavery and the other to freedom. . . . In 1940 it was not the cry of nationalization that made this country rise up and fight totalitarianism. It was the cry of freedom and liberty.

During the first Dartford campaign, she met businessman Denis Thatcher, heir to a prosperous chemical business. Sharing a political philosophy and more, they married on December 13, 1951. Thatcher maintained her job at J. Lyons long enough to study for the bar in the evenings and began to practice law in January 1954, one month after giving birth to twins. Thanks to her husband's income, she could afford to hire the assistance she needed to bring up her children, freeing her to pursue a full-time career. "Marriages are made in heaven, but it is better if the money is made on earth," she said at the time. In 1956, she decided to reenter politics, but it took her another two years to win the party nomination for a Conservative seat in Finchley, a prosperous middle-class slice of North London. Her campaign added important planks to her political agenda, including a call for low taxes as an incentive to create wealth, and an attack on government bailouts for troubled industries as being morally wrong.

The Convinced Don't Need a Consensus

At the age of thirty-four, she won easily in Finchley and her career took off. Thatcher made her first speech in Parliament on February 5, 1960, introducing a bill that would widen media access to local government meetings. Speaking for thirty minutes and citing reams of facts and figures without notes, her energetic and aggressive performance was a template for her future speeches in Parliament.

Starting in the summer of 1961, a series of government posts followed. She served as junior minister of pensions and national insurance under Prime Minister Harold Macmillan, seeing firsthand the complex bureaucracy dealing with welfare and other programs relating to the social safety net. In the shadow cabinet of Prime Minister Harold Wilson, starting in 1964, she held junior posts in pensions, finance, housing, transport, and education and science, developing an exceptional mastery of the details of programs and governance. Preparing to debate a Labour tax bill, Thatcher studied every budget speech and finance bill of the past decade and came into the House of Commons with a grip on the issue so superior to her opponent, she derided him publicly for understanding his government's own policy less well than she did. Writes biographer John Campbell, "The force which transformed British politics over the next twenty years was Mrs. Thatcher's belief that politics was an arena of conflict between fundamentally opposed philosophies, her contempt for faint hearts and her ruthless view that a party with a clear philosophy needed only a 'sufficient' majority—not an inclusive 'consensus'—to drive through its program."

The Conservatives won the election in 1970 and were poised to return to power with Edward Heath as prime minister, but his four-year term would prove to be a disaster that set the stage for a split in the party and the beginning of the Thatcher revolution. Heath came into office promising a clean break with the past and a turn to true conservative policies. Like President Richard Nixon, he pledged to reduce government but ended up expanding it. He vowed to slash public spending, to clip the power of labor unions, and to cut government subsidies for failing industries, but he broke each of these pledges; in fact, he attempted to reflate the economy with more deficit spending, he bailed out sick industries, he caved to union demands, and he imposed wage and price guidelines in a vain attempt to control inflation. Heath seemed more comfortable managing the system better rather than changing it. The media came to describe one Heath move after another as a "U-Turn," a term that infuriated the right wing of his Conservative Party, none more than Thatcher. She deeply resented Heath for his failure to take a strong stand against socialism and for his inability to promote clear conservative values.

Gathering around a think tank called the Institute of Economic Affairs, some Conservatives were now demanding a reexamination of the basic tenets of economic policy. Foremost among these voices was Keith Joseph, a member of Heath's leadership team, who gave a series of high-profile speeches outlining a coherent new set of policies. Said Joseph, "We [in Britain] are now more socialist in many ways than any other developed country outside the Communist bloc, in the size of the public sector, in the extent of government controls, etc. . . . We have inherited a mixed economy which has become increasingly muddled, as we tried our best to make semi-socialism work." Though Heath was the target of this withering criticism, he seemed willing to listen to his critics and asked Joseph to establish a new research center, the Centre for Policy Studies. Joseph asked Thatcher to be the vice chair and she accepted—the first overt sign that she was inclined to challenge the prime minister, who was by now her direct boss.

Heath had chosen Thatcher as his secretary of education and science, and she was consumed by her new cabinet responsibilities, driving subordinates crazy with her micromanaging and domineering style. She insisted on reading and editing every document that came before her, and she came down hard on staff. One official recalled coming out of a meeting with her feeling like a "peeled banana." Another colleague said, "She's a killer. She makes up her mind about someone in ten seconds and rarely changes it. As soon as you open your mouth you are categorized." She was determined always to win every argument. If she was losing the point, she would shift to another. On the other hand, she could be gracious and charming, making coffee for everyone or even cooking supper when they worked especially late.

During nearly four years as a cabinet minister, Thatcher's toughest lessons came in a scandal that erupted when she proposed that schoolchildren should no longer receive free milk at school. Even though Thatcher was implementing the plans of other men—Heath's plan to cut costs, and her Labour predecessor's plan to save some 8 million pounds sterling each year by charging schoolchildren age eight to eleven for milk—her foes quickly exploited the image of a mother denying milk to children. One Labour MP said that "[Thatcher] was to British education what Attila the Hun was to Western civilization." The *Sun* called her the most unpopular woman in Britain, as headline writers cracked jokes about "Thatcher the Milk Snatcher." The outcry grew so loud that her husband, Denis, suggested to Margaret that she should consider leaving politics. She would have none of that. "Iron entered my soul," she later recalled. The scandal would prepare Thatcher for the far larger controversies she would kick up as prime minister.

Thatcher survived in large part because Prime Minister Heath publicly stood by her. But the Heath government was mortally wounded by its own U-turns, all the more so as Britain was battered by rising unemployment, labor strikes, oil price spikes, and inflation. Even after the Conservatives lost the national election in March 1974, and despite the failure of his centrist policies, Heath refused to step down as party leader. Nevertheless, the right wing of his party would not stand behind him for much longer.

"You Haven't Got a Hope"

The first blow to Heath came not from Thatcher but from Keith Joseph, who made a decisive break in a speech saying the British government should drop the priority it had placed for decades on fighting unemployment. Haunted by the Depression era with its long lines of desperate job seekers, every postwar government, whether Labour or Conservative, had followed the prescriptions of economist John Maynard Keynes, who argued that in hard times the government should work to stimulate growth and create jobs, even if it had to borrow heavily to do so. Joseph rejected Keynes, saying that government spending would only drive up the public deficit, drawing down the pool of capital available for loans and pushing up interest rates for private business—and business was the real source of new jobs, he said. Worse, government bailouts would backfire by propping up inefficient businesses, which would lower productivity and therefore stimulate inflation, undermining economic growth, and making it even more difficult for slumping businesses to pay back their loans. The new approach, said Joseph, would be to fight inflation by controlling the money supply and lowering government borrowing and deficits, freeing up capital to fund private-sector growth. A more competitive private sector would flourish and generate jobs in this dynamic new economic environment, created by conservative policy, he promised.

When Joseph decided to challenge Heath for the party leadership, Thatcher agreed to manage his campaign, reinforcing and elevating her open opposition to her boss. A few weeks later Joseph delivered a speech about birthrates among adolescent mothers. It was so politically incorrect that the public reaction forced him to withdraw from the election. Thatcher told her husband that she was thinking of stepping in to directly dislodge Heath herself, to which Denis replied: "You must be out of your mind. You haven't got a hope." On November 25, 1974, Thatcher nevertheless visited Heath to say she would challenge him.

Like Denis Thatcher, many Conservatives thought Thatcher had no chance and began working to block her candidacy. They dredged up an old interview—in which Thatcher said she kept stores of canned food in case of emergencies and advised other housewives to do the same—and twisted the context to imply that Thatcher was hoarding food. The media piled on, reporting false accusations that Thatcher was buying up large quantities of sugar, which was in short supply at the time. It was a vicious campaign reminiscent of the "Milk Snatcher" scandal, and it died out only when it was discovered that the store at which Thatcher was allegedly buying all the sugar did not even exist. Thatcher later recalled being bitterly upset, sometimes near tears or shaking with anger, but said she had told a friend: "I saw how they destroyed Keith [Joseph]. Well, they are not going to destroy me.' "

In early January 1975, she launched her campaign against Heath with a pitch pared to the essentials. She said the British people had come to believe that "too many Conservatives have become socialists already" and had little reason to "support a party that seems to have the courage of no convictions." She went on to highlight her basic convictions: concern for individual freedom; opposition to excessive state power; the right of hardworking and thrifty people to reap the rewards of success and pass them on to their children; the right to work without oppression by either employer or trade union boss. Thatcher also went to great lengths to soften her image. She presented herself on TV and in the press as a commonsense housewife. In meetings with her parliamentary colleagues, she started asking for their advice and listening patiently to their views. On February 11, 1975, Thatcher scored an upset victory over Heath and became the first woman to lead a major Western political party. The *Daily Telegraph* spoke for many in the stunned political establishment when it wrote:

> What kind of leadership Mrs. Thatcher will provide remains to be seen. . . . But she believes in the ethic of hard work and big rewards for success. She has risen from humble origins by effort and ability and courage. She owes nothing to inherited wealth or privilege. She ought not to suffer, therefore, from the fatal and characteristic defect of guilt about wealth. . . . This is one reason why Britain has traveled so far down the collectivist road. What Mrs. Thatcher ought to be able to offer is the missing *moral* dimension of the Tory attack on socialism.

That is precisely what she tried to do.

As shadow prime minister, Margaret Thatcher led the Conservative Party in opposition as the economy spiraled toward a collapse. The environment made her warnings about the evils of socialism appear all too real, as critics were comparing Britain to the gray mediocrity and corporatist state of East Germany; as inflation approached 25 percent; as labor strikes were mounting with increasing frequency and severity; and as marginal tax rates hit 98 percent for the richest Britons, stifling individual initiative. Thatcher was making a connection with millions of British citizens who were growing tired of the suffocating bureaucracy and the politics of compromise, and who were increasingly open to Thatcher's claim that Britain was in decline because it had embraced the immorality of socialism. At one point, she walked into a meeting on how the Conservative Party needed to find a centrist path through the usual British compromises. She loudly demurred, slamming down a copy of Hayek's *The Constitution of Liberty*—which presented a hard-core view of conservative philosophy—and proclaimed, "This is what we [Conservatives] believe."

Thatcher was also forceful on the world stage. As shadow prime minister between 1975 and 1979 she visited the United States twice, garnering extensive publicity as a supporter of the "special relationship" between Britain and the United States that had grown out of their World War II alliance, and as an outspoken advocate of the idea that the main goal of the allies should now be to eliminate (and not merely contain) the insidious threat of Marxist socialism at its source, the Soviet Union. On January 19, 1976, she condemned the expansionist global designs of Soviet totalitarianism on Eastern Europe and on developing nations in Asia and Latin America. In response to this speech, the Soviet army newspaper labeled her "the Iron Lady," a name that was to remain with her throughout her public life, delighting most Britons and erasing any suggestion that she might prove weak because she was a woman. The Soviets "never did me a greater favor," she later recalled.

Thatcher assembled a coterie of sophisticated image makers in order to present herself to the media as tough yet not threatening, a leader who was taking control of the Conservative Party agenda and could govern when the time came, but who was also a housewife in touch with the concerns of ordinary people. To craft her speeches, Thatcher hired an accomplished playwright, Ronald Millar, who at their first encounter gave her some quotes from Abraham Lincoln:

> You cannot strengthen the weak by weakening the strong.
> You cannot bring about prosperity by discouraging thrift.
> You cannot help the wage-earner by pulling down the
> wage payer.

Thatcher reached into her bag and pulled out a yellowed newsprint with the same words. "It goes wherever I go," she said.

Her advisers began choreographing her public appearances with the aim of landing a good sound bite on the evening news. They helped spread the story of the grocer's daughter, the housewife who shopped for the family and made breakfast for her husband every morning, the woman of modest means who worked her way up the social and political ladder. She picked fashion advisers to select her wardrobe and took speech lessons to take the crust off the diction she had picked up at Oxford. These modern elements of a perpetual campaign, although known in the United States, were entirely new to the stodgy political environment of Great Britain at the time, and they paid off handsomely. The tabloids started calling her "Maggie" as well as the Iron Lady.

Britain was lurching from one crisis to another under Labour prime minister James Callaghan. In 1976, inflation and unemployment helped cause a severe currency crisis that forced Callaghan into the humiliating position of asking the International Monetary

Fund—known primarily for supplying financial lifelines to bankrupt developing nations—for a loan to support the value of the British pound. Within two years, Callaghan had to call again on the IMF, but this time the quid pro quo for a loan was a severe cutback in public spending and a tightening of the money supply, the same program that right-wing Conservatives had been preaching. By late 1978, interest rates stood at 16 percent, inflation at 20 percent, British productivity had stalled, and unemployment topped one million.

Worse was coming in what came to be known as the "winter of discontent," when Britain was paralyzed by the most extensive labor strikes in half a century. So many professions stopped working—the hospital workers, the garbage collectors, the truckers, the dockworkers, the train engineers, even the grave diggers—that it looked like Labour had lost control of the country. An outraged citizenry turned to the Conservatives to save Britain from what looked like terminal decline, and in April 1979, they handed Thatcher and her party a landslide victory that would change the course of the nation. "You know," said Callaghan in defeat, "there are times, perhaps every thirty years, [when] there is a sea change in politics. It then doesn't matter what you say or do. There is a shift in what the public wants and what it approves. I suspect there is now such a sea change—and it is for Mrs. Thatcher."

"Don't Tell Me *What* to Do, Because I Know That. Just Tell Me *How*."

Thatcher knew that economic revival was a necessary step to achieving her most fundamental ambition, restoring Britain's global status. Economic strength was the requisite foundation to support Britain's role as Washington's senior European partner, as a bulwark against Soviet expansion, and as a major player in the European Community (which was the successor to the Common Market and eventually became the European Union). Britain entered the European Community in 1973, and although Thatcher supported that move, she always wanted to limit Europe's powers and maximize the scope of Britain's sovereign decision making. Despite the nation's dismal economic condition, Thatcher believed that the clout to achieve all these ambitions was within reach. Britain had recently discovered oil in the North Sea, and Thatcher saw that this bonanza would reduce the need to import petroleum while lowering energy costs for British manufacturers, which would make their goods more competitive in global markets. Equally significant, she felt that the nation would respond to policies that went in a totally new and unambiguous direction. "Having previously exaggerated our power," she wrote in her memoirs, "we now exaggerated our impotence."

In the 1970s most of the Western world was suffering from the same economic shocks that had threatened to derail Britain. In 1971, Washington took the dollar off the gold standard, and the value of all major currencies began to swing sharply, creating uncertainty across the world. Two years later the Arab oil producers mounted an embargo and then quadrupled oil prices, undermining manufacturing industries, increasing unemployment, and leading to extreme volatility in financial markets. The price pressures came at a time when inflation was already creeping up across the industrial nations, due to growing wage demands by unions and increased government spending on welfare programs. In Western Europe, welfare expenditures were growing at 7 percent a year, or twice as fast as average GDP growth.

Over the previous five decades, in fact, the role of government had been expanding in the industrial economies of Western Europe, North America, and Japan. In these nations, total state spending had represented about 10 percent of GDP before World War I but had climbed steadily to 43 percent by 1980. Government programs required higher taxes or higher debt, and they came with increasing regulation, putting the squeeze on the individuals and entrepreneurs who were the heroes of Thatcher's world. Entering office with a clear mandate from voters demanding change, Thatcher pressed her advisers on how to turn the tide, saying: "Don't tell me *what* to do, because I know that. Just tell me *how*."

In short, Thatcher was attacking the Socialist impulse to use the hand of the state to control market chaos—an impulse hardly confined to Britain. To contain inflation, London had imposed wage and price controls, but so did the administration of President Nixon. England nationalized industries, but so did France and many other nations even outside the Communist bloc. Not long after Thatcher became prime minister a number of Latin American countries, including Mexico, would default on their sovereign debts, which they had wrung up to pay for bloated public sectors and inefficient state-owned enterprises.

It took a uniquely decisive and powerful figure to give people the confidence to imagine a government that would fight the creeping expansion of its own influence in the economy. The critical policy moves were to reduce government spending, control inflation, remove exchange controls, and let the British pound find its own level in the global market; to slash tariffs and open the country to foreign investment; to deregulate the economy and privatize state-owned companies; and to reduce the influence of trade unions. Later, Thatcher recalled both the magnitude of the moment and her own inner confidence. "[The eighteenth-century statesman and prime minister] Chatham famously remarked: 'I know that I can save this country and that no one else can.' . . . I must admit that my exhilaration came from a similar conviction."

From the beginning of her reign as prime minister, Thatcher dodged the compromises of consensus politics by working through an inner circle within the cabinet or just one-on-one with the key official responsible for a major policy. She hated the very notion of consensus, calling it "the process of abandoning all beliefs, principles, values and policies in search of something in which no one believes, but to which no one objects." She rarely put issues to a cabinet vote, and, even on major economic decisions, often bypassed even a cabinet discussion—nearly unprecedented for British politics. She trusted few of her colleagues, often demeaned one official in front of another. She relished confrontation.

"The Lady's Not for Turning"

From her first day in office Thatcher began implementing the radical ideas of Keith Joseph in ways that reversed the trends of the past four decades by jettisoning targets for employment and economic growth, and by pushing through a budget that ruthlessly cut public spending, including subsidies to state companies, and reduced government borrowing. She simultaneously lowered direct income taxes and, in order to plug the resulting shortfall in revenue, raised sales taxes (which, of course, fell hardest on lower-income citizens). She abolished the controls on the British currency and let the pound sterling trade freely in world markets. Thatcher abolished the Price Commission, which had been setting wages and prices in a futile effort to contain inflation. She attacked the Civil Service, proposing significant cuts in both its size and in its influence on shaping policy. From here on out, *she* would set the policy and the job of the Civil Service would be to execute it.

Within six months Thatcher established her dominance, but the initial effect proved to be disastrous. By the start of 1981, the economy was in worse shape than when she took office. Inflation was at 18 percent and unemployment at three million. Economic growth and business competitiveness were declining. Most projections indicated that the worst was yet to come, in part because a global recession was beginning and in part because the pound was getting stronger. The result was a double whammy for British exports—the global recession would constrict demand for British products, and the strong pound would make those products more expensive in overseas markets. Public approval for the Conservative Party fell to 30 percent and for Thatcher herself to just 23 percent, one of the worse ratings for any government in decades. As members of her own cabinet grew uncomfortable with her direction, Thatcher delivered one of the lines that would define her forever. "To those waiting with bated breath for that favorite media catchphrase, the 'U-Turn,' I have only one thing to say. You turn if you want to. The lady's not for turning."

Then came one of her most daring and controversial moves: In March 1981 Thatcher presented a budget that ignored her popular opposition and doubled down on her defiant conservatism with more cuts in public borrowing and spending. She capped child benefits at a rate that failed to keep up with inflation, imposed higher taxes on alcohol, tobacco, and gasoline, and created new taxes on government benefits for the sick and unemployed. Proposing a deflationary budget in a recession was a radical departure from the established Keynesian approach—which was to spend generously in a recession—and it was greeted as heresy even by most Conservative leaders. In a joint letter to the *Times* that March, 364 of Britain's leading economists wrote that the budget would "deepen the depression, erode the industrial base of our economy and threaten its social and political stability."

By the summer, riots broke out in several major cities, sparked by unemployment and racial tensions. Thatcher stood her ground, comparing Britain to a patient that would have to take its medicine if it was to be cured of its "disease" and vowing that "the government will have the guts to see it through." Even as the budget debate raged, it also sparked a certain admiration for

Thatcher. "She has reasserted her political dominance and restated her faith in her own policies," said the *Times*. "If she succeeds it will be a remarkable personal triumph. If she fails the fault will be at her door, though the damage and the casualties will be spread wide through the political and economic landscape."

As late as May 1983, disaster was still knocking at that door, with unemployment stuck at 3 million, having risen from 1.2 million in 1979. During the same period, the economic growth rate had declined 3 percent and industrial output had fallen by 15 percent. In addition, the tax burden during these four years rose sharply, from 34 percent to 40 percent. True, there was a global recession, but Great Britain was doing relatively worse than most other Western industrial nations, even with its North Sea oil. Then fate intervened, elevating Thatcher's stature to a height no one could have imagined.

War to the Rescue

On April 2, 1982, Argentine dictator Leopoldo Galtieri launched a surprise invasion of the Falkland Islands, a British territory three hundred miles off the coast of Argentina with a population of just two thousand, the majority being British nationals. Wrote biographer Charles Moore, "The humiliation of Britain was sudden, and complete. Unless it could be reversed Mrs. Thatcher could not expect to survive as prime minister." Thatcher met with her commanders, who assured her that they could recapture the islands, and she took the country to war. She knew the stakes: victory would erase the anger over the 1981 budget and deliver on her promise to begin restoring Britain's "greatness," while failure would almost certainly bring down her government in humiliation. She formed a war cabinet and for ten straight weeks met with them every morning at 9:30 a.m. She became a high-profile commander in chief, heir to the tradition of her idol, Winston Churchill.

As British naval vessels steamed eight thousand miles to the southwestern Atlantic, Thatcher formed close ties with her commanders, who greatly appreciated her ability to listen, absorb information, and make clear decisions. She oversaw intense diplomatic efforts to get the support of the United States and other European countries and to isolate Argentina in the United Nations. Her allies assumed that a peace could be struck once it was clear the British armada was serious, but Thatcher demanded that Argentina withdraw from the Falklands before she would even consider talks. In the event, the naval battles erupted before any negotiation could be arranged, and fighting raged for three weeks. By the end of the ordeal, Britain had lost six ships, thirty-four aircraft, and 255 lives, with another 777 wounded; and Argentina lost two ships, seventy-five aircraft, and 650 lives. On June 14, seventy-two days after the invasion, Argentina surrendered.

The military victory made Thatcher a global star, the great female warrior, bringing even closer international attention to what she was doing at home. In Britain, Thatcher's standing rose on a wave of nationalist pride, as the British discovered in the Falklands saga a new respect for her enduring trademarks—single-mindedness, clarity, a sense of moral righteousness, and a sense of Britain's destiny. When she returned to 10 Downing Street on the day of victory, a crowd sang "Rule Britannia." On October 12, 1982, Thatcher held a dinner for 120 officials whom she considered critical contributors to the victory, all of them men. Because of space constraints, spouses were invited only for post-dinner drinks. At the end of the dinner and following her remarks and subsequent toasts, Thatcher rose. "Gentlemen," she said, "shall we join the ladies?" Wrote biographer Moore, "It may well have been the happiest moment of her life."

For Thatcher it was a personal victory of monumental proportions. "No transformation in modern British history had been swifter, or more complete," wrote Moore. "She now had command of the whole field."

Winds in Her Sails

Thus emboldened, the prime minister began to press her agenda with even more determination. During her 1983 campaign for re-election, she echoed almost verbatim the themes of her 1950 run for office; once again she was fighting a "historic election" offering a choice between "two totally different ways of life" and "the chance to banish from our land the dark, diverse clouds of Marxist socialism." She went on to win in a landslide, her threats carrying a new credibility after the war. "I had a very tough time the first three years, a very, very tough time," she later wrote. "But after the Falklands War, people understood that we were going to do what we said we were going to do."

Thatcher's second term produced better economic results—lower inflation; higher industrial production; soaring exports; the fastest economic growth in Europe; soaring imports; a stock market boom. Whereas between 1980 and 1981 there were sixteen months of negative growth, in the 1981–1989 period, GDP growth averaged 3.2 percent. In addition, the cost of servicing the government debt came down. Government spending held steady as a share of the growing economy, even as it rose in absolute terms.

One consequence of her policies was that the British economy was dividing into two distinct tracks. Skilled workers prospered, often in services such as finance, but those losing jobs in the manufacturing sector were entering an ever-enlarging underclass. Wages were rising for those in the workforce, but unemployment remained at record highs, widening the gap between the haves and have-nots. The goal of social equality, so important to previous postwar British governments, was replaced by a Darwinian society that closely resembled what Ronald Reagan's America would become.

For the most part, the political opposition to Thatcher was divided and inept. Historian John Campbell described Thatcher's second term, from 1983 to 1987, as "the moment when the hundred-year-old political argument between capitalism and socialism seemed to have been decisively resolved in favor of the former. The moral and practical superiority of the market as an engine of wealth creation and the efficient delivery of public services was incontestably established." With her success, Thatcher became even more of an autocrat, in total control of the government agenda. She reorganized her cabinet to include more loyalists, and the leadership team acted increasingly as a rubber stamp.

Selling Off the State

Thatcher's next big move was to focus on selling off government-owned companies. There was no significant precedent for privatization in Britain or any other country. Political journalist Peter Riddell described it as a "striking and internationally influential policy innovation" that transformed the British industrial structure and the global business culture itself. Co-authors Daniel Yergin and Joseph Stanislaw said it was "the most decisive element of Thatcherism, and the one—along with the philosophy itself—that would have the greatest impact around the world."

The background was that after World War II British governments had begun nationalizing key industries so that they could harness production for the public interest. In the 1950s and '60s, British politicians had come to believe that the results of competition among private companies were often perverse: short-term gains for managers and shareholders, lowest possible wages for laborers. Owned by the state, so the theory went, companies would act in the national interest, treating workers better, investing in long-term projects, and preventing foreign companies from taking over important British economic assets. As it turned out, these Socialist ideals were rarely fulfilled.

Thatcher envisioned a new kind of capitalism that gave individuals a larger ownership stake in the country—ownership of homes, of businesses, of property. In 1985, her chancellor of the exchequer, Nigel Lawson, captured Thatcher's goals in a speech that called the ownership of private property "crucial to the survival of freedom and democracy" because it gives a citizen "a vital sense of identification with the society" as well as "a stake in the future—and indeed, equally important, in the present."

The sell-off began slowly but picked up pace in Thatcher's second term, under the principle laid down by Lawson that "no industry should remain under State ownership unless there is a positive and overwhelming case for it so doing." Once the momentum began, it fed on itself. Between 1979 and 1991, the British government sold two out of every five companies that it had owned. It sold more than fifty large enterprises, employing some nine hundred thousand people, including Britoil, Associated British Ports, Enterprise Oil, Jaguar, British Telecom, British Gas, British Airways, Rolls-Royce, and British Airports Authority. During this period the amount of money that the British Treasury took in on privatization deals rose by a factor of ten, reaching 5 billion pounds per year. Most of the companies were sold to the public in stock market offerings. The proportion of the adult population owning stock shares shot up from 7 percent in 1979 to 25 percent ten years later.

Some of these deals broke world records for the size of initial public offerings, but their scale was matched by the resulting controversies. Critics said that the assets were sold too cheaply and that London's brokers made unseemly profits. Labour politicians, joined by many academics, charged that the government was simply placing state monopolies in private hands, without the anti-monopoly regulation needed to protect the public. They also complained that the government was booking the sales as current income and spending it on the needs of the moment, rather than investing the proceeds in long-term assets, such as better roads, bridges, and rail services. "In advance of every significant privatization, public opinion was invariably hostile to the idea, and there was no way it could be won round except by the Government going ahead and doing it," recalled Lawson in his memoirs.

Breaking the Unions

Thatcher's next big battle was with the labor unions, which had mortally wounded both Labour and Conservative governments in the 1970s and had brought the nation to a standstill in the 1979 winter of discontent. In the mid-1980s, in a showdown with the most militant of the major unions, the National Union of Mineworkers (NUM), Thatcher set out to demonstrate that Britain was governable, after all. Nationalized in 1947, the mines had been losing more than $1 billion a year and were simply too big for Britain's needs. The only rational answer from London's standpoint was to close uneconomic mines and drastically reduce the size of the workforce. Nevertheless, NUM leader Arthur Scargill refused to acknowledge the problem of overcapacity and argued that the only reason for closing a mine should be safety or geological exhaustion.

Thatcher had avoided a direct confrontation with the miners in her first term, instead moving slowly to curtail their room for maneuver. She worked to limit union power to close shops to nonunion workers, to dodge liability for damage caused in protests, and to employ violent tactics in work stoppages. Her government was also creating a unified command center for local police forces, and stockpiling coal to keep key industries running in the event of a nationwide strike.

That came in 1984, when the National Coal Board, an official agency in charge of managing the mines, struggling under heavy debts, announced plans to close twenty uneconomic pits and cut forty-four thousand mine jobs. Scargill responded by calling a strike that would run for a year, escalating into bloody clashes between strikers and police, and between strikers and miners who stayed on the job. In her memoirs, Thatcher vividly recalled the violence against the "scabs," including the beating of one miner in his own home and the death of another, when a three-foot concrete block was thrown off a bridge and through his windshield as he drove to work. But the violence would work against Scargill, who made a series of blunders that made him look like a radical thug. For example, he failed to hold a vote to authorize the strike, as union rules required, and he solicited financial support from Libyan dictator Colonel Mu'ammar Gadhafi, from French Communists, and even from Soviet sources—all of which came out in the British tabloids.

To Thatcher, the battle with the miners was like the Falklands War all over again, and she was determined to achieve an unconditional surrender. She set up the equivalent of a war cabinet and began secretly guiding the moves of the National Coal Board, which by statute was supposed to be an independent agency. Behind the scenes, she and her inner circle befriended or otherwise bought off other unions who might have supported the miners, coached the coal board to offer miners large bonuses to return to their jobs, worked to prevent the board from compromising with the union, and made sure that union members who violated the law would face criminal and not civil penalties.

Thatcher defined the confrontation in her typically apocalyptic way, especially after she escaped unscathed from the blast of a bomb planted in her hotel, presumably by the Irish Republican Army. "At one end of the spectrum are terrorist gangs within our borders and terrorist states which finance and arm them," she said. "At the other end are the hard left operating inside our system, conspiring to use union power . . . to defy and subvert our laws." She left no doubt she was intent on crushing both.

The strike lost momentum over time, in part because Scargill called it when coal reserves were already high and demand was falling with the end of the winter heating season. The government stockpiles helped keep factories humming, and the centralized police command helped contain violence. On March 3, 1985, the union's governing assembly, against Scargill's advice, voted to end the strike. Thatcher stood outside 10 Downing Street and declared a victory for the miners who had stayed on the job. The outcome shattered the labor movement, and by 1990 union membership had fallen by 25 percent and the number of officially recorded strikes had fallen to a fifty-five-year low. The result was a change in the basic relationship of labor, management, and government. Thatcher put it this way: "What the strike's defeat established was that Britain could not be made ungovernable by the Fascist left. . . . They failed, and in doing so demonstrated just how mutually dependent the free economy and a free society are. It is a lesson no one should forget."

Freeing Up the City of London

The series of victories had generated extraordinary political momentum for Thatcherism, and her ministers ran with it, nowhere more dramatically than in the campaign to cut red tape that would lead in 1986 to the "big bang" deregulation of the City of London.

Great Britain had been the center of global finance for most of the nineteenth century and well into the twentieth century. Even as the United States eclipsed Britain as an economic power after World War I, the City of London held on to its global financial preeminence for decades, if only because Wall Street remained relatively provincial, focused on the booming American domestic market. The City remained an important pillar of the British economy, with financial services accounting for 10 percent of GDP and also 10 percent of its employment in 1985.

When Thatcher came into office, her economic team was concerned that the City could be threatened by rival financial centers. The United States, Switzerland, Germany, and Japan were lifting regulations that had kept financiers closeted at home. Investors and companies were expanding across the globe, listing their shares in multiple countries, looking for the most efficient places to do business. The OPEC (Organization of Petroleum Exporting Countries) price increases in the mid-1970s put vast wealth in the hands of oil-producing nations, which were looking for places to invest outside their own tiny domestic markets. As the globalization of finance gathered force, governments were engaged in competitive deregulation to grab a share of the foreign money flows, and top British politicians worried that the City was losing this race.

Indeed, the City was a clubby old community dominated by Oxford and Cambridge graduates and governed by rules that were outdated in a competitive new era. The Thatcher government set out to abolish regulations that rigidly and unnecessarily separated various activities within the London Stock Exchange, fixed commissions on the sales of securities like stocks and bonds, and barred foreign firms from buying major British banks and brokerages. The government was also ready to set higher standards for professional behavior, disclosure of information, and detection of fraud.

The impact on London was dramatic. In just a few years most of the major British firms were bought up by American, Swiss, and Japanese companies with global reach and more modern financial acumen. London began to attract some of the best talent in Europe and became the world capital for euro trading, even though Frankfurt was the seat of the European Central Bank (which issues the euro). The City became an innovator in finance and in financial derivatives and hedge funds, and it emerged as the center for initial public offerings from emerging markets such as Russia, Hungary, and Poland. It dominated the $3 to $4 trillion daily market for foreign exchange trading. Prosperity in London soared with a new generation of highly paid bankers, followed by lawyers and accountants, new office buildings, and fancier restaurants. The City expanded from its square mile in London to Canary Wharf on the outskirts, and developed a comprehensive set of new regulations that became a twenty-first-century model. New York developed as a major global financial center, too, but two decades after the big bang, a major debate would rage in the United States about how London had pulled ahead and what Washington and Wall Street could do about it.

Exhaustion before the Fall

In 1987, Thatcher was again reelected, entrenching her revolution deep in the British structure and psyche. But she had become too autocratic, and she started making ill-advised moves. One was the replacement of certain property taxes with a flat-rate, regressive poll tax. Another was her extreme hostility to signing on to new European Union financial arrangements that her own advisers, loyal to her as they were, adamantly favored joining.

The economy also began to sour. Inflation returned. It was clear that even the Iron Lady couldn't repeal the business cycle. In 1989, ten years after she took office, polls showed that only 37 percent of Britons thought that their country had become a "better place" under Thatcher. Her approval ratings sank to her lowest since 1981, and as low as any prime minister since Neville Chamberlain made his infamous deal with Hitler at Munich in 1938. In November 1990, when a challenge to her leadership arose from within her own party, Thatcher, sensing imminent defeat, decided to step down. "All Britons remember where they were when Margaret Thatcher resigned in 1990," wrote Daniel Hannan, a Conservative member of Parliament. "It was our equivalent of the Kennedy assassination."

Legacy

Prime Minister Thatcher became Baroness Thatcher, lived to eighty-seven, and died of a stroke on April 8, 2013. The many international dignitaries who attended her funeral included Queen Elizabeth, who had not attended a funeral for any prime minister since Churchill.

The arguments still rage over her legacy. The basic statistics on GDP growth, inflation, and unemployment oscillated wildly during her long tenure, so it is easy to paint very different pictures with these numbers, depending on which period is selected. There are disputes over the human cost of her policies, and over how much any improvement in the economy was a credit to her policies at home rather than the result of changes for the better in the global economy. It is possible England would have turned itself around with or without Thatcher.

There are, however, a few basic achievements on which many will agree.

When Thatcher took office, growth was negative, and during her time as prime minister moderate growth was the norm. Government spending as a percentage of the total economy was substantially reduced, from 23 percent to 20 percent; public debt declined from 50 percent of GDP to about 30 percent. Thatcher presided over a dramatic decrease in inflation, from 11 percent to

7 percent, and a very substantial *increase* in unemployment. Between 1955 and 1979, unemployment in Britain averaged 3.3 percent; between 1980 and 1995, the figure was 9.7 percent. The percentage of industry owned by the state was substantially reduced. Thousands of businesses went bankrupt. Driven by growth, increased consumerism, and a strong exchange rate, Britain's trade deficit grew. Manufacturing and mining were hard-hit by labor unrest, but ultimately productivity per person increased. Financial services expanded, but a huge swath of it was acquired by foreign banks. Tax rates on income were dramatically simplified and reduced, with some top rates cut by half.

It is fair to say that the Thatcher revolution created a more competitive nation and also a much harsher one. She was said to have administered "shock therapy," but as John Cassidy of the *New Yorker* wrote, it was all shock and no therapy. In relying so heavily on market forces to generate growth, Thatcher opened the floodgates to a version of winner-take-all globalization in which those lacking advantages in education, skills, or social connections suffer greatly. Thatcher had no interest in helping the poor and the middle class level the playing field, and she was very blunt about it. In 1987 she told an interviewer:

> . . . too many children and people have been given to understand "I have a problem, it is the government's job to cope with it!" . . . and so they are casting their problems on society and who is society? There is no such thing! There are individual men and women and there are families and no government can do anything except through people and people look to themselves first.

When she died, commentary around the world reflected the division over her legacy. Thatcherism was described both as the "most popular and successful way of running a country" of her era and as a "cult of greed" that did more to shackle than to free the human spirit. Whatever the verdict, no one doubted that she had been the strongest prime minister since Churchill and had reshaped her nation's culture and temperament for generations to come. Perhaps her official biographer, Charles Moore, summarized it best: "Many think she saved Britain, many that she destroyed it."

By the time Thatcher became prime minister, socialism—with its doctrine that government should own many industries, regulation should be all-encompassing, and unions should be central to society—had been building for several decades. Its momentum derived from a number of events and trends: the Russian Revolution of 1917 that ushered in communism, an extreme form of socialism; the enhanced role that governments were forced to play in their collapsing economies during the Great Depression; the necessity of governments' directing all parts of their economy during World War II and the accompanying popular acceptance of that role; and, except in the United States, the need for governments to orchestrate a recovery from the devastation of World War II. This long period of heavy state intervention in national economies resulted in a philosophy that led to widespread government involvement in what would otherwise be much less constrained commerce. It also led to societies that were intensely inward-looking. Thatcher dismantled these government barriers, releasing a gusher of trade and money that moved across borders and linked Britain with other countries as it had not been for generations. By reversing the course of socialism in an iconic nation, she established the first clear proof that it could be done, at a time when the prevailing belief was that the continued growth of government over the private sector was inevitable. Many countries around the world that had also embraced socialism took note and followed suit.

Thatcher was not an originator of ideas; her gift was an ability to grasp important concepts and, above all, to simplify them in language that could mobilize an entire society to try radical reform. By the late 1980s, Thatcher's conviction and clarity helped inspire a deep backlash against Socialist policies across Western Europe; in Mexico, Brazil, and Argentina; in Japan, India, and throughout much of Asia; and in Hungary, Poland, and others behind the former Iron Curtain. Most notably, in Russia, Gorbachev's efforts at perestroika were no doubt influenced by Thatcherism. And even in China, Deng Xiaoping was opening the country to the winds of global markets during the same decade as the Iron Lady was in control.

It is notable, for example, that the Thatcher years of shrinking government were accompanied by similar trends elsewhere. One study of forty-three major developing nations shows that between 1980 and 1998, total government expenditure fell as a share of GDP, from 19.3 percent to 16.3 percent. In addition, virtually every other country that had nationalized industry after World War II would eventually reverse course and copy Thatcher's model. By the 1990s, Russia, India, Japan, Malaysia, Mexico, New Zealand, the Philippines, Sri Lanka, Singapore, Poland, and Turkey, among others, had launched privatization programs. To describe it another way, in the fifteen years after the great British privatizations, other governments sold more than $100 billion worth of state assets; France sold Renault, Germany sold Lufthansa, Italy sold its oil company, ENI. "Whether they managed well or badly," wrote the *Economist*, "all of them looked to the British example."

By the mid-1990s, the entire package of free market policies that had been pushed in their purest form by Thatcher were relabeled the "Washington Consensus," and aimed at the wide swath of countries being labeled "emerging market nations." The Clinton administration became a relentless evangelist for this approach in the developing world, including the former satellite states of the Soviet Union. It pressed these nations to balance budgets, deregulate their industries, sell off state assets, reduce rigid work rules supported by unions, and free up their currencies. Many countries came to believe that the role of the state had to be vastly curtailed, and that a smaller state created more space for the markets to work and for individual enterprise to flourish. They bought into Thatcher's focus on economic growth rather than economic redistribution; the goal was to distribute the slices of an economy that was expanding rather than one that remained static. They endorsed Thatcher's philosophy that the formula for shrinking the state involved lower trade barriers, less regulation, and less government ownership, all of which made commercial dealings within and among countries easier. In fostering such thinking, Thatcher added steroids to globalization.

In this new century, the gale winds of globalization are creating a host of new problems for which existing answers seem totally inadequate—growing inequality, deteriorating public services and social safety nets, financial systems subject to serial crises, monumental environmental challenges—to take but a few examples. All of these destabilizing trends create opportunities for the type of interventionist regimes and international institutions that Thatcher detested. Today, for example, China, Russia, India, Brazil, and the Persian Gulf oil states all have political cultures that are comfortable with state ownership of companies, banks, and investment firms. In response to the global financial crisis of 2008–9, all major countries, including the United States, the UK, and Japan, have instituted extensive new regulations on banks. In the face of recession, these same nations have turned to government for economic stimulus, unemployment benefits, and the like. The big question now, therefore, is, will Thatcherism be reversed?

My answer is that because the Thatcher legacy runs so deep, its edges can be blunted, but it is difficult to imagine a fundamentally new long-term direction. Just compare the world today to what it was on the eve of the Iron Lady's election as prime minister. Then most national currencies were not freely traded; today, no one is considering going back to the status quo ante. Then the unions were powerful forces in the governing of companies; today, there is little resurgence. Then there was no international authority to stop nations from blocking trade with sky-high tariffs or subsidies; today, we have a credible international policeman in the World Trade Organization.

Still, it is worth taking a minute to reflect not just on the nature of the revolution that Thatcher wrought but also on its ultimate impact. It is conventional wisdom that the lowering of barriers to all kinds of commerce enhanced the connections among countries that characterize exactly what globalization has become. It is an article of faith among mainstream economists that these new channels on balance enhanced global economic growth, expanded individual opportunities, and facilitated many attendant benefits. That these trends also contribute to stress on employment and widening income disparities is also widely accepted. What is now becoming more apparent, moreover, is that the free market economy may also account for the growing vulnerabilities of the global system itself. Markets alone are not self-correcting, as we saw in the 2008–9 crash. Markets may create sleek, efficient supply chains, but these same chains are highly susceptible to snapping if there is a disruption from natural or man-made catastrophes, as we saw when floods in Thailand in 2011 jeopardized the global computer industry. Indeed, the very connectedness that Thatcher's policies encouraged has created a counter-requirement for more cushions, more redundancies, more advance planning to avert catastrophes—in other words, more government involvement.

Here, then, is a big paradox of our time: the really big threats—such as global warming, global financial, economic and humanitarian crises—suggest that Jean Monnet was right about the need for more supranational authority. No national government has the resources or clout to successfully meet these challenges. At the same time, the most pressing domestic economic issues for most advanced nations now are high levels of indebtedness, effective delivery of essential services, and, above all, remunerative jobs. And a large number of national governments cannot operate effectively enough to deal with those issues, either, which involve explosive political conflicts over allocating resources among segments of the population, including among different generations. As Thatcher liked to say, "there is no alternative" to shrinking the size of government and letting markets flourish.

How to deal with these two sets of seemingly irreconcilable imperatives? Thatcher offered a philosophy of governance that responded coherently to the crises of her day. Her approach could even now form a springboard for what comes next, especially in an era of such rapid technological change where speed and agility will be so important, and where government has such difficulty keeping up with the market. But Thatcherism alone won't be enough. The future of globalization belongs to a judicious balance between the private and the public, between competition and collaboration, between the market and the state, between what nations can do on their own and what they must do together—in other words, between Thatcher and Monnet. This much is certain: given the economic and social pressures that have arisen around the world—such as slow growth, income inequality, and high public-sector debt—a major debate is now taking place concerning the size, scope, and function of the public sector. These issues are at the heart of what managing globalization is all about, and the ideas and policies of Margaret Thatcher will never be far from the center of those heated discussions.

AP PHOTO/MARTY LEDERHANDLER

Chapter IX

ANDREW GROVE

The Man behind the Third Industrial Revolution

1936–Present

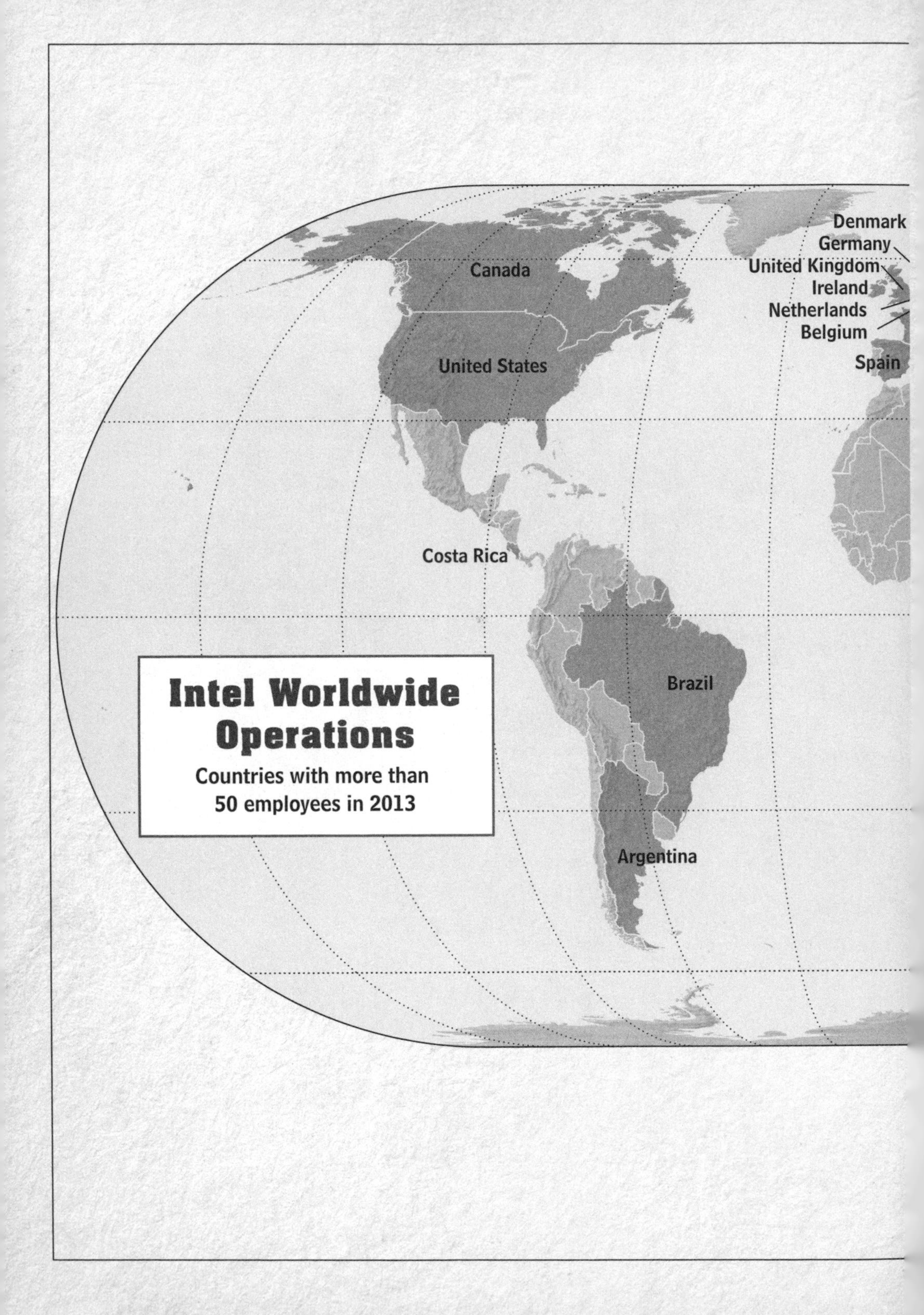
Intel Worldwide Operations
Countries with more than 50 employees in 2013
Canada
United States
Costa Rica
Brazil
Argentina
Denmark
Germany
United Kingdom
Ireland
Netherlands
Belgium
Spain

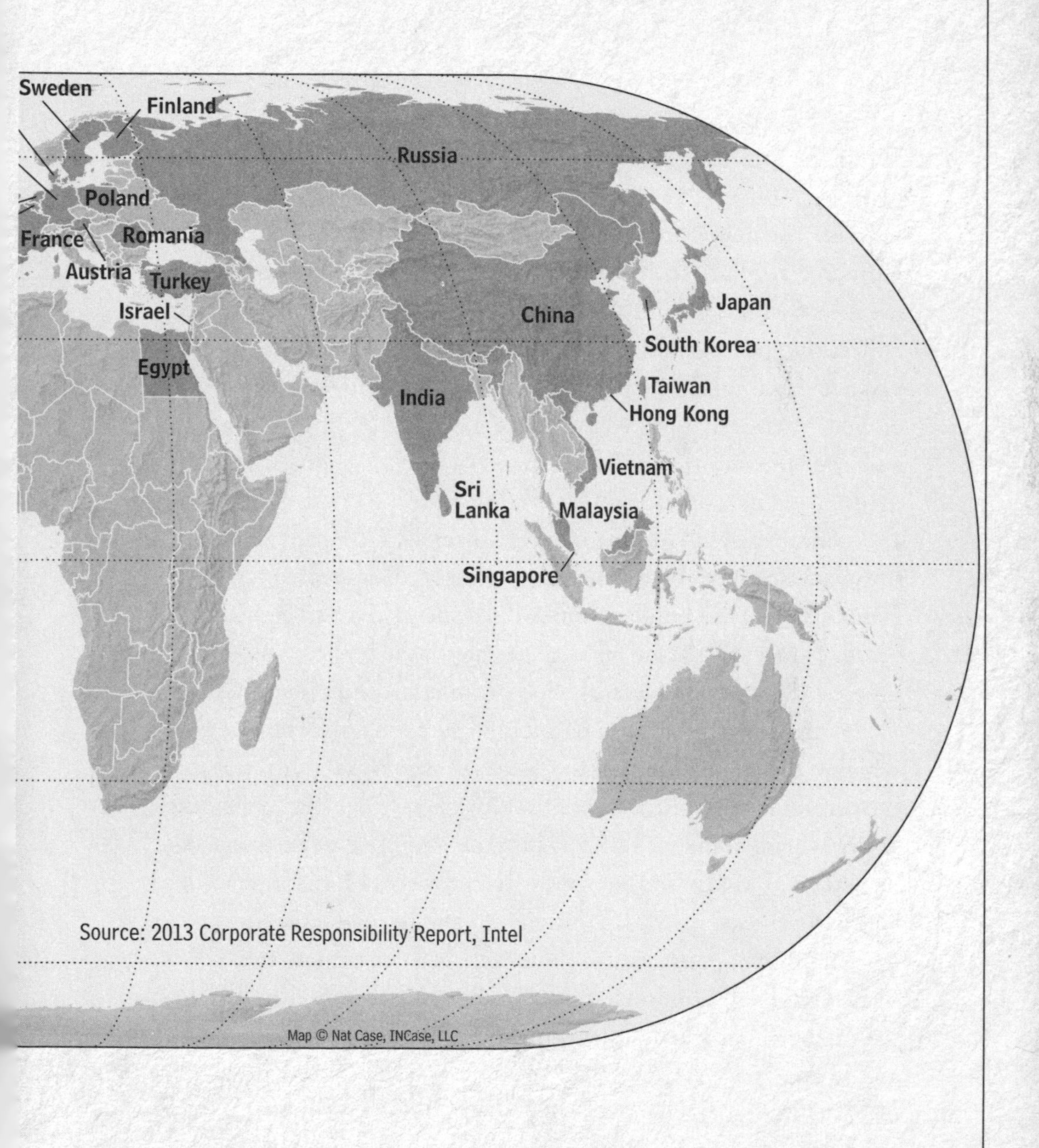
Sweden
Finland
Russia
Poland
France
Romania
Austria
Turkey
Israel
Egypt
China
Japan
South Korea
Taiwan
Hong Kong
India
Vietnam
Sri
Lanka
Malaysia
Singapore
Source: 2013 Corporate Responsibility Report, Intel
Map © Nat Case, INCase, LLC

Thanks in large part to the global channels that each of the individuals in this book opened, today the new industrial revolution, driven by digital technology and the Internet, is unfolding much faster than the earlier industrial revolutions. One manifestation of this phenomenon is the advances we see in robots, artificial intelligence, massive data collection and analysis, nanotechnology, and bioscience—to cite just a few examples. Another aspect is the technological capacity to embed Internet-connected computers in anything, from clothes and appliances to the individual parts of aircraft engines, opening up dizzying new opportunities to command and control our living and commercial environments. McKinsey & Company says that the change brought by the digital revolution is happening ten times faster and at three hundred times the scale of the late-eighteenth-century mechanization of the textile industry or the early-nineteenth-century advent of mass production and assembly lines.

At the heart of what's happening is a tiny microprocessor, which is the closest thing to the brains of a computer. It is not just the output of microprocessors that looms so large in the way the entire world is linked, and not just the exquisite technology behind them, but also the very industrial process of making the devices that has provided such a strong underpinning to contemporary globalization. Michael Dell, chairman and CEO of Dell Corporation, once said, "Ideas are a commodity. Execution is not." Excellent execution means exquisite management, and in large part the computer age arrived as a result of a revolution in the management of high-technology industries. Andy Grove was the leader of that revolution.

The Origins of Paranoia

Andy Grove became famous for urging his staff to maintain an attitude of acute paranoia toward Intel's rivals and for tracing his own natural anxiety to his experience as a child. He was born András Gróf on September 2, 1936, an inauspicious year to be born as a Jew in Hungary, or in any country within striking distance of Nazi Germany. That was the same year that Adolf Hitler reopened the arms factories of the German Rhineland and sent his armies into Austria, soon to invade Czechoslovakia and Poland as well. The campaign to round up and slaughter millions of Jews was imminent, though its full force would not hit Hungary until 1944, when András was eight years old and the German armies marched on his hometown of Budapest. One of Hitler's most notorious henchmen, Adolf Eichmann, took personal charge of the extermination campaign in Hungary, and he succeeded in killing two out of every three Jews in a population of 445,000.

For all practical purposes, however, the war came to Hungary in 1941. That year Hitler invaded Stalin's Soviet Union and Hungary joined the Nazi cause. In 1942, András's father, George Gróf, a partner in a small dairy business, was conscripted by the fascist Hungarian government who sent him to the Russian front, further shattering the comfortably middle-class existence he had carved out for his family. Maria, András's mother, an energetic woman who had finished a college preparatory academy (a rare feat for women in those days), was left to keep András alive as best she could. For years she shuttled András between their apartment and a friend's house in the countryside, trying to avoid the war between the Germans and the Russians taking place on Hungarian soil, not to mention the German search to round up Jews for eventual extermination. "Things were gloomy," Grove would recall years later. "But my mother was there, and that made me feel there was something warm and normal at the center of this strange existence." Still, Grove lived in fear—fear of separation from his mother, of capture by the Germans, and of Hungary's own fascist gangs who roamed the streets and murdered Jews.

At the end of the war in May 1945, the Russian Communists took control in Hungary. George Gróf returned from the war and found his place in the new state-owned economy, imposed on Hungary by the Communists. He worked first in a department store and then as director of a livestock breeding company. Maria took a job as a bookkeeper in a heating oil business. George was keen that his son make up for lost time; he insisted that András study English and take piano lessons, and he hired a math tutor for his son. Maria sold her jewelry to pay for lessons. The family was deeply concerned about András, who, five years prior at the age of four, had contracted scarlet fever, which permanently damaged his hearing. To compensate, András learned to lip-read and would always sit in the front of the class. Over the next twenty years, he underwent five reconstructive ear operations.

By the early 1950s, Budapest had gotten much of its old bustle back. Grove had developed into a good student, and at sixteen he was eager to express "what I thought of as my rich and multifaceted self in writing." He reveled in his position as a student journalist documenting his daily experiences for an obscure newspaper. But Hungary was a satellite of the Soviet Union and this foray into the world of imagination would not last. By 1952, the Soviets began clamping down on free expression, and in that year George Gróf was fired from his management job for hiring a "bourgeois person." The Gróf family standard of living collapsed and, as additional retaliation against George, András's editors began to reject his articles. His interest thus turned to chemistry, a profession less susceptible than journalism to capricious interference from Communist mandarins.

When Stalin died in March 1953, Moscow began easing its grip on Eastern Europe, lifting controls on free speech and releasing hundreds of thousands of political prisoners. Given a taste of freedom, Hungarians let loose a tide of anger against Russian rule and began demanding change. By 1956, what became known as "the Hungarian revolution" had begun. András was happy to see the Communist regime under pressure, but he feared the consequences. Sure enough, Soviet tanks rolled into Budapest to crush the revolution. The experience of the earlier Russian invasion seemed to repeat itself as the family took refuge in the cellar just before mortar rounds hit their building. Russians moved back into the Grófs' apartment. Budapest had again become a war zone.

By December, Russians were rounding up Hungarian youths and putting them in covered military trucks for shipment to political prison camps. András's aunt, an Auschwitz survivor, urged her nephew to escape immediately. George gave him the name of a cousin in the United States, and András, with nothing but the clothes he was wearing and a knapsack, set out with two friends for Brooklyn, New York.

First came a long day's ride on a train out of Budapest, then two days of furtive walking in the countryside, evading ubiquitous Russian patrols, finding sympathetic villagers who gave the three young men shelter at night in farmhouses with dirt floors. At every stop, people demanded money to protect the fugitives. Tired and desperate, they moved from one village to another, searching for the border. In his memoirs, Grove recalled the final minutes of their escape:

> I lost track of time. After a while we emerged from the woods. I could see some faint lights far across an open field. "Those lights are Austria," [the farmer who was guiding us] said. "Head towards them and don't take your eyes off them," he said. . . . After what seemed like miles and miles, the lights finally came close. Had we made it? We snuck up to the first house that we could see. Dogs immediately started

> barking in the dark. We again threw ourselves to the ground. A man came out of the house holding a kerosene lantern over his head, and called out—in Hungarian—"Who is there?" My heart stopped. . . . "Who is there?" he repeated. We hesitantly picked ourselves up from the ground and forced ourselves to approach. When he saw us, he smiled a big, warm smile and said, "Relax, you're in Austria."

They headed for Vienna and found it was filled with international relief agencies trying to help escapees. András applied to the US consulate for political asylum, and after a few weeks he received permission to go to America.

His long nightmare was over. Many years later Grove summed it up this way:

> By the time I was twenty, I had lived through a Hungarian Fascist dictatorship, German military occupation, the Nazis' "Final Solution," the siege of Budapest by the Soviet Red Army, a period of chaotic democracy in the years immediately after the war, a variety of repressive Communist regimes, and a popular uprising put down at gunpoint.

András crossed the Atlantic in a rusty US troop carrier and moved in with his cousins, who had emigrated from Hungary two decades earlier and ended up in Brooklyn, New York. With his poor and highly accented English, he entered Brooklyn College, a tuition-free school. Because of his exceptional scholarship and his growing interest in chemical engineering, he was urged by his professors to transfer to City College of New York, also a free university, where the academic standards were higher and the curriculum broader.

He changed his name to Andrew Grove and soon everyone called him Andy. He graduated number one in his chemical engineering class. He married Eva, an immigrant from Austria, who came to America after living many years in Bolivia. After earning a PhD at the University of California at Berkeley in 1963, he took a job at Fairchild Semiconductor in the town of Mountain View, south of San Francisco.

Grove and the Roots of the IT Revolution

Fairchild was a product of the dog-eat-dog entrepreneurial spirit that shaped the early days of what was to become Silicon Valley. This spirit drove the advances in technology that would ultimately shrink the mainframe computer to a device we could carry in our palms. The first mainframe computer, which emerged from military-sponsored research in 1945, was called ENIAC (Electronic Numerical Integrator and Computer). It comprised eighteen thousand vacuum tubes, housed in a vast machine one hundred feet long, ten feet high, and three feet deep, with huge fans to keep the tubes from overheating. In thirty seconds it could do calculations that would have taken a human being twenty years. It was housed at the University of Pennsylvania, and lore has it that it used so much power that the lights in western Philadelphia went dim when the mainframe was turned on.

Over the next thirty years, the story of the computer revolution was in large part the story of making the device that transferred electronic signals within the computer's brain ever smaller, lighter, more reliable, and faster. It started with replacing the vacuum tube in the original mainframe. The first replacement emerged in 1947 at Bell Telephone Laboratories, where a team of scientists, including William Shockley, John Bardeen, and Walter H. Brattain, invented the transistor. A tiny slab of metal one hundred times smaller than the vacuum tube, the transistor was more powerful, yet it generated less heat. Researchers at Bell began a process that would make transistors smaller and smaller, creating room for millions of transistors on a silicon chip of one square centimeter. When transistors are in close proximity, the electric signals move faster between them, boosting the chip's computing speed. Yet because the price of the chip stayed the same, the cost of this rising power would decline sharply. Any company that set the standard for packing the most transistors onto a chip would set the pace of the computer revolution.

Ultimately, that company was Intel, an organization that grew out of a period of corporate infighting and turmoil that is now legendary in the tech industry. It all began in 1954, when William Shockley, a disgruntled member of the Bell team, set up his own semiconductor lab in an old shed outside of Palo Alto and recruited some of the best minds from around the country, including Robert Noyce and Gordon Moore. But the company was plagued by problems. Customers like the Department of Defense and IBM required a highly reliable and repeatable process for mass-producing transistors to increasingly tiny specifications. That became Shockley's big problem because of a disconnect between his team's great ideas and its spotty execution—between research and development on the one hand, and manufacturing on the other. He and those around him could create pathbreaking technology, but they could not produce reliable devices in large quantities.

A second problem was that although Shockley was great at spotting talent, he was also a micromanager who refused to give creative latitude to his brilliant staff. His autocratic and narcissistic temperament tended to ignore or reject any proposal that was not his own. As a result, Noyce and Moore started looking to break away and start another company. The search for funding took them to Sherman Fairchild, an eccentric, wealthy playboy and entrepreneur whose Fairchild Camera and Instrument Corporation made photographic and aviation equipment, based in Syosset, New York. Fairchild agreed to bankroll the defectors and establish Fairchild Semiconductor, which set up shop in Mountain View, California, about two blocks from Shockley's operation. The defectors from Shockley came to be known in Silicon Valley as the "Traitorous Eight," but in an admiring way. They are seen as the true fathers of the technological revolution that unfolded in California over the next seven decades.

It was an ideal time to start a new business. In 1957, the space race between the United States and the Soviet Union had elevated the microelectronics business to national prominence. The government needed computers that were smaller than mainframes. The many engineers that the United States was churning out were drawn to the wealth of opportunities in the Mountain View area. Acre by acre, fruit trees were uprooted in favor of factories and small office buildings. General Electric, IBM, and start-ups like Hewlett-Packard were moving in. Fairchild began to achieve pioneering breakthroughs, including the discovery of a process that could produce complex microelectronic devices far more cheaply, and radical advances in the operation of transistors. Working separately and unbeknownst to one another, Jack Kilby of Texas Instruments and Robert Noyce both invented what became known as the integrated circuit, a silicon chip that replaced first hundreds and then millions of transistors. Instead of using a separate transistor for memory, another for calculation, and another for other functions, and then wiring the transistors together, each integrated circuit could perform all the functions on one chip, with no wires.

Boosted by Noyce's work on the integrated circuit, Fairchild took a big lead in the race to make chips smaller and faster. Then in 1959, Sherman Fairchild bought out Noyce and Moore, making them wealthy men. They stayed on at Fairchild for a while. Early on, the corporate culture had deliberately rejected the command-and-control, perk-ridden culture of big East Coast corporations. There was little or no hierarchy and dress was casual. The seating plan was open and the walls separating work spaces were low, a design intended to encourage conversation and unleash creative energy. But as Fairchild grew, the culture began to change, much to the dismay of Noyce and Moore.

What happened was that Fairchild became the largest semiconductor company in the world, with eleven thousand employees, sales of over $150 million per year, and operations around the United States and as far away as Hong Kong. At the same time, much of its profits were being transferred to Fairchild's corporate headquarters in New York, and an East Coast–type bureaucracy was being imposed on the company. Noyce was being forced to assume a senior management role that he did not want or enjoy. In addition, Fairchild had not solved the quality problem that was endemic at Shockley. The lack of close collaboration among key parts of the organization led to flawed products. Years later, Andy Grove would recall, "The research lab and the manufacturing location were seven miles apart. Those seven miles, from the standpoint of collaboration, could have been seven thousand miles."

The critical importance of these organizational flaws began to come into focus after April 1965, when Gordon Moore presented a paper in *Electronics* magazine that described what others would later call Moore's law. Its essence was that the number of transistors that can be placed on an integrated circuit could double at regular intervals—every eighteen months to two years. The implications were mind-blowing, for it amounted to a forecast of unprecedented speed, unprecedented miniaturization, and unprecedented precision.

The "law"—it wasn't really a law but a framework for thinking about the business model for the semiconductor industry—allowed companies to envision the capabilities of computer chips five to ten years into the future. It pointed to the opportunity—or perhaps the inevitability—of sustained exponential growth of computer technology. This in turn opened up possibilities and imperatives for research agendas, fabrication capacity, and investment programs. If you subscribed to Moore's law, then you could be working on several generations of chips at the same time because you knew how the future would unfold in terms of the size, power, and cost of chips. Beyond that, you would be ensuring that devices in the distant future would be compatible with their predecessors—so-called upward compatibility. The "law" became a driving force for the entire technological revolution of the late twentieth and early twenty-first centuries. But to sustain the pace of progress that Moore described would require a company that combined the freewheeling open-plan creativity of Fairchild's early years with a level of organizational discipline that had never been achieved in any company in the transistor world.

The Simple Plan for Intel

In 1968, Noyce and Moore decided to leave Fairchild for two reasons: because it was too bureaucratic for their tastes and because it was not organized enough to solve the Shockley problem of being unable to manage the transition from flawless design to flawless manufacturing of microscopic parts. They wrote a three-page business plan, describing their intention to build one corner of the transistor business—the one focused on computer memory—into an industry. Their reputation was such that within forty-eight hours they had raised $2.5 million on the phone.

A month after leaving Fairchild, Noyce and Moore established Integrated Electronics—Intel for short—in a half-abandoned thirty-thousand-square-foot concrete building, one hour south of San Francisco. At the time, big mainframe computers were storing information in crude devices called magnetic cores, which Noyce wanted to replace with tiny transistors that could store more information in much less space, accelerating the speed of the entire computer by allowing different parts to communicate more quickly. Rather than go for low margin and high volume, as Fairchild did, Intel wanted to get so far ahead of the competition that it could sell its products in high volumes for high margins. The market was driven by America's powerful defense establishment, but Noyce and Moore also saw the rapidly growing opportunities in consumer markets.

They recognized that commercial success depended on cutting the lag between great ideas and mass production to a time interval no semiconductor company had ever achieved. This would require managerial talent rare among tech engineers, most of whom wanted to build things, not supervise people, and who were motivated more by the thrill of creation than by commercial incentives. At the time, the number of transistors that worked relative to the number produced was often well under 20 percent—an obscenely low proportion that had to be dramatically increased. Even making a small batch was complicated enough, a task that has been aptly compared to doing surgery on the head of a pin, in circumstances where the slightest impurity in the air or on the material would kill the patient. Workers could not eat, smoke, or even wear cosmetics on the job. They had to wear protective suits with gloves and booties. There were highly precise procedures for lifting a tool or attaching a wire. Noyce and Moore had to find a tough manager who could do all this plus oversee an organization that would have to be preeminent in research and development, marketing, and after-sales service—all the while being ruthlessly competitive. They chose Andy Grove, with whom they had worked closely at Fairchild.

He was a surprise choice for director of operations at Intel, to say the least, since he was more of a physicist than an engineer and more of a professor than a businessman. His English was heavily accented and his cumbersome hearing aid looked like it had been made behind the Iron Curtain. Nevertheless, he clearly had the necessary toughness. Whereas Noyce and Moore could articulate goals, Grove was riveted on achieving them. Noyce and Moore could explain where the train should be heading and when it should arrive; Grove had the ability and desire to get it there on time, in good condition. He relished removing all obstacles in its way and in taking any action necessary to do so. Over the next four decades Andy Grove was the person most responsible for putting Moore's law into practice.

If Noyce and Moore led the way toward a nonhierarchical management culture in Silicon Valley—the opposite of the corporate environment of the East Coast and a major factor in the creativity of the US technology industry—that alone would not have been enough. What was also required was decisive and highly disciplined management. Andy Grove's great and distinctive contribution was to achieve both.

A key difference between Grove and his bosses was how he dealt with colleagues. In the early days at Fairchild, Grove had been an assistant director of research and development, very much part of the creative culture and yet a bit of an odd man out, with a reputation for being direct, extremely well organized, and sometimes abrasive. While Noyce and Moore were gracious and low-key, Grove could yell, pound the table, and intimidate anyone.

But there was a deeper difference. The two bosses would give instructions and assume they would be followed. There were no penalties for ignoring them. Not so with Grove. "In Intel's first decade, this was Andy's unwritten role," wrote journalist and historian Michael S. Malone. "He kept the company aligned on a daily basis with the long-term goals set by Noyce and Moore. He imposed consequences on every employee and action in the company. And he ruthlessly enforced cost accountability on every office at Intel—Grove did not accept excuses for a failure to hit one's numbers." As biographer Tim Jackson wrote, Grove was so disciplined that he made people wonder "how much he had been unwittingly influenced by the totalitarian regime he had been so keen to escape."

Grove became Intel employee No. 3. Unlike Noyce and Moore, he neither participated in nor profited from their two previous start-ups, and he did not identify himself as a self-starting, job-hopping entrepreneur. Of course, he wanted independence and control of his professional life, but because he had a wife and was supporting his two parents in Hungary, moving into a risky venture was a big gamble. "It was terrifying," he later recalled.

Grove quickly found the secret to solving Shockley's quality problem. He taught himself the manufacturing techniques that would dominate the computer age. It came down to shaping and inspiring a workforce that functions and adapts smoothly and swiftly enough to keep up with the accelerating speed of the computer chip.

In 1969, Intel introduced its first chip, which could store 64 numbers (and was called a 64-bit DRAM, for "dynamic random-access memory"). Within a year, Intel chips could store 256 numbers, and within two years they could store 1,024 in a chip (called the 1103) that was smaller and more energy-efficient than its predecessor. Due to Grove's relentless refining of the manufacturing process, the 1103 became the answer to the problem Intel had been created to solve—replacing the magnetic core that was the bulky memory center of the mainframe computer. Within two years, the 1103 was the biggest-selling semiconductor in the world, making Intel the largest global producer of memory semiconductors.

Grove later wrote about the lessons he learned in these early days in his widely read book, *High Output Management*. His key ideas were pathbreaking for their time, especially for a young industry dominated by creative but often undisciplined scientists and engineers.

First, Grove wrote, every person at Intel, whether they worked on the manufacturing line, in the marketing office, or the R&D lab, was responsible for attaining specific targets and was held accountable for that output. Second, output was measured by the team, and the critical role of a manager was to increase the output of his or her teams. Third, a responsible organization had to shed management layers. Supervisors and subordinates had to be in direct and constant communication. These ideas may seem commonplace today, but at the time they broke critical ground in the science and practice of management. All of this was simple to understand but difficult to execute. And executing policy with the bare minimum of errors was Grove's passion.

Early on Grove kept a journal to record his thoughts on management. One of his more striking entries details the passage of a new product from the design phase to the manufacturing plant, and identifies all the steps in between where something could go wrong. He underlined the importance of having managers who acted as generalists and had the sole objective of facilitating handoffs from one stage of the process to another. This was a critical insight, a recognition that it was not within each department, but in the links between them, that mistakes were most often made. The solution lay not so much in more technology but in better management to coordinate the entire process.

Grove was inventing the management process that would make Moore's law work in the real world. That was no small feat, for that law was at the heart of the revolution in semiconductors, computers, and communications since the late twentieth century. It was a central force behind the emergence of personal computers, mobile phones, Web servers, network routers—and so much more. In his writings Grove mused about the balance required to keep the whole of Intel inspired and ahead of its rivals. How does a manager best deal with a complex problem when a number of specialists must be involved? How fast can an organization grow and stay highly productive? How to be persuasive and not commanding? He was not only preoccupied with these fundamental issues of managing, but he was defining them clearly, debating them with colleagues, writing about them, and exploring them with students at Stanford University, where he became a part-time professor. By 1971, the year a local paper coined the term "Silicon Valley," Intel was becoming a polestar of the tech industry and Grove was becoming the axis around which Intel turned. He would often run morale-building seminars with Intel employees, one of which went like this:

> ***Grove:*** "How would you sum up the Intel approach?"
> ***Employee:*** "At Intel you don't wait for someone else to do it. You take the ball yourself and you run with it."
> ***Grove:*** "Wrong. At Intel you take the ball yourself and you let the air out and you fold the ball up and put it in your pocket. Then you take another ball and run with it and when you've crossed the goal you take the second ball out of your pocket and reinflate it and score twelve points instead of six."

The Cult of Management

Grove came to embrace what he called a culture of "constructive confrontation" as the best means to coax maximum performance out of his teams. Fiercely argumentative and well prepared, he could be brutal in challenging the less well prepared, grilling subordinates to the point of demoralizing them, often shouting at colleagues who sought to challenge him. Craig Barrett, Grove's longtime number two and his eventual successor, later told the *Washington Post*, "Occasionally we . . . suggest [to Grove] there may be an alternative to grabbing someone and slamming them over the head with a sledgehammer."

In 1976, Grove became chief operating officer of Intel and in 1979 he became president as well. In these combined posts he subjected every production process and every administrative process to numerical measurement. Each employee had to make exceedingly detailed budget projections, establish targets for their work, prepare constant updates, and explain discrepancies. He would ask how many functioning integrated circuits a section produced, and how long it took. How many recruits were interviewed, and what was the yield? How many lawsuits were filed to protect Intel's patents? Of janitors he would ask, how many square feet could you clean in an eight-hour period? Obsessed with cleanliness, Grove and his assistants would make surprise inspections of bathrooms, janitors' closets, and offices, and would criticize staffers for having too many papers on their desks.

Once, when Grove criticized a senior manager for failing to fix a hole in his wall, the manager responded, "Andy, I've got bigger problems to worry about than that." To which Grove replied, "There are no bigger problems. There are just problems."

Grove also insisted people come to work on time and established a sign-in process for all employees from the CEO on down who arrived more than five minutes after 8:00 a.m.—an unpopular requirement that provoked pouting and ridicule. Under Grove, Intel became a company in which everyone was clearly accountable for every aspect of their performance. Grove was the exact opposite of the leaders he saw at Fairchild Semiconductor who couldn't translate ideas into products. He was all about high-quality commercial output.

The Computer on a Chip

While Intel was enjoying the success of the 1103, it was also quietly developing one of the most important technological breakthroughs of the century, though even Grove didn't realize its significance at the time. In 1969, the Busicom calculator company from Japan asked Intel to design specialized chips for printing, display, calculations, and other functions—an advanced calculator. The assignment was given to a young Intel engineer, Ted

Hoff, who concluded that it would be cheaper to create one single device that did it all.* His idea was to build one calculator that would include some two thousand electronic elements, ranging from memory chips to a clock for synchronizing operations. No bigger than an index card, this device packed the same computing power as did the room-size mainframe in 1947.

Busicom was nonplused by the all-in-one chip that Intel was proposing. It had not requested all these features and thought it was paying for more than it needed. In 1971, it demanded a discounted price. Intel agreed to accept a lower payment if it could keep the rights to sell the brainy chip for non-calculator applications to other customers. On that basis a deal was struck. In his book *The Microprocessor: A Biography*, Michael S. Malone wrote that Busicom had just made one of the worst business decisions of the century.

The new device would soon be called a microprocessor. Whereas the prevailing computers at the time were gigantic mainframes, the microprocessor was, literally, a "computer on a chip." It would become the brains of the personal computer and so many other devices that we use today, from tablet computers to smartphones. By the first decade of the twenty-first century, it could be embedded in machine tools, cars, appliances, medical devices, robots—virtually *anything*.

To be sure, while the first smoothly functioning microprocessor didn't emerge until the early 1970s, the full recognition of its utility would take another decade. Microprocessors started to appear in smart traffic lights and car brakes, coffeemakers and refrigerators, elevators and medical devices. Its greatest impact, arguably, was on PCs.

---

* As in almost all such breakthroughs, more than one person deserves credit for the achievement. In this case, another Intel engineer, Federico Faggin, was key to the project, as were several others.

The microprocessor was initially a niche product, used in a few types of products such as cash registers, microwave ovens, and sophisticated machine tools. But Intel kept tinkering, and in 1974 the company built and introduced a more advanced microprocessor, called the 8080. "History may well recognize [the 8080] as the most important single product of the 20th century," wrote Malone. Soon after, a team of engineers concocted a keyboard and monitor, which allowed a person to control the 8080, and proudly displayed their creation to Gordon Moore. "What's it good for?" he asked. Housewives can store recipes, was the response. Moore wasn't interested. Neither Moore nor Grove realized that they were looking at a key tool of the digital age, nor were they aware that, not far away in Silicon Valley, two unknowns named Steve Wozniak and Steve Jobs were building in their garage the first Apple computer, the first computer designed for ordinary people, the first you could lift and move around. The PC industry would become the chief market for Intel's microprocessors.

For the first few years, Intel and several of its rivals, including Motorola and Texas Instruments, were engaged in a fierce race to design, manufacture, and acquire customers for the new microprocessor, with Intel eventually emerging as the leader. Intel's edge was the organization Grove had created, which could not only design and build microprocessors with unprecedented efficiency but could also provide an unmatched package of training and services to customers. Intel prevailed in large part because of Andy Grove's obsessive drive to excel over rivals in every one of these dimensions.

The 1970s ushered in the first cycles of shortages and gluts in the tech industry, which Intel also learned to manage with unusual discipline. Here Grove's insights and talents were critical, and they constituted another major advance in high-technology manufacturing. The cyclical downturns were devastating to the industry, and the natural response of most companies was to cut back on *all* spending, including R&D. Grove thought differently. He was following Moore's law, not the business cycle. Alone among the industry leaders, Grove had the courage to respond to industry slumps by cutting budgets, cutting jobs, and forcing staff to work longer for less money. At the same time, however, he pushed Intel to become the first major company to *expand* R&D during downturns. This required Grove to ignore screaming shareholders—who wanted spending cuts to protect quarterly earnings—in order to come out of recessions much stronger than Intel's rivals.

The power of this insight cannot be underestimated. As the world economy has become more integrated, the advanced technology industries have become more vulnerable to amplified booms and busts, but so, too, have other industries—from oil to banking. It has always been the conventional wisdom among business managers to cut back in bad times and gun the investment engines when the dark skies start to clear. In doing the opposite, Grove understood at a visceral level the changing nature of competition, especially the importance of investing in innovation and talent when others were doing the opposite. Given the natural antipathy of Wall Street toward outlier companies, especially those whose spending seemed unnecessarily extravagant, Grove's actions were not just highly prophetic, they took enormous courage.

The strategy worked. During the recession of 1974, when Grove cut staff but dramatically boosted R&D, Intel's stock dropped 80 percent. Two years later, when the cyclical recovery came, the company's stock quadrupled in value from $21 per share to $88, as earnings rose 65 percent from 1975, and the payroll nearly doubled. Another case: in the economic nosedive of 1981, Grove cut expenses and staff, required the remaining workers to put in 25 percent longer hours for no additional pay, and reinvested in R&D. Intel came roaring out of the recession with four successful new products. Much later Grove recalled, "You can never come out of a downturn with the same products you had when you went into it. Leaders have to understand at a gut level the new possibilities that are out there."

As Intel's fame and clout grew, so did Grove's. In the 1970s, the company saw yearly revenues grow from $9 million (with a profit of $1 million) in 1970 to $854.2 million (with a profit of $96.7 million) by the end of the decade. It was in the 1970s, too, that Intel expanded operations around the United States and the world, eventually to have facilities in several parts of California, in Oregon and Arizona, and in Malaysia.

Crush the Competition

By the end of the 1970s, Intel faced serious rivals for technological leadership in memory chips, particularly the Motorola 68000, which many experts decreed superior to Intel's latest model, the 8086. Determined not to relinquish Intel's global lead in the memory business, Grove launched Operation Crush.

It was as if he were General Eisenhower planning the D-day invasion. He mobilized the sales and marketing force, offering rich bonuses to every staffer who could keep an Intel customer—or potential customer—from leaving for Motorola. He oversaw an aggressive publicity plan that included articles in the press extolling Intel's products over Motorola's. He kept in touch with people in the field, receiving a steady stream of progress reports. Grove whipped up such enthusiasm for victory that many at Intel were focused not just on outselling the Motorola 68000 but on destroying Motorola itself. After a year of trench warfare, Intel had won, preserving its lead and reputation in the field. Operation Crush "was the perfect expression of [Grove's] conception of business as a contact sport," said biographer Richard S. Tedlow.

During the 1980s, Grove and Intel continued to prosper, all the while living on the edge of violent business cycles and warp-speed technological change. By the beginning of the decade Intel was a regular on *Fortune*'s list of the world's five hundred largest companies, but it would soon be confronted by two monumental challenges—the dawn of the era of personal computers and a new competitive threat from Japanese companies such as Fujitsu and Hitachi. Under Grove's leadership, Intel emerged to stand alongside Microsoft as the standard setter in the personal computer industry and to dominate the microprocessor sector.

Japan and Other Challenges

By 1980, IBM was the behemoth of the computer world, which was still dominated by large mainframes and by vertically integrated companies that tried to make every part of the mainframe in-house. But Steve Jobs and his new Apple computer were a growing challenge, which IBM decided to quash by building its own desktop computer. Unable to build the microprocessor necessary to drive its machine, IBM turned to Intel, and its personal computer took off. By 1984, IBM's PC division would have been No. 25 on the *Fortune* 500 had it been a stand-alone company.

The IBM PC led to a very rapid scaling up of the personal computer industry, but it also spelled the beginning of the end for the vertically integrated model that defined computer companies like itself. IBM relied on Intel for its PC microprocessor and on Microsoft for the operating system, and from then on the industry would increasingly be organized horizontally, with outside manufacturers supplying most or all of the computer parts. Intel microprocessors soon became the PC industry standard, and the makers of operating systems and hardware began to design their products for Intel's microprocessor. (Apple was the one exception, as Steve Jobs bought microprocessors from Motorola.)

In the early 1980s, the semiconductor business was becoming a global industry as everything from electronic games, computer-intensive car engines, digital control patterns, and the Internet was spreading. Intel kept expanding, too. By now it had a factory in Singapore and one in Hong Kong, and at least eighty-seven sales offices in seventeen countries. Meanwhile, Japanese rivals were rising to challenge Intel in its bread-and-butter business, the memory chip.

At first Grove did not pay much attention to the competition from Asia; it was conventional wisdom in the US semiconductor business that the Japanese manufacturers were good copiers but not innovators. Few realized that Japanese manufacturers were improving quickly, investing heavily in R&D, and capitalizing on tight relationships between Japan's manufacturers and suppliers to improve efficiency and innovation. By the early 1980s, Japan was better at making memory chips than America, with high-quality yields of 80 percent, compared to American yields on the order of 50 percent. When recession hit the US economy in the early 1980s, Japanese companies used their superior efficiency to lower prices and increase market share—setting off a global trade battle over Japanese dumping of products in the United States. (Among the reasons they could do this so successfully was the fat profits they made at home with higher prices.)

Then, in one of the great turnarounds for a global company, Intel simply changed the contours of the battlefield. As the company's profits fell from $198 million in 1984 to less than $2 million in 1985, Grove and Moore held a quiet but intense discussion about the dire situation. "If we got kicked out and the board brought in a new CEO, what do you think he would do?" Grove asked. Without hesitation, Moore replied, "He would get us out of memories" and focus on microprocessors. Grove felt numb but then recovered. "Why shouldn't you and I walk out the door, come back and do it ourselves?" That's what they did: Grove led Intel out of memory chips and into microprocessors, a move that required firing some eight thousand people and spending over $180 million to rebuild the company around a new core business.

To the naked eye, it might appear that building memory chips and microprocessors are more or less the same. But the skills and the thinking were dramatically different. It was as if the Caterpillar Corporation decided to abandon tractors and start making sleek sports cars, requiring a wholesale retooling of the company, including philosophy, culture, and specific operations. Grove had to replace the prevailing mind-set that was at the core of Intel's DNA, to decide who to keep, who to retrain, who to fire, not to mention which assembly lines to restructure or to close down. He had to do all of this in the space of one year—1986—when Intel had its first financial loss since 1970. It was transformative leadership in its purest and most decisive form.

Intel's response to the Japanese onslaught illustrates a theme that pervaded Grove's career—the ability to recover from setbacks. Time and again Intel faced a near life-threatening situation. At every juncture it bounced back. This was life in the hypercompetitive high-technology arena where speed and risk taking were essential to success; where skilled scientists and technicians often jumped ship to competitors, taking with them commercial secrets; and where the only road map for the future was the dizzying framework of Moore's law.

Grove's come-from-behind agility became legendary. Grove, said Michael Malone, possessed an acute survival instinct. He could pick up Intel and himself from severe setbacks, learn from the experience, and come out on top. "For more than thirty years," Malone wrote, "whenever Intel fell (sometimes when it was Andy's own fault), it was he who, through sheer force of will, pulled the company back to its feet and told it in which direction to change."

Building a Company That Can See the Future

Years later, Grove wrote about the shift from memory semiconductors to microprocessors in his second major book, *Only the Paranoid Survive* (1996). His reflections redefined the way Intel, and ultimately many large corporations, dealt with crisis. Grove underlined the idea of an "inflection point"—a moment, or a period of time, when a set of forces are so overwhelming that they compel a fundamental change in the rules of the game for a company or an industry. This happened in banking, he writes, when cash machines replaced human tellers. It happened in the retail business when Walmart put so many mom-and-pop shops out of business. And it was the same story in memory chips, when the Japanese took on America. "It's like sailing a boat when the winds shift on you, but for some reason, maybe it's because you are down below, you don't even sense that the wind has changed until the boat suddenly keels over. What worked before doesn't work anymore," Grove wrote. He noted that the inflection point compares to ordinary change the way Class VI rapids, so turbulent and dangerous they routinely capsize even professional rafters, compare to a calmly flowing river.

In his book Grove admits he failed to see the Japanese challenge coming and counsels other business leaders to remain vigilant—even paranoid—to the inevitability of inflection points, so that they can quickly respond or, better yet, get ahead of the curve. He talks about how to spot the signs of an inflection point, particularly by involving employees who are closest to the market—salesmen, middle management. He discusses how to understand what is happening in the minds of customers—in the battlefield—and what to do about it. All this required an organizational structure that makes it possible for information and advice to travel quickly from the field to the top officials, but then empowers leaders to fully mobilize the company behind a chosen plan of attack. You have to have feelers out in the market in many different ways, so you have to allow your employees to have multiple touch points. But amid all that unstructured interaction, you have to be organized, too, he explained. "Allow chaos," he advised, "then rein it in."

The concept of inflection points is one of Grove's most profound contributions to the theory of management in the unpredictable environment wrought by technology and globalization. His insights point to the way to navigate radical disruptions. When inflection points come, he warns, be ready to drop all previous assumptions and to start from scratch. Be ready to open your mind to multiple sources of information and to advice that is frank, even confrontational and contrary to what you want to hear. Grove's genius was recognizing the critical importance of this process and identifying specific steps to take. Today the best leaders in all sectors better understand the need to anticipate sudden economic, political, or commercial headwinds, but they have built on much of what Grove learned and wrote about Intel's battles to survive.

By 1986, Grove saw that he and Intel were in the middle of a revolutionary inflection point, the dawn of the PC age. With PC sales picking up quickly, Grove went to IBM with the new 386, a chip that contained 275,000 interconnected transistors. IBM was interested but wanted a second supplier, a backup supplier, which was then a standard demand in the industry, because semiconductor customers wanted a hedge against unreliable manufacturing and against monopoly pricing.

Jealous as ever of company secrets, Grove said that Intel would not hire a second company as its backup manufacturer but instead would allay at least some of IBM's fears by manufacturing chips at several of its own facilities, thereby creating security of supply that comes with diversified sources. When IBM balked, Intel then triggered another inflection point by selling the 386 to emerging rivals of IBM that were entering the PC business such as Dell, Zenith, and Tandy. After several months, IBM finally agreed to buy the 386, but it was too late; IBM had surrendered its position in the PC wars, and Intel had established itself as the sole source provider of microprocessors for nearly the entire industry (again, except for Apple, which stayed with Motorola).

For Grove, it was a game-changing victory. Establishing Intel as the industry standard meant that manufacturers of other computer products—software, keyboards, sound systems—had to make their products compatible with Intel's microprocessors. The company had become a de facto monopoly with profit margins of 90 percent. As demand exceeded supply, shortages arose and Intel was in the pivotal position of deciding who received the microprocessors and when, not to mention having exceptional leverage over pricing.

Grove's challenge now lay in fighting off competitors, many of whom resented his win-at-all-costs campaign to establish Intel's dominance. One market research firm summed up the industry view: "Intel has never been a company to be altruistic or even fair. Greed, avarice and paranoia are in its corporate culture." It was a widely shared view. Grove would open his customer presentations with a drawing of Intel rendered as a giant castle, harboring inside the gleaming jewel of the 386 microprocessor, but set upon by attackers firing all manner of weapons—lawsuits, new products—in an effort to steal or destroy the jewel. In these presentations, Grove had a chart showing how he fired back with countless lawsuits and other legal maneuvers to keep competitors such as Advanced Micro Devices, Motorola, Texas Instruments, and Cypress Semiconductor Corporation at bay. Indeed, Intel mounted ferocious legal attacks against rivals, accusing them of patent infringement and of engaging in conspiracies to weaken Intel. Enmeshing others in long, costly lawsuits was a central part of Grove's strategy to intimidate and even annihilate competitors. "We believe Intel views all competitors, both existing and emerging, as credible and serious," wrote a Wall Street analyst in 1996. "It is the paranoia that is ingrained in Intel culture that has motivated the company to attack its own product line before its competitors get a chance."

A Decade to Remember

In 1987, Andy Grove became CEO, presiding over a golden decade in his corporate castle. Intel had contributed more than any other company to the creation of the microprocessor, which made possible the PC, which in the 1990s became the gateway to the Internet. The Internet would open a new age of communication and collaboration, afford opportunities for other upstarts to challenge incumbents, and destroy and create billions in shareholder wealth. The 1990s would see the founding of pioneering Internet companies like Yahoo!, Amazon, eBay, and Google, all of which fed the demand for PCs powered by Intel. Nevertheless, success did not have a calming effect on Grove, who had been stalked by fear as a child and had learned to use that fear to his advantage as a leader and manager, always on the watch for new inflection points. "I worry about products getting screwed up, and I worry about products getting introduced prematurely. I worry about factories not performing well and I worry about having too many factories," he admitted in 1996, at the pinnacle of his career. "I worry about hiring the right people and I worry about morale slacking off. And of course I worry about competitors."

A year earlier, Grove had encountered the most personal of inflection points: he was diagnosed with prostate cancer. He attacked it with the same approach he applied to invaders in Intel's market. Listening to his doctor, he saw quickly that medical science was divided on the best course of treatment, so he decided to investigate the alternatives himself. Over the course of eight months, he read countless books and articles and delved into the medical literature, burying himself in technical research papers. He contacted physicians from different specialties around the country. He talked to former patients. He plotted the information he gathered on charts that correlated various treatments with outcomes. Ultimately he selected high-dose radiation treatment as opposed to surgery or a number of other options. The cancer went into remission.

Not so with Intel's competition. More major competitors had emerged from Taiwan and South Korea. But still, during Grove's tenure at the helm, the total value of Intel's stock market share had grown from $4.3 billion to $114.7 billion and it rose from No. 200 on the *Fortune* 500 list to No. 38. Sales grew from $1.9 billion to $26.3 billion, profits from $246 million to $6.1 billion. The company was doubling in size every two years. In some ways, it seemed like its own growth was following Moore's law. Intel even became one of the leading sources of venture capital for new start-ups. By August 31, 2000, when its stock reached $78 per share, Intel, now valued at nearly $500 billion, had become the most valuable manufacturing company in the world, worth more than all of the US automakers combined.

In the 1990s, the concept of the global company was changing, and, under Grove, Intel was at the forefront. To be global was much more than setting up more subsidiaries or hiring more foreign employees, or earning more revenue in foreign markets. It also meant setting the international quality standards for products and services, developing the most talented workforce in many countries, and figuring out the best way to manage worldwide supply chains. In every one of these arenas, Grove drove Intel to the top ranks of global corporations.

At the beginning of the 1990s, for example, Intel already earned more than half of its revenues from abroad. By the year

2000, that number was 63 percent, and by 2008 it was 85 percent. In 1990, it had operations in six other nations—Malaysia, the Philippines, Singapore, the West Indies, Japan, and Israel. By 2008, it had added dozens more, including India, China, Vietnam, and Brazil. It employed approximately eighty-two thousand people around the world. Grove, for all his paranoia about Intel secrets, was one of the first American high-tech CEOs to establish R&D labs outside the United States. In moving investment in research abroad, Intel was at once trying to become more integrated in local markets and looking to tap into the best minds wherever they were. It was a trend that many of America's most dynamic multinational companies were following. Grove was at once an internationalist and also a fervent proponent of the idea that the US government should help level the global playing field when it was tilted against American companies, especially Intel itself. Thus, at the same time that he was pushing Intel to locate abroad, he was urging Washington to take vigorous trade action against Japan, which he accused of subverting free trade with heavy government support for Japanese exporters, including throwing up obstacles against Intel and others in the Japanese market.

In 1998, Grove stepped down as CEO, having established himself as one of the three giants of the Internet age along with Bill Gates at Microsoft and Steve Jobs at Apple. He became chairman of Intel, pulling back from day-to-day operations to manage the board of directors, and became a sage public voice for a nation that seemed to be losing its way on management issues. When the financial crisis of 1998 erupted following the exposure of major fraud at technology companies like Enron and WorldCom, Grove weighed in as a forceful public advocate for an overhaul of corporate governance.

He also became a strong proponent of using electronic communication to create a database of personal medical records in order to modernize and improve the management of the complex health care system, including reducing mistakes in diagnosis and medical procedures. In 2000, he was diagnosed with Parkinson's disease, an incurable condition, but with the use of medication he kept the tremors under control and continued to follow a full professional and personal life. Even after he stepped down as chairman in 2005, Grove not only stayed active in the national health care debate but also started pressing Silicon Valley to apply its technological expertise to developing clean energy, particularly batteries to power automobiles.

Legacy

Andy Grove was not a pathbreaking scientist like Robert Noyce. He did not author anything so important as the law associated with Gordon Moore. He was never a household name like Bill Gates. Unlike Steve Jobs, he was not a design genius, and he did not have the same intuition for consumer sentiment. He never achieved the personal wealth of Silicon Valley icons like Larry Ellison of Oracle. But with the test of time, I believe he will be seen as exceptionally pivotal to the age. He was not just a visionary CEO, he was a leader who created an organization set up to anticipate change and respond quickly to inflection points. He thus captured the essence of what it takes for a high-technology company—or perhaps any ambitious company—to succeed in our era of extraordinary change and volatility.

As much as anyone, Grove embodied the zeitgeist of the late twentieth century, when a burst of technological creativity and innovation was reshaping the world. The Cold War and the space race had produced a steady burst of technology. The free market evangelism of Margaret Thatcher, the collapse of the Berlin Wall, the opening of China under Deng Xiaoping (as we will see in the next chapter), all signaled an era of fantastic new possibilities.

There was Andy Grove, an East bloc disciplinarian with the modish sideburns and the clunky earpiece, whipping a motley crew of early industry pioneers into the world's most important and global technology company. He built the products—the semiconductors, the transistors, the integrated circuits, the microprocessors—that drove the consumer products revolution that has transformed global society in the twenty-first century. He showed the world what speed was, what precision was, what risk taking was, what take-no-prisoners competition was, how you should use recessions to your advantage by accelerating investments, how to learn from your mistakes and bounce back even stronger, how to rapidly scale up, and how to spread your presence to the corners of the earth. He gave us a vivid picture of how to survive and thrive in business when the only constant is mind-bending change. Grove's Intel became a symbol for the age.

Intel stands out not only because its product sits at the core of the computer revolution but because Grove did so much to define and spread the management ideals throughout America, Europe, and Asia that are now at the core of many other technology companies. For at least a century, the companies around the world have sought out advice from American management gurus like Frederick Taylor, who made monumental contributions to industrial efficiency with his time and motion studies, or Peter Drucker, with his insights into the way people should organize themselves for effective management. Others have studied the management practices of industrial giants like Henry Ford, who revolutionized mass manufacturing for consumer consumption with his assembly lines. But those men were either thinkers (Taylor) or doers (Ford). Andy Grove was both. And his influence came not just from his business achievements but also from his capacity to communicate his thoughts and experiences via his teaching and writing. He helped create a corporate culture that cultivated individualism, egalitarianism, innovation, and—miraculously in light of all of that—exquisite teamwork.

Moreover, Grove defined a management process that continues to generate the production of computer chips that are increasingly smaller, more powerful, and cheaper. Between 1971 and 2011, Intel increased the number of transistors on a microprocessor by one million times. Today you can fit more than six million transistors into a space the size of the period at the end of this sentence. In the same period, the price of a transistor dropped by a factor of about fifty thousand. The amount of energy consumed by each transistor dropped by a factor of about five thousand, and their speed increased by about four thousand times. By 2013, Intel was producing six billion transistors per second, or twenty million per year for every person on the planet.

Here's another reference point: the computing power of a 1975 supercomputer costing $5 million at the time is matched today by an iPhone costing about $400. This is what is making possible the arrival of the "Internet everywhere." There is no letup. Fearful that they will soon have crammed as many transistors into a computer chip as will be possible, or that at a minimum Moore's law will be undermined by the physical limits of miniaturization, Intel and its competitors, including IBM and HP, are now racing to create technology that uses lasers and light to transmit data faster than electronic current can, something that holds the potential for breakthroughs of the same magnitude as the transistor or the integrated circuit delivered in years past.

Like the other leaders in this book, Grove did not create the forces that shaped his era, but he did exploit them with exceptional effectiveness. He entered the information technology industry when the trajectory of innovation was straight up, and the wide-open environment for entrepreneurial talent in Silicon Valley was unique in the world, which is why there has been so much appetite for Grove's ideas in countries such as Taiwan, South

Korea, Singapore, China, Israel, and the UK, all of which are trying desperately to create their own zones of innovation and entrepreneurialism.

Today the world is entering a third industrial revolution. The first began in late-eighteenth-century England with the mechanization of the textile industry. The second took off in early-twentieth-century America with innovations such as the assembly line and mass production. The third is powered by modern communications technology, particularly the Internet and digitalization, which makes possible robots, artificial intelligence, digital sensors, 3-D printers, big data, new heat-resistant materials based on nanotechnology, and much more. Indeed, our era promises to be one that experiences a dramatic transformation in what is produced, how it is produced, and where it is produced. Three-D printers can already build aircraft parts, but in the future they may be able to create human organs, one layer at a time. There are plans for ingestible sensors that can transmit diagnostic information from inside our bodies. MIT professor Andrew McAfee describes the digital era this way: "The [previous] Industrial Revolution was when humans overcame the limitations of our muscle power. We're now in the early stages of doing the same thing to our mental capacity—infinitely multiplying it by virtue of digital technologies."

The information gathered, the array of new activities they will create, the connections they will foster all over the world, and the enhancement of the human condition—all are incalculable. Toward the end of 2014, half the world had access to the Internet, but within another five years it is likely that two-thirds will be online. In 2014, Cisco Systems, Inc., estimated that some 13.5 billion devices were connected to the Internet, but by 2020 the number could climb to 50 billion. "The things that are—or will be—connected aren't just traditional devices, such as computers, tablets, and phones," write two top Cisco executives, "but also parking spaces and alarm clocks, railroad tracks, streetlights, garbage cans, and components of jet engines."

None of this would be possible without trillions of microprocessors, which historian John Steele Gordon has called the most fundamental new technology since the steam engine. To dramatize the impact of the microprocessor, Gordon proposes this thought experiment:

> Imagine it's 1970 and someone pushes a button causing every computer in the world to stop working. The average man in the street won't have noticed anything amiss until his bank statement failed to come in at the end of the month. Push that button today and civilization collapses in seconds.

Cars would not run, he says, planes would be grounded, and industry would be paralyzed. All communications would break down.

Grove's critical contribution to globalization was to spread both the technology and the process by which transistors and microprocessors could be commercially manufactured and distributed to all corners of the earth. He is thus a pioneer of the modern communications age. As the story of globalization has progressed, information has traveled farther and faster, and in greater volumes. It was carried on foot, horse (Genghis Khan), ship (Prince Henry), telegraph (Cyrus Field), telephone, radio, auto, and airplane. Now globalization is propelled by—and cannot be separated from—information technology. At the heart of that industry has been Silicon Valley. At the heart of the valley has been Intel. And at the heart of Intel was Andy Grove.

NEW CHINA PICTURES, MAGNUM PHOTOS

Chapter X

DENG XIAOPING

The Pragmatist Who Relaunched China

1904–1997

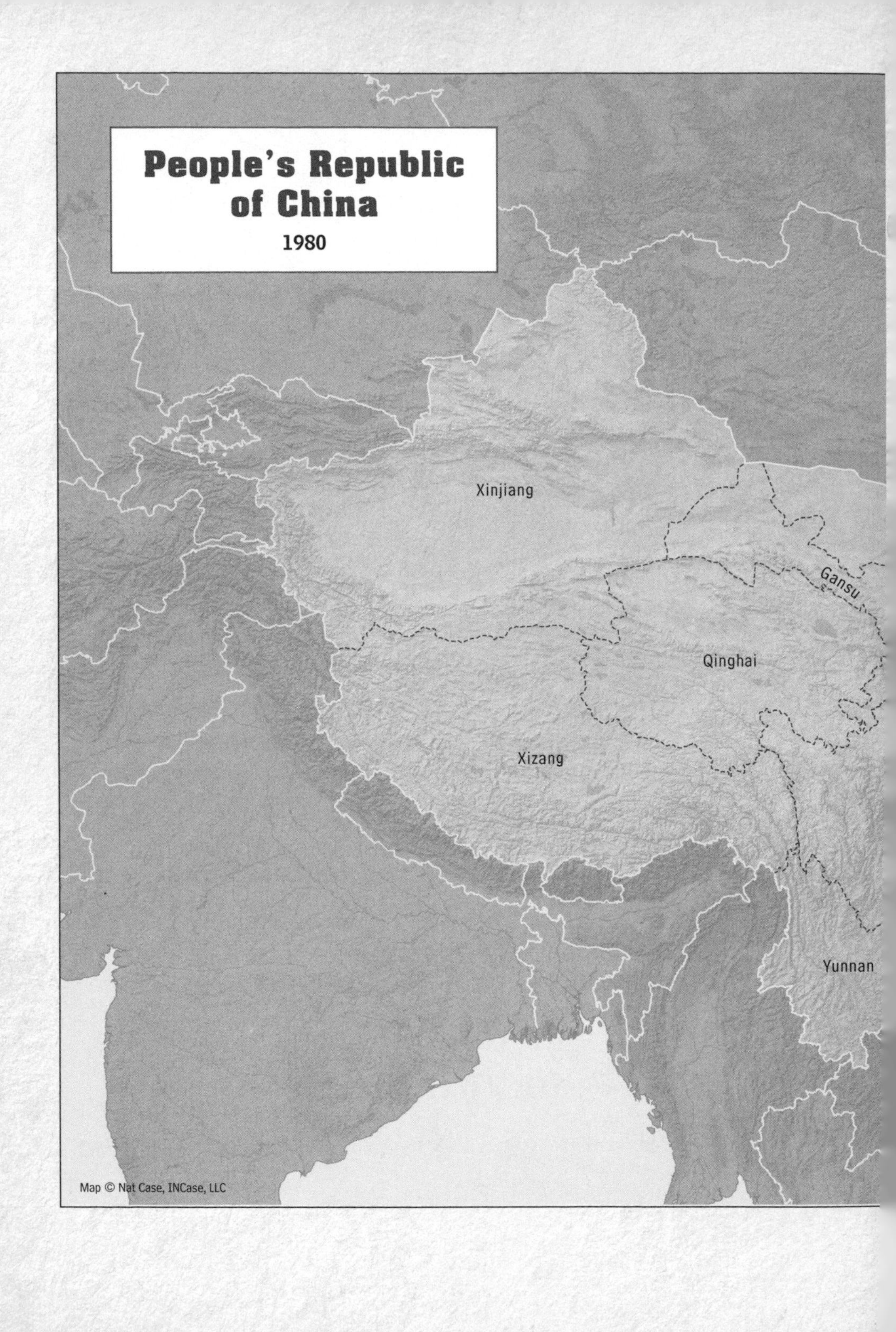
People's Republic of China
1980
Xinjiang
Gansu
Qinghai
Xizang
Yunnan
Map © Nat Case, INCase, LLC

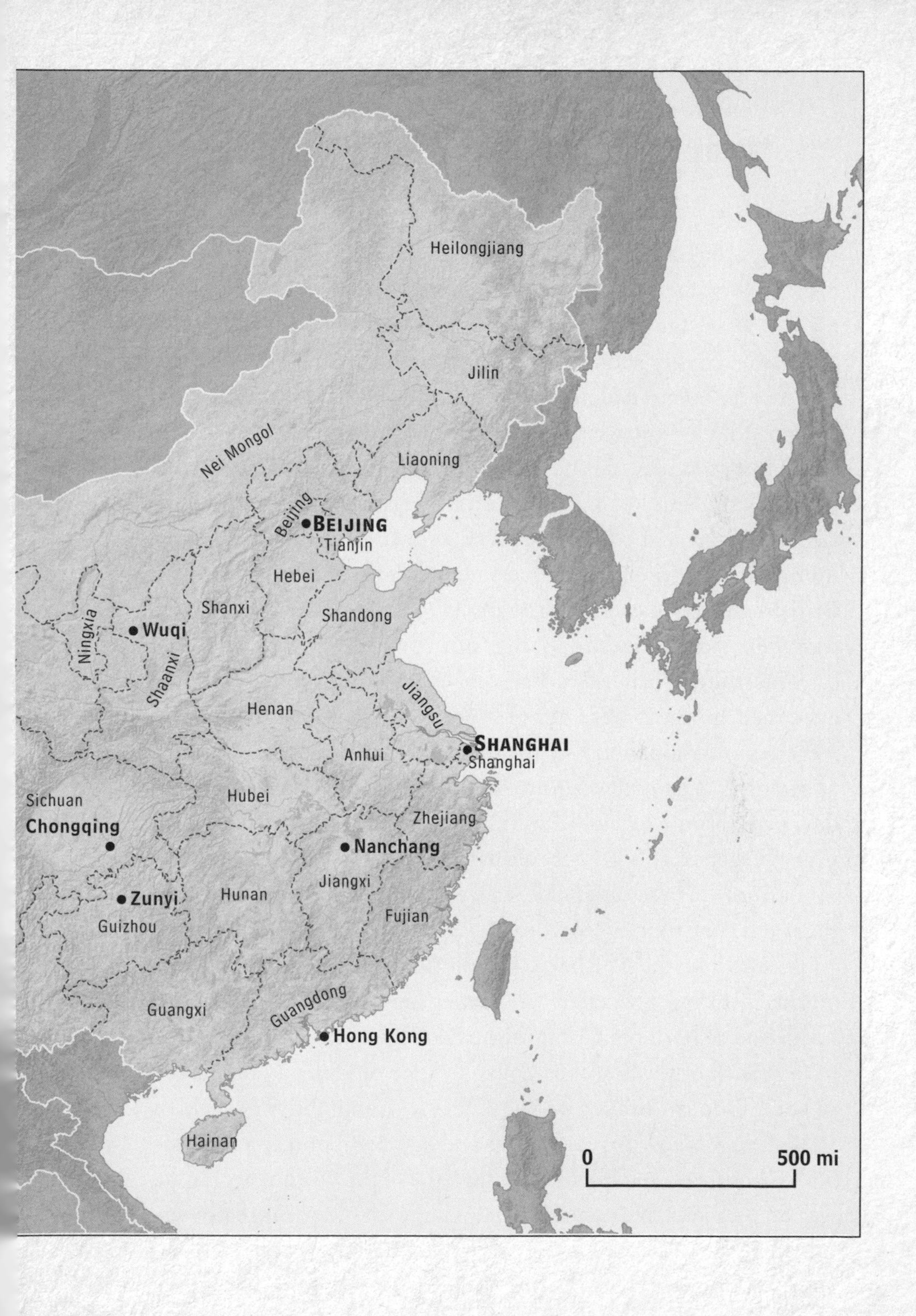

Heilongjiang
Jilin
Liaoning
Nei Mongol
Beijing
BEIJING
Tianjin
Hebei
Shanxi
Shandong
Ningxia
Wuqi
Shaanxi
Henan
Jiangsu
SHANGHAI
Shanghai
Anhui
Sichuan
Hubei
Chongqing
Zhejiang
Nanchang
Jiangxi
Zunyi
Hunan
Guizhou
Fujian
Guangxi
Guangdong
Hong Kong
Hainan
0
500 mi

Deng Xiaoping found his calling as a Communist leader while living as a student in Paris. Born in 1904 in a tiny village deep in Sichuan Province, he was the son of the county sheriff, a respected landlord who—despite the fact that his village had no post office and got its news from traveling peddlers—knew enough about the world to want his son to study in France. At the age of sixteen, Deng joined some sixteen hundred students on a work-study program in France, which China had created to help modernize a society that was being humiliated by foreign occupiers. Deng left for Paris from the British-controlled enclave in Shanghai, where he had to pass by a sign that read, "Chinese and dogs not allowed."

The students landed in Paris in 1920. Within months, however, their sponsors were out of money, and Deng and his cohort were out of school, fending for themselves. Living with as many as twenty students in a room, they often got by on one meal a day, perhaps just a glass of milk and a croissant, or a small portion of vegetables. With only a smattering of French, Deng picked up odd jobs where he could—making rubber shoes and bicycle tires, and later running machinery in a Renault factory.

Meanwhile, the Bolshevik Revolution was spilling across the world, inspiring a workers' movement in France as well as the formation of both the Communist Party of China and the European branch of the Socialist Youth League of China (a precursor to the European branch of the Chinese Communist Party). In 1922, Deng joined this last group and was soon immersed in rallies and lectures, socializing late into the night. Within a year, he was on the executive committee, mixing with older, more powerful colleagues. Still just a teen and standing barely five feet tall, Deng developed the skill of looking tough and confident while forging ties to men who would become prominent leaders in the Chinese Communist Party. Foremost among these was Zhou Enlai, destined to be Mao's right-hand man for nearly three decades, much of that time as premier.

Deng helped Zhou to run *Red Light*, a fortnightly newspaper that published Marxist texts and reports on the Communist movement. Anger among Chinese students in Paris erupted in May 1925, when British troops brutally suppressed a strike at a British factory in Shanghai. In June of that year the Chinese students, including Deng, blockaded the gate to the Chinese embassy in Paris, demanding that the Chinese ambassador lodge official protests against British behavior. Over the next few months, as student violence mounted, the Paris police began closing in on foreign students. Sensing personal danger, in January, Deng and many of his friends fled to Moscow, the center of the growing worldwide Communist movement. The timing was exquisite, for shortly afterward the French authorities raided his room in an attempt to arrest him. Paris had exposed Deng to the modern achievements and glaring inequities of Western capitalism, to the internal battles of a fractious Communist Party, and to the risk of arrest and prison. He was ready for the highly combative politics of revolution in China.

The French experience was hard and tumultuous, but it was an important foundation for Deng. Although he would spend several decades in the shadow of Mao Zedong, in the third quarter of the twentieth century he would emerge as China's supreme leader. He would proceed in astonishingly rapid order to undo parts of Mao's Communist Revolution, opening the path for China to move to a position of extraordinary influence in global politics and economics, positioning it to be the hub of global manufacturing, and unleashing hundreds of millions of workers into the worldwide pool of labor, with equal numbers of global consumers sure to emerge in their wake. In joining China to the world and vice versa, Deng made globalization larger and deeper than ever, and he unleashed new forces that will move the center of global commerce and political influence from the West toward the East.

The Clandestine Operative

After fleeing to Moscow, where he spent most of 1926 studying and later teaching Communist doctrine, Deng returned to southern China to join up with Zhou Enlai. The region was in chaos. With the Qing dynasty having expired, the new central government was weak, and warlords often held sway in the provinces. Russia, Japan, and many of the big European powers had seized commercial outposts on the southern coast, and the Communists had formed a tense, temporary alliance with the rival Nationalist Party, hoping to stave off further foreign advances. But the alliance of antagonists could not hold, and soon the Nationalists were pursuing a brutal campaign to eliminate the Communists.

Under pressure from the Nationalists, Zhou, Deng, and their comrades relocated from the city of Wuhan to Shanghai, where Deng became a key player in the clandestine apparatus of the Communist Party. He relayed instructions to party cells throughout the city, guarded and dispensed funds, kept minutes of critical meetings, and coordinated with equally secretive party branches across China. To stay a step ahead of the Nationalist police, the Communist Party frequently moved its headquarters. Members often assumed false identities, some wearing fancy clothes on the street to look like members of the bourgeoisie. Deng put up a variety of fronts, first as a small grocer, then as an antiques dealer.

In the summer of 1929, when he was twenty-five, the party transferred Deng from the clandestine offices of Shanghai to the emerging battlefield of Guangxi Province, in the mountainous terrain of southern China. He learned how to recruit and organize peasants by the thousands, and he led guerrilla attacks against the Nationalists. Although his troops often had little more than pitchforks and poles, he created an enthusiastic fighting force with ties to local peasants and warlords. He established a propaganda arm, stole weapons from government arsenals, and made pacts with local warlords to support the Communists. He encouraged peasants to kill landlords on the theory that once they had blood on their hands, they would be committed to the revolution. Either they would believe in the cause or they would keep fighting to prevent retribution from the enemy. In the end, however, his efforts failed. The bases he established were overrun, and many of the leaders he befriended were killed.

Deng was nevertheless promoted to political commissioner in one of the Communist Party armies, a position that made him partner to the commanding general—if not the superior of the two. He operated in the southeastern province of Jiangxi, a Communist stronghold where the young Mao Zedong was on the rise. Though the rebellion's Russian advisers urged them to attack the opposition forces in urban areas, Mao was pushing instead a strategy of agrarian revolution that avoided meeting superior military force head-on.

The anti-Maoists accused Deng of "deviations" from the party line and forced him to make a public self-criticism. They stripped Deng of all his titles and responsibilities, and exiled him to a lowly post as a party inspector in a small town. To add to his humiliation, Deng's second wife left him for his accuser. (His first wife had died in childbirth in 1930.)

The Communist Party was forever reexamining its leaders for ideological purity, forcing them to undergo withering public humiliation and self-criticism. If the authorities were satisfied with the self-criticism, they might sentence the subject to jail or exile, ending with a declaration that the subject had been rehabilitated; if not, the process could end in execution. In this case, Deng was rehabilitated within weeks—probably through the intervention of Zhou Enlai, and possibly Mao—and appointed editor of a military publication called *Red Star*. Deng now had the support of the two men who would eventually rule China.

The Long March

By early 1934, Nationalist troops under their leader, General Chiang Kai-shek, were encircling Mao's troops, threatening the survival of the Communist Party. Once again the Communist leadership was forced to uproot. On October 10, they set out on what became known as the Long March, a yearlong trek covering six thousand miles, and ending in the northwest province of Shaanxi. Deng was part of the march. Carrying everything from food to guns and office supplies on their backs or on mules and horses, the caravan made an easy target for Nationalist raiders and artillery. The marchers were engulfed with sickness, hunger, combat wounds, and sheer fatigue as they slogged across harsh mountains, windswept plains, and treacherous swamps. When horses died, the marchers often would eat their flesh. Deng contracted typhoid and nearly lost his life. Of the eighty thousand members who set out on the march, less than 10 percent survived to reach the end.

In January 1935, the weary group stopped at Zunyi, a city in Guizhou Province, to take stock of all that had happened and to plan for what they thought might come next. Zhou Enlai asked Deng to prepare the final summary of a historic debate, in which the Communist Party ratified Mao's military leadership and set him on the path to become China's supreme leader. As the chief recorder, Deng saw firsthand the internal party warfare and the clash of personalities and ideologies. He was later made head of propaganda in one of the armies, an assignment that entailed his explaining Mao's policies to the troops. Many veterans of the Long March, Deng among them, became the aristocracy of the Chinese Revolution.

Communist Victory and Early Experiments in Pragmatic Reform

Back in 1931, when Deng was on the run in Guangxi, Japanese forces had attacked northern China and occupied Shanghai the next year. By 1937, the Japanese advance had forced a new alliance of necessity between the Communists and Nationalists, both of whom knew the compact would last only so long as they had a common foe in Japan. Immediately after Japan surrendered in September 1945, China's civil war broke out again, and fighting between the Communists and Nationalists raged for another four years. Deng emerged from the wars against the Japanese and the Nationalists as a revered military leader and trusted aide to Mao, but also as a reformer who had laid the seed of ideas that would reverse a good deal of Maoism.

Deng became known for his assiduous study of enemy forces and an ability to exploit their weakness. As a political commissar in the army—the senior civilian officer in the military unit—he became an expert in recruiting and organizing peasants, indoctrinating them into the army, and mobilizing them for battle. One of his roles was to identify and recruit promising young organizers, some of whom he would one day rely on to run China, such as Zhao Ziyang. He set up training schools to teach guerrilla tactics inspired by Mao and experimented with ways to improve the lives and win the loyalty of towns under his control.

He earned his military reputation in many battles, but most notably in one of the largest ones in human history—the Huai-hai Campaign in 1948, a battle against the Nationalists for control of central China. As party secretary for the command structure, Deng coordinated more than half a million men to defeat Chiang's armies, putting them on the defensive for the rest of the war. It was a key turning point for the Communist Revolution, and Deng's reputation as a military leader was cemented for the rest of his life.

Even at this stage Deng was showing his "capitalist leanings," stretching Mao's guidance on how to run a socialist economy. He allowed farmers to sell their crops with no tax on the profits. He permitted property owners to rent out fields so farmers could expand production. He experimented with financial incentives, arguing that economic benefits were the key to winning the population's loyalty. Perfecting the Socialist ideal—a society without profits and greed—would come only gradually, in his view, and change would come through effective organization and management.

On January 1, 1949, the Communist forces entered Beijing and within months established the People's Republic, uniting China and annulling all unequal treaties with Western powers. Mao told the crowd, "We, the four hundred and seventy-five million people of China, have now stood up." But after forty years of war, China was in a catastrophic state, plagued by inflation, food shortages, and famine, with schools and roads in disrepair and millions of people homeless. Looting and corruption were rampant. Foreign threats loomed: the Nationalists had retreated to the island haven of Taiwan, backed by the United States, while the British reclaimed Hong Kong and Tibet moved toward independence. In these circumstances, Mao dispatched army leaders to run many of the provinces, assigning Deng to oversee the southwest region, which included Sichuan, Guizhou, Yunnan, and Xikang, an abjectly poor and widely diverse region that included one-third of China's territory and one-third of its population.

Over the next several years, Mao gave Deng a series of increasingly important posts. By 1956, Deng was considered one of the six most powerful men in China. With his third wife and five children, he settled in Zhongnanhai, Beijing's protected enclave for top Chinese leaders. In the many internal power struggles swirling around Mao, Deng showed himself to be a shrewd bureaucrat. The "Chairman" was capricious, quick to twist facts to suit his ideology, and paranoid, eventually taking revenge on anyone who challenged him. In this potentially deadly political environment, in which every official was hypersensitive to his own status with Mao, Deng proved a deft organizer of major party events and policy decisions, picking speakers, coordinating opinions, and summing up meetings to accommodate clashing views. At this point Mao saw Deng as a superb political organizer, someone who knew how to execute decisions, not an ambitious firebrand who would challenge him. Deng kept the trains running. He knew how to pick people and where to assign them. He could force corrections when policies deviated from the central party line. These skills would fuel his ascent to the top of China's power structure.

Nevertheless, while many top officials were concerned about the rigid cult of personality that had developed around Mao, and the retribution he would extract on those who disagreed with him, Deng was one of the few who did not hide his views. He argued for more effective public administration, especially for clearer separation of the Communist Party and the government in policy making. He favored stimulating rapid economic growth and downplayed the potential for inflation. He was fond of experimenting with different economic incentives on a small scale to see what worked and what could be scaled up. While exalting the importance of "Mao thought," he also talked about the need to apply Mao's doctrines to specific situations with some flexibility, depending on the circumstances at the moment.

Deng Pushes Too Fast, Not for the Last Time

By late 1956, Mao decided to release political frustrations in China of the kind he saw building up in Eastern Europe against

Soviet rule. Relaxing constraints on public expression, he proclaimed a new policy encompassed by the expression "Let one hundred flowers bloom, let one hundred schools of thought contend." Mao was so shocked by the subsequent outpouring of criticism from intellectuals that within two months he had unleashed a vicious countercampaign, tarring millions of intellectuals, officials, and even workers and peasants as "rightists," and throwing tens of thousands in jail. In contrast to his willingness to experiment with economics, Deng was doctrinaire when it came to politics. An opponent of the "Hundred Flowers" opening from the beginning, he was a zealous supporter of the crackdown. In fact, he had no practical use for democracy and saw the jailing of dissidents as just deserts for arrogant intellectuals who were getting in the way of hardworking government agents.

Deng was also an early supporter of Mao's next big ideological campaign, the Great Leap Forward, launched in 1958 as a crash program to bring industrial production up to the level in the United States and Europe. Mao set wildly ambitious and totally unattainable targets for increasing the production of steel, coal, and electricity, and soon the Great Leap was morphing into a megalomaniacal attempt to mold a perfect Socialist society. Mao attempted to abolish money and to force all peasants into large cooperatives where no one owned property. Instead they all lived together in dormitories, wearing the same blue tunics and trousers. Before long the industrial strategy was backfiring spectacularly; for example, when Mao ordered communes to ramp up steel output in backyard smelters, peasants started to ignore the harvest and melt down their hoes and plows to meet the targets.

The crazier the Great Leap Forward became, the more difficult it was for practical leaders like Deng to continue supporting it. When Deng and a number of other senior officials traveled to rural areas to see the barren fields and abject hunger for themselves, they were appalled. By 1962, Mao conceded that the Great Leap was not working, and he turned to Deng and Liu Shaoqi, then president of China, to put the economy back on track. Mao promised to remove himself from daily affairs, and Deng proceeded to reverse Mao's approach. He started by demanding that officials deliver honest data, rather than pretending to overfill their targets, and he abandoned the ideology of mass mobilization in favor of virtually any practical step to raise production. "It doesn't matter if the cat is black or white, so long as it catches the mouse it is a good cat," Deng told the party. He pushed to dismantle large communes and restore some private farms. He imported large amounts of food, using scarce foreign exchange that Mao had hoarded. He cut subsidies to money-losing heavy industries. He attacked the corruption that had infected the party, restored some leaders demoted during the Hundred Flowers crackdown, and worked to restore the shattered morale of the party.

Watching Deng's policies take hold and his status rise, Mao began to question the loyalty and revolutionary authenticity of his once-trusted aide. As soon as the economy was brought back under control, Mao began angling to reimpose his vision of the ideal Socialist society. It was clear to Mao that Deng, with his willingness to follow any experiment that worked, would not get that job done. What China needed, in Mao's view, was to rediscover its original ideological zeal. To that end, in 1966, Mao launched the Great Proletarian Cultural Revolution, which would lead to another decade of economic chaos, international isolation, and mass jailings of Mao's perceived foes, this time including Deng.

House Arrest, Exile, Isolation

The first shot of the Cultural Revolution came in a party circular calling for a purge of "reactionary bourgeois thinking" in education, news media, literature, and art. Soon the party was calling for the same cleansing at the highest levels of government, the military, and the Communist Party itself. Starting in Beijing, students began taking to the street, destroying property, criticizing and assaulting teachers and principals, and plastering campuses and cities with posters attacking authority figures. Unbeknownst to Deng and others, Mao had created a Central Cultural Revolution Group comprised of Mao's wife, Jiang Qing, and three others, who would eventually be called the "Gang of Four," with broad license to isolate and punish any authority figure, in any field, for lack of Maoist enthusiasm. Failing to appreciate the power of this group, Deng and Liu tried to send party representatives into the schools to negotiate a peaceful resolution with the marauding students, who had taken the name "Red Guards." Deng, in particular, failed to show respect to Jiang Qing, whom he detested. For example, he refused to attend the preachy political operas that Jiang sponsored, and when he did go, he often ostentatiously fell asleep.

Before the summer of 1966 was over, Mao denounced both Liu and Deng for trying to quiet rather than fire up the Red Guards. Vilified as a "capitalist roader," Deng was stripped of his titles and publicly humiliated at a mass rally against him in Tiananmen Square. When the time came to issue his self-criticism, Deng said what he needed to say in order to save his party membership, and his life. "Whenever I think of the damages caused by my mistakes and crimes to the revolution I cannot help but feel guilty, shameful, regretful, and self-hateful," he told the party. "No punishment is too much for a man like me." It did not help him.

The Red Guards were compared by one journalist to jackals attacking a crippled deer. Their campaign to tear down bourgeois authority ended up paralyzing the national rail network, shuttering schools, factories, and farms, and destroying the lives of millions accused of being capitalist roaders. The Guards were even allowed to enter Zhongnanhai, dragging officials out of their homes, beating them in front of their families, sometimes filming these attacks as a lesson to the broader public. No one was immune—not men, women, children, or the elderly. After Mao dismissed Liu Shaoqi as president in 1967, Guards hounded him and his family, ultimately killing Liu and two of his sons. In all, the Red Guards killed an estimated four hundred thousand people and maimed many more in torture sessions.

The government held Deng and his wife under house arrest in Zhongnanhai, where the media set up loudspeakers to blast insults at him twenty-four hours a day. In one instance, the Red Guards hauled Deng and his wife out of their home and forced them to assume the airplane position—hands extended outward, head bowed—for hours. Then they spat on the couple and beat them in front of their children. At another location the harassment of Deng's family peaked when Red Guards beat the eldest son, Pufang, and then dumped him in a room where he was told the only exit was the open window. Pufang exited—whether he jumped or was thrown is still unknown—and was paralyzed from the waist down for life.

At the height of the revolution, the Gang of Four and their violent hordes gained a momentum that even Mao seemed unable or unwilling to slow. But the Chairman did create a limited cordon of protection around Deng, who never went to jail, and had a special back channel to communicate directly with Mao through a trusted aide. In October 1969, Deng and his wife were moved from Zhongnanhai to the distant city of Nanchang in Jiangxi Province in southeast China where he had once fought the Nationalists, and where they had minimally decent housing and jobs in a nearby tractor factory.

Now sixty-five, Deng had to chop his own wood or break up coal for heat, and mop floors while his wife washed and mended clothes. The summers were so hot that during the night the Dengs had to get up repeatedly to wet down their bamboo beds. But at least they were isolated from the Red Guards, and later reunited with two of their children, including Pufang, now confined to a wheelchair.

The Dengs attended to Pufang, washing him, seating him on the lavatory, putting him to bed. Deng walked some three to four miles each day, creating a well-worn path around his little house, seemingly lost in contemplation about China and what he would do if he ever returned to government. No one can know why Mao gave Deng partial protection from the Cultural Revolution, but it is possible that he also was looking to the day when Beijing would need Deng again, given his proven record in government, party, and military affairs.

Comeback

That opening came in 1970, when Mao turned on his designated heir, Lin Biao, who was accused of plotting a coup. By the next year, Lin was dead, killed when his plane crashed under mysterious circumstances in the mountains of Mongolia. With Lin out of the way, Deng knew that the succession to Mao was again in play. In November, he wrote Mao a pandering note praising the Chairman's brilliant leadership in uncovering Lin's plot. In the same note he asked that one day he be allowed to work again for the party, perhaps in some technical capacity in order to pay back what he owed to the country.

Although Deng didn't receive an answer, his living conditions began to improve: he was allowed to send Pufang to a top hospital in Beijing, and his guards seemed to disappear, giving him more freedom to move around. In the spring of 1972, Zhou Enlai wrote Deng that if he penned a short self-criticism and sent it to Mao, he might be called back to Beijing. Deng did just that, stating his unreserved support for the Cultural Revolution and recanting his previous "deviations."

In early 1973, Mao allowed Deng to return to Beijing, which infuriated the Gang of Four and particularly Mao's wife, Jiang Qing. But by now Mao was at political odds with his wife, and Zhou Enlai was running the government. So many top leaders had been purged that Zhou clearly needed the talents of Deng, his former protégé. In addition, Deng still had many strong supporters in the military, clearing the way for his rehabilitation. In restoring Deng as deputy premier that February, Mao told the party's Central Committee, "Deng is a rare talent. He is known in both military and civilian circles for this. He is like a needle wrapped in cotton. He can deal with different problems with responsibility."

With Mao increasingly enfeebled by Parkinson's disease, and Zhou Enlai succumbed to cancer, the battle to succeed Mao would in the coming years come down to Deng against the Gang of Four. Both Mao and Zhou were thrusting more and more responsibility on Deng—for the economy, for reform of the military, for rejuvenation of a demoralized party, even for substituting for the ailing Zhou at diplomatic receptions. Deng gained political strength from being seen with Mao at international events. His status was also enhanced when he represented China on his own, including a 1974 trip to the United Nations General Assembly, his first trip to the West since his student days in Paris and his first look at the vitality of New York City and the economic and social gaps between China and America. Just a few years later, he encouraged his youngest son, Deng Zhifang, to study in the United States. "You will see what the modern world looks like," he said.

In the vacuum into which he had been thrust, Deng moved fast, exactly as he had after the Great Leap Forward, risking another backlash from the Chairman. Indeed, Mao's sweeping talk of China's advance toward Socialist perfection, and the constant purges to cleanse the party of ideological impurities, gave way to Deng's roll-up-the-sleeves pragmatism. Mao's cult of personality was eclipsed by Deng's focus on building professional and technical competence in the party apparatus. Where Mao glorified workers toiling in the fields, Deng would glorify entrepreneurs who built something new in the fields.

By the beginning of 1975, the government had launched "the

Four Modernizations," a blueprint for progress in agriculture, industry, military, and science and technology, which Deng would execute with his characteristic get-it-done approach. He provided material incentives for workers and peasants to raise production. He dropped Mao's bias toward protecting homegrown industry in favor of promoting foreign trade. He was silent on several Maoist themes—like the revolution of the masses and egalitarianism—warning instead that the revolution was an empty vessel without economic progress. Under Deng's oversight, roads and railroads were repaired, factories reopened, farm output started to climb, and China began to emerge from isolation. In 1975, Deng led the first state visit of a Chinese Communist leader to a Western country—France.

Deng's pragmatism would ultimately win out, but only after yet another fall into exile. By the middle of 1975, the Gang of Four had launched a campaign attacking Deng for putting China back on the capitalist road. Mercilessly criticized at high-level meetings, in the media, and on billboards, Deng defied the Gang at every turn. He wore hearing aids and would turn them off when Jiang Qing and others berated him, allowing him to retain his composure as if nothing happened. His fall came when he presided at the funeral of Zhou Enlai in January 1976, which Mao and the Gang of Four refused to attend on the grounds that Zhou had deviated from the path of revolution. (Mao may have been sick at the time, as well.) Knowing this act of respect for an accused deviant could land him in jail or worse, after the funeral Deng fled to Guangdong Province with the help of an old ally from his army years, Marshal Ye Jianying.

Mao had appointed Hua Guofeng as his successor, seemingly a safe choice to preserve his legacy. Deng waited and watched, but not for long. Six months later Mao died, and a scramble to rule in his wake ensued. Marshal Ye convinced President Hua that the Gang were a threat to the country and to Hua personally, and he gained the support he needed to arrest and imprison Jiang Qing and her cohorts. Celebrations broke out all over China, demonstrating mass relief that the chaos of the Mao era was over as well as a deep base of support for more pragmatic policies.

A few days later, writing from his secret location, Deng sent Hua a flowery letter of congratulations, support, and loyalty, while at the same time admitting to his own failures. Shortly after, he asked Hua if he could be rehabilitated. Hua saw Deng as a challenge to his own preeminence and to the survival of the Maoist policies he supported, but with strong lobbying help from Marshal Ye and other old allies, Deng was back in government by the summer of 1977.

The Deng Era Begins

Deng was now in his early seventies, fit, energetic, and unrepentant. Even as President Hua held the top positions, Deng maneuvered to place his own people in key posts and orchestrated the rehabilitation of thousands of people disgraced in the Cultural Revolution. Soon Deng had all but repudiated the Cultural Revolution, an exceptionally delicate task that required his condemning Mao's policies without discrediting Mao himself. In December 1978, Deng described a 180-degree change of course, saying that China needed "large numbers of pathbreakers who dare to think, explore new ways and generate new ideas." China's resentful insularity would have to go. "One must learn from those who are more advanced before one can catch up with and surpass them," said Deng. "Independence does not mean shutting the door on the world, nor does self-reliance mean blind opposition to everything foreign."

Over the course of 1979 Deng, still serving under Hua, reassumed many of his former roles. He also volunteered to take responsibility for science, technology, and education, telling colleagues that people who worked with their minds should be considered respectable members of the working class. One year later, Deng was also overseeing foreign affairs, traveling throughout Southeast Asia and going to Japan, and making a trip to the United States to meet with President Jimmy Carter, where he visited the New York Stock Exchange, famously donned a ten-gallon cowboy hat in Texas, and worked the controls of the space shuttle simulator. It was Deng who completed the process of normalizing relations with the United States that had begun with Nixon's visit to Beijing in 1972.

To be sure, Mao's successor, Hua, was a modernizer as well, sending missions abroad to study foreign management practices and starting a discussion of special zones where foreigners would be free to invest in China, but his dynamism was no match for Deng's. Bombarded by criticism of his ineffectiveness—criticism Deng did not discourage—Hua resigned in 1980 and moved quietly to a lower post in the politburo. There were no anti-Hua billboards and there was no public humiliation, no self-criticism, and no brutal period of exile. The Maoists were gone, and Deng was in charge. He did not do to the Maoists what they had done to him.

The Paramount Leader

By 1980, the Age of Reform—the reign of Deng—was already unfolding. Deng, the new emperor of China, was leading the transformation of the Middle Kingdom, fundamentally changing the course of the global economy and global politics. Rejecting any semblance of a cult of personality, Deng's titles—vice premier and chairman of the party Central Military Committee—at first vastly understated his power. It would take thirty months for Deng to be publicly recognized as "the Paramount Leader," the giant among supposed equals in the communal leadership. "In the annals of world political history," wrote biographer Ezra Vogel, "it would be difficult to find another case where a person became top leader of a major nation without formal public recognition of the succession."

In Chinese tradition omnipotence lies in the mystery that invisibility invokes, and everything about Deng seemed to undermine his visibility. He entered a room surrounded by assistants who were a head taller than he was. He greeted guests with a limp handshake. He conveyed his thoughts perfunctorily and with no emotion. He slurred words together. He backed away when asked for clarification of what he said. "It was not just that he discouraged any suggestion of a cult of personality," wrote China scholar Lucian W. Pye, but "he actually seemed to work at creating anonymity." On the other hand, Deng radiated a desire to get to the heart of the matter with no fanfare. Henry Kissinger recalled his own experience with Deng. "Having grown accustomed to Mao's philosophical disquisitions and indirect allusions and to Zhou's elegant professionalism, I needed some time to adjust to Deng's acerbic, no nonsense style, his occasional sarcastic interjections, and his disdain of the philosophical in favor of the eminently practical," Kissinger wrote years later. "Compact and wiry, he entered a room as if propelled by some invisible force, ready for business."

Deng had also developed a view of the world that was less Sino-centric than that of his colleagues. Every previous Chinese leader had blamed China's problems on the depredations of foreign invaders and emphasized the superiority of Chinese ways. Deng came to exactly the opposite conclusion on his trips abroad: China had to stop blaming foreigners for its problems and to learn all it could from the world, as soon as possible. In the 1960s, much of Asia was still an economic backwater, but by the 1970s, the Japanese miracle was taking off and spreading to Taiwan and South Korea. To Deng the message was clear: China had to get moving, too. And so by word and deed, public and private, Deng turned China's focus away from nursing old grudges and toward a future he felt could be promising.

Over the next two decades, Deng put economic development at the heart of all his efforts, with an emphasis on achieving a high rate of growth. He reduced state intervention in China's economic policies and supported the idea that it was okay for some people to get rich faster than others—which would have been heresy in Mao's China. He took another radical step by shifting scarce resources from the military to the civilian sectors. His government introduced market reforms in agriculture, abolishing the large communes and giving peasants the right to sell more of their products on the open market. It cut the price of critical farm supplies like pesticides and raised the price the government paid for grain. Deng also allowed peasants to expand into new enterprises, unleashing a wave of entrepreneurship that created new businesses in light industries such as textiles and toys.

As a result of these reforms, farm production rose along with the living standard of farmers. From 1978 to 1982, peasant income roughly doubled. And agricultural productivity—defined as output of grain per head—increased by 3.8 percent annually between 1978 and 1984, compared to 0.2 percent from 1957 to 1977. Rural consumption per head increased 300 percent between 1978 and 1986, compared to 30 percent over the previous twelve years. In 1981, China was the world's fourth-largest importer of cotton; four years later it was a net exporter.

The state still ran every industry, but Deng pushed to give business managers more flexibility. He eased rules that made it difficult for workers to migrate to new jobs within China. He brought to life the special economic zones, which eased taxes and regulations on foreign investment, with an eye to building export companies. Based on Hua's policies, the first special zones had been identified as early as 1979 in the Pearl River delta province of Guangdong and also in Fujian Province; Deng would open these and more. As investors, particularly from Hong Kong and Taiwan, poured in money, the special zones started to prosper and to adopt foreign management practices and machinery. Some of Deng's colleagues still feared that the foreign influence would weaken China, but Deng saw this threat as an inevitable price of opening the economy. "When you open the windows for fresh air," he once said, "some flies get in."

While Deng created the political environment for reform, he never pretended to be an economist. "I am just a layman in the field of economics" he said in 1984. "Sometimes I make a few remarks in this field, but all from a political vantage point. I have proposed China's economic policy of openness to the outside world, but as for the details on specifics of how to implement it, I have to say I myself know very little indeed." In contrast to Mao, who managed by keeping underlings off balance, refusing to give clear guidance on big decisions, and then intervening in details, Deng delegated clear responsibility to carefully chosen subordinates and left them free to experiment, no longer constrained by Maoist ideology. In fact, many of his specific economic reforms were designed by others, especially his two key lieutenants, Zhao Ziyang and Hu Yaobang. He made it clear he believed in some key concepts—modernizing management, the power of incentives—but there was no overarching economic blueprint. He was fond of saying: "In order to cross a large river, you feel the stones along the way," meaning that in periods of great uncertainty extreme pragmatism—taking one step at a time and adjusting as you go along—is itself the best strategy. In fact, a similar notion guided both Jean Monnet and Andy Grove.

Deng was obsessed with science, technology, and education. Even before he became paramount leader, Deng developed plans to establish an independent Chinese Academy of Social Sciences. He supported increased budgets and better working conditions for researchers. In a sharp departure from Mao, he encouraged more Chinese students to study abroad, and more contact between Chinese and Western scientists. To make his modernization priorities clear, Deng built trips to the West around visits to companies such as Boeing and IBM.

Deng knew that he needed a long period of peace with the outside world in order to keep the door open to Western technology and markets. He thus reversed a quarter century of economic and political isolationism by opening China to the world economy and normalizing relations with the United States, the Soviet Union, and all of Asia. He worked to peacefully reclaim Hong Kong from its British rulers under an ingenious formula called "one country, two systems," designed to allow Hong Kong to remain an enclave of Western-style capitalism, under Chinese Communist rule.

In all this, Deng unleashed the long-repressed ingenuity and commercial drive of the Chinese people. Suddenly everyone was an entrepreneur, every institution an enterprise, and even provincial and metropolitan governments started businesses ranging from food processing to real estate. The heroes of the working class were replaced by new icons such as the man who turned a small business making iced lollipops into a billion-dollar soft drink empire, or the person who sold all his belongings to start a pig-feeding operation, which grew into a giant company with over one hundred factories. With few commercial laws, no notion of property rights, and no clear accounting procedures, this was capitalism of the Wild West, capitalism without rules, but Deng acted as if he was unafraid of the resulting chaos. In time, Deng said, the laws and the enforcement authority would catch up with the ferocious new competition in business.

Deng never broke with Mao on the critical importance of the Communist Party monopoly on political power. He aimed to build a market-oriented economy, but within an unquestioned dictatorship of the party. He believed that the party should secure popular backing not through the ballot box but by effectively delivering economic goods. Thus, he was obsessed with streamlining both the party and the government bureaucracy, and recruiting talent that was younger, better educated, and more technically competent than the older generations.

The dichotomy between his openness to experimenting with capitalism and his determination to maintain disciplined rule under an all-powerful Communist Party was perhaps the single most important feature of his philosophy and his manner of ruling. This paradox would raise one of the biggest questions facing China—and, given its importance, the world as well—in the twenty-first century. Until this point, Deng had been consistently progressive on economic issues and consistently ruthless in resisting challenges to hard-line party doctrine. He orchestrated Mao's campaign to punish intellectuals after the Hundred Flowers movement in 1957. In 1975, he called in the People's Liberation Army to quell demands for religious freedom in Shadian, and more than sixteen hundred villagers were killed in the fighting. In 1979, he imprisoned human rights leaders such as Wei Jingsheng, and in 1990, he exiled dissidents such as Fang Lizhi who had been agitating for democratic reform. Deng especially viewed the Cultural Revolution as the result of too much democracy leading to anarchy, violence, and demagoguery. As China scholars Orville Schell and John Delury write, Deng's reforms were a mixture of Vladimir Lenin's prescription for a tightly organized state and Milton Friedman's infatuation with free markets.

Massacre at Tiananmen

By the mid-1980s, Deng's view was starting to clash with the demands of a rising middle class, which expected at least some political openness to follow from economic progress and resented the special privileges accorded to the families of top Chinese officials. In the first six months of 1988, for example, rallies took place on seventy-seven campuses in twenty-five cities. Deng kept admonishing his colleagues that firm action was needed to counter these demonstrations, which he found unacceptable.

On April 15, 1989, with the death of the liberal former party head Hu Yaobang, students began to gather in Tiananmen, demanding the rehabilitation of the reputation of the once-demoted Hu, as well as more political freedom and more funding for education. On April 26, the *People's Daily*, the official government publication, roundly criticized the students, triggering on the next day what was perhaps the largest spontaneous student protest in modern Chinese history. About a month later, Deng ordered the military to prepare to impose martial law, a decision that divided the politburo between younger reformers like Zhao, who saw China emerging as a more democratic nation, and older hard-liners like Deng, who had spent their whole lives fighting to strengthen the party and state and were not about to let it fall to student radicals. Once Deng decided to call in the army, the previously unthinkable was now a possibility: the People's Liberation Army, the troops that Deng once recruited from the masses, would confront the citizenry in street battles.

On June 3, tanks rolled down the streets of Beijing,* killing and wounding civilians who attempted to halt their progress toward Tiananmen Square. The official death toll was three hundred, although independent observers put it at ten times higher. By the time the tanks reached Tiananmen, the crowds had cleared out, and for five days Deng remained silent. Then on June 9 he went on TV to praise the military and to say that nothing would stand in the way of China's economic reform. Foreign reactions were harsh. The United States in particular levied a variety of sanctions on technology transfer and trade. But it did not take long for most of China's foreign relations to return to normal; the lure of the increasingly large and open Chinese market was just too great. By early fall, Deng announced that he was offi-

* For a complete picture of what happened, the following two sources provide excellent background: Zhang Liang, compiler, *The Tiananmen Papers*, ed. Andrew J. Nathan and Perry Link (New York: PublicAffairs, 2001); Gideon Rose, ed., *Tiananmen and After*, from the *Foreign Affairs* Anthology Series (New York: Council on Foreign Relations, 2014), electronic edition.

cially retiring, with one proviso: "If I have any useful ideas and suggestions I will gladly convey them to the new leadership."

"Be Bolder in Reform"

Tiananmen occurred in the same year that the Berlin Wall came down. Deng concluded from the collapse of Communism in Eastern Europe and the pressures in his own country that the party could survive only if it continued to deliver prosperity. He calculated that if Beijing kept improving the material lives of the people, then they would not object to tight political control. By the early 1990s, however, Deng's successors seemed to have lost his enthusiasm for more economic reforms. Some leaders were legitimately concerned about inflation and grew fearful at the breakneck pace of economic expansion. A number worried that the infusion of foreign capital was bringing with it new opportunities for corruption and the destruction of Chinese culture. Some even wanted to re-collectivize agriculture. From retirement, Deng objected to any rollback of economic liberalization but felt his concerns were not being heard. He had always worked through the party and government apparatus before, but now he decided to take his case straight to the public through the media.

On January 8, 1992, at the age of eighty-seven, he embarked on a tour of southern and eastern China, where growth was exploding in the special economic zones. At various stops in Wuhan, Shenzhen, Zhuhai, and Shanghai he publicly proclaimed that the counterreaction to reforms had gone too far and that conditions were ripe for even faster economic growth. In Zhuhai, Deng took aim at his successor, Jiang Zemin, saying, "Whoever is opposed to reform must leave office." Surrounded by the press at every stop, and accompanied by comrades from the army who were previously identified with forces opposed to further liberalization, Deng's words were broadcast around the world and had a catalytic effect all over the Middle Kingdom. In a raw demonstration of his hold on the government and his understanding of China's yearnings, he turned the tide of public opinion in favor of reform, lighting a fire under top officials. By October, Party Chief Jiang Zemin was singing Deng's praises and strongly endorsing his vision. On the day Deng returned to Beijing, the editorial in the *People's Daily* was entitled "Be Bolder in Reform." Reforms were accelerated and broadened. Foreign investment poured in. The growth rate shot up.

It was only a short time later that China opened up negotiations to join the World Trade Organization (WTO), a monumental step for China and the world. Until Beijing became a member, its economic reforms followed only the dictates of Deng and his administration. But when it eventually joined the WTO, Beijing signed a multilateral treaty obligating it to reach new international standards of openness that had already been set by the Western members of the WTO, or face the threat of multilateral sanctions. Deng effectively bound the Middle Kingdom to the rules of an open world economy.

Slowed in his later years by Parkinson's disease and lung infections, Deng passed away on February 19, 1997, at the age of ninety-two. In accordance with his will, his organs were donated to science and his body was cremated and spread around the sea. Unlike the terrifying uncertainty that followed Mao's death, Deng had put in place a smooth succession plan and had left an unequivocal policy path. An obituary in the *New York Times* captured the mood in Beijing.

> As the sun rose on Beijing this chilly morning, many residents learned the news on radio and television broadcasts before they set off for work and school, and the reaction of those on the street at dawn was restrained and, in some cases, nonchalant. It is part of Mr. Deng's legacy that ordinary people no longer hang on daily reminders of their leaders' every pronouncement.

Ten thousand guests attended the funeral under tight security. No foreigners were invited. Wrote Ezra Vogel, "The structural changes that took place under Deng's leadership rank among the most basic changes since the Chinese empire took shape in the Han Dynasty two millennia ago." Summing up the impact of the Deng era, former US ambassador to China J. Stapleton Roy said, "If you look at the 150 years of modern China's history since the Opium Wars, then you can't avoid the conclusion that the last 15 years are the best 15 years in China's modern history."

Legacy

Deng led China out of the wilderness. He rescued it from being an isolated, backward nation in a world moving quickly toward advanced technological modernization and unprecedented prosperity, and he gave it the opportunity to become a central and vibrant part of a globalized society. He wasn't the only official in China who wanted to break with Mao and change direction, but he was the leader who commanded the influence, respect, experience, relationships, and political skills to do it. It is a sure thing, in my view, that in the future China's progress will confront some significant setbacks, as has befallen every other country in history, but Deng gave it such a promising start and such strong momentum that the long-term prospects for such an inherently energetic and talented people to constitute an extremely powerful political, economic, and social force in the world should not be in doubt.

Deng had totally transformed the prospects of a country that is home to 20 percent of all humanity. After the tumultuous years of Mao, he restored a sense of normality. He injected competence and dynamism into the Communist Party and into a government that had proved utterly incapable of meeting the challenges of the modern world. "Perhaps never in human history has an established society gone through such a total transformation, without a war, violent revolution or economic collapse, as did China with the ending of Mao Zedong's reign and the emergence of Deng Xiaoping as paramount ruler," wrote Lucian W. Pye. Gideon Rose, editor of *Foreign Affairs*, agreed when he wrote, "The scale and speed of the transformation rank it as one of the great events in human history."

Deng initiated reforms that led to thirty years of growth averaging more than 9 percent annually—twice the average of other developing countries, and three times the average of developed nations. When he took the reins of power, China's trade with the world totaled $10 billion. Three decades later, it had expanded one hundredfold. In his early days as leader, Deng asked the United States to take a few hundred Chinese students. By 2007, some 1.4 million Chinese students were studying abroad. He helped lift over 200 million Chinese out of poverty with policies that dramatically freed ordinary Chinese to open their own businesses and to make choices about how and where they would live and work. Under Deng, China had accomplished in two decades what took other nations a century or more to do. When he left the scene an inevitable question loomed for a nation that had experienced tumultuous volatility throughout its history: could China continue to prosper without him?

During 2012, in the third peaceful transfer of power since Deng died, a new leadership took over in China, headed by President Xi Jinping. This was the first generation of leaders that had not been in some way handpicked for power by Deng himself, and yet they continue to identify themselves with his name. In his first trip outside of Beijing as president and Communist Party chief, Xi Jinping set out to establish his free market credentials by visiting Shenzhen, one of Deng's early special economic zones, and by laying a wreath at a bronze statue of Deng nearby. In 2014, top Chinese officials, again evoking Deng's opening of Shenzhen, inaugurated a new special zone in Shanghai to promote a more open financial system and attract foreign investment in related services—echoing Deng's approach to economic liberalization. On the 110th anniversary of Deng's birth in August 2014, the Chinese government went to great lengths to laud the former leader with a forty-eight-episode TV series, the release of an official biography, and extensive ceremonies in Beijing as well as at Deng's birthplace. President Xi took the occasion to not-so-subtly hitch himself to Deng in terms of the nature of his mission and his leadership qualities, a link reinforced heavily by official commentary from the media. "Great times make great men [and] Comrade Deng Xiaoping is a great man arising from the great struggle of the Chinese people," Xi said at a high-profile ceremony. He continued, saying that Deng "proved during the practice of a lifetime to be a visionary thinker, statesman, and strategist, but [also] someone who was realistic, pragmatic and practical . . . his whole being was filled with reform and innovation, indomitable courage, creative thinking . . ." In the years since Deng died, Xi said, the Communist Party has adhered to his principles and policies and made notable progress along the way.

It is not surprising that China's leaders would model themselves after their most successful predecessor in centuries. But they know, too, that China has changed since Deng's time. It is a much more powerful player in the world now, much more enmeshed in geopolitics and in the global economy. Global enthusiasm for China's entrance on the world stage has given way to the competitive realities of a fast-rising world power bent on recapturing its central role in the world. China now has a more prosperous population, but a more demanding one, too, with huge segments being left behind economically, resulting in widening income and wealth disparities. Its growth model, based as it has been on labor-intensive exports and massive investment in industry and construction, needs to change to generating more economic stimulus from within China itself. It needs to build industries that serve consumers in order to make life better for a broader variety of citizens, industries providing better health care, financial services, and education, for example. The banking and credit system needs reform. The social safety net is seriously inadequate. Environmental degradation is acute. Widespread corruption at all levels of society has become corrosive. The Xi government is well aware of these challenges, and has an elaborate program of pragmatic reform to address them. Indeed, it never tires of articulating its ambitious agenda for reform in these and other arenas.

The central Communist Party remains supreme, and the repressive conditions for human rights and free expression have not changed. If anything, President Xi is strengthening the authoritarian nature of the regime, even reestablishing a cult of personality more reminiscent of Mao than of Deng. It is becoming increasingly dangerous to criticize the state, with dissenters and their lawyers being rounded up and often imprisoned, and journalists being muzzled. Minority populations are being severely repressed. The activities of nongovernmental organizations are being dramatically cut back. The government is expanding its control over the Internet and social media. A sweeping new law on the definition of national security is giving Beijing new powers to clamp down on a broad range of its citizens' activities. All this is amplified by the drumming up of nationalist fervor. It's no exaggeration to say that economics and politics are moving in opposite directions—the former in a relatively progressive way, and the latter toward ever tougher authoritarianism.

The Chinese leadership has also become much more assertive on the international stage than Deng ever was. Perhaps no one should be surprised. Except for the last few centuries, the Middle Kingdom had been a major power on the world scene, and now that its economic clout is restored, China has every right to speak with a louder voice in international organizations, every right to assert its interests. Still, its aggressive stance toward territorial ownership of the East and South China Seas, combined with its sustained expansion of naval forces, leaves little doubt that Beijing has left behind Deng's policy of hiding Chinese power and intentions in the geopolitical realm in order not to alarm other nations.

When it comes to its ties to the United States, perhaps the most important bilateral relationship in the world, and one on which the future of global peace and prosperity could revolve, the environment has changed since the Deng era. On the one hand, economic interdependence has increased dramatically as China has become a major trading power and as common interests in arenas such as climate change have been the subject of increasing cooperation. But on the other, Washington now recognizes that economic ties do not translate into common political and security interests. In this regard, a host of big problems between the two countries have not gone away, such as violations of human rights or intellectual property rights. New sources of tension have also arisen, such as cyber theft and cyber espionage. The notion that China could be a true partner with the United States is giving way to a hardening attitude in America that the US-China relationship could easily become a cold peace—or worse. We don't know whether China's toughening political stance at home and abroad constitutes Beijing's effort to distract its citizens' attention from problems at home, such as much slower economic growth, and whether Deng would have steered the Middle Kingdom to the same place. But this much is for certain: President Xi faces an America much more suspicious of China's character and intentions than it was as recently as a decade or so ago.

No one can know how these trends will play out. There are a number of potential narratives. For example, China's rise could be peaceful and it could take its place as a leader in the existing global order, the very order that has until now served it so well. A second scenario is that military conflict between China and the United

States is inevitable, in the tradition going back through at least two millennia in which rising powers and incumbent leaders have always clashed. Another possibility is that the confluence of economic opening and extreme centralized political control will result in severe internal or international tensions that could lead to a cataclysmic crisis and paralyze China in years to come. That would be based on the notion that the Chinese economic miracle, so central to China's broader power and status in the world, is fading, and that the usual blunt policy levers that Beijing has employed in the past will be less effective now that the Chinese economy is larger, more complex, and more deeply tied to the global market. In this situation, the Chinese leadership may not be skillful enough to continue growing GDP at anywhere near the pace in which it has expanded the last few decades. In that case popular confidence in the Communist Party could erode, thereby undermining the political legitimacy of the government itself. That in turn could lead to destabilizing social unrest, Beijing's ultimate nightmare. Such an outcome would have negative repercussions around the world, to say the least. Perhaps what everyone can agree on is that China is undergoing several transitions at once—in economic and social policy at home, in its economic and political relationships with other countries, and in its own view of what kind of nation it wants to be in the future, all against a background of slower growth within the country and in the global economy. The challenge of balancing all these transitions in a successful way will test even the most exceptional government officials.

Deng had formidable challenges, to be sure, but he was able to manage them during a reign that lasted more than two decades. He was the right person for his time. Amid dramatically different circumstances, let's hope his successors are equally worthy. The stakes for China and our interconnected world couldn't be higher.

Chapter XI

THE BEST IS YET TO COME

Writing this book took me on a long trip through history. Somewhat to my surprise, this journey left me considerably more optimistic about the future, because my ten protagonists shared certain qualities that allowed them to move mountains in moments of crisis and periods of turmoil. And it's not terribly hard to imagine individuals of similar caliber stepping forward to face down the modern crises of globalization. So what are some of the common traits of those about whom I have written?

They Are Hedgehogs

The ancient Greek poet Archilochus famously said, "The fox knows many things, but the hedgehog knows one big thing." Everyone in this book was a hedgehog, relentlessly pursuing one big idea. They had a missionary-like faith in what they were attempting to do, and they were driven, day by day, year by year, decade by decade, to that end. Their power to focus on one big idea for so long engendered a simplicity and clarity to the complex maneuvers that were required to execute their ideas. Hedgehogs refuse to accept failure and demonstrate resilience in the face of daunting setbacks. Their fierce determination overwhelms any sense of despair; indeed, they seem incapable of sustained discouragement. They are bold risk takers when the specific risks are critical to their singular goals. By definition, the risks are more educated than they would otherwise be, because they are within the relatively confined field of ever-accumulating experience and expertise.

Robert Clive never took his eye off conquering more territory for the East India Company and Great Britain. For six decades, Jean Monnet concentrated all his energies on reducing the barriers to deep-seated cooperation among sovereign countries. For all his professional life Andrew Grove was riveted on making semiconductors smaller, faster, and cheaper and to make their production of flawless quality.

If someone has just one overriding obsession, it's more likely that he or she would reject failure and bounce back from major obstacles encountered along the way. The option to do otherwise carries dire implications. They bet the farm and they cannot afford to lose it. Thus, when Cyrus Field's *fourth* try at laying the transatlantic cable failed—in front of a worldwide audience—he immediately told the crew, "This thing is to be done," and he disappeared belowdecks to plan the next mission. When Margaret Thatcher faced excruciating political pressure to change her austerity policies in her first few years in office—not just pressure from the political opposition but within her own party and from the public—she not only refused to modify her approach but redoubled her efforts. "The lady's not for turning," she said. Moreover, hedgehogs seem impervious to humiliation or even tragedy. Prince Henry was not stopped even after he had to surrender his younger brother to death by torture rather than give up his newly conquered city of Ceuta. The indignities of virulent anti-Semitism did not deter Mayer Amschel Rothschild.

Hedgehogs who are highly successful take big risks because

in the end their obsessions give them no choice. Warriors like Genghis Khan and Robert Clive time and again put their physical survival in jeopardy. Others, such as Jean Monnet, Margaret Thatcher, and Deng Xiaoping, gambled that their revolutionary ideas would not backfire and send them into political oblivion.

As hedgehogs, they not only focused on a single goal but also started their efforts at a very early age and never abandoned them. No one could say they had their one second of fame, or that they were wunderkinder, or that they peaked early and went into obscurity. To the contrary, they started early and their accomplishments built with time: Robert Clive, the teenager who went to India and built his career step by step, battle by battle; or John D. Rockefeller, who began as a clerk in the commodities business and ended up controlling the world's most important commodity; or Margaret Thatcher, who started out working for Conservative campaigns as a girl and became the Iron Lady, espousing the same beliefs all along the way; or Deng Xiaoping, who before twenty fell in with the future leaders of the Chinese Revolution and rose to the commanding heights of the Middle Kingdom. Long-term efforts, long-term horizons, all of them.

They Rode the Tides of History

It is tempting to say that the essence of great achievement is to change the course of history, but that idea doesn't describe the ten people in this book. Each of them capitalized on fundamental trends that were already in train. They swam with the stream rather than against it. In all cases, someone else might have done what each of them did. But our protagonists got there first. Was it because they were smarter and quicker and more obsessed than their would-be competitors? Perhaps. Was luck involved? Of course. But whatever the reason, they and not some others succeeded in being close to first movers.

In the end, the characters in this book do not validate the great man theory of history, as articulated by nineteenth-century historian Thomas Carlyle, who said, "The history of the world is but the biography of great men." Political philosopher Herbert Spencer was closer to the mark when he focused on the way historical trends and events have shaped the careers of great leaders. "Before he can remake his society," Spencer wrote, "his society must make him." It is the case that the individuals here helped to create globalization, but equally so that globalization had a role in creating them. The people in this book were of their time and they made their time. They steered history only insofar as they seized the opportunity that contemporary circumstances afforded them.

Shakespeare said it well:

There is a tide in the affairs of men.
Which, taken at the flood, leads on to fortune;
Omitted, all the voyage of their life
Is bound in shallows and in miseries.
On such a full sea are we now afloat,
And we must take the current when it serves,
Or lose our ventures.

Consciously or not, each of the ten picked their moment "at the flood."

Thus, Prince Henry came to the fore at a time when the West was awakening, the spirit of exploration was growing, and the European hunger to find a sea route to Asia was becoming an obsession. Robert Clive rode the momentum of rising British industrial and naval power and British appetite to acquire colonies. Mayer Amschel Rothschild took advantage of the Napoleonic revolution to create a new financial order out of the chaos of war and its aftermath. Cyrus Field had the strong tailwinds of Samuel Morse and others who had created the telegraph, and he lived at a time when other titans were demonstrating seemingly limitless opportunities for grand projects. John D. Rockefeller discovered oil at a time when America's industrial revolution was gathering force and petroleum was in rising demand around the world. Jean Monnet's push to dissolve borders would not have stood a chance if Europe and the United States had not been desperate to unite the countries on the Continent to mount a consolidated recovery out of the devastation of World War II, and to ward off another war involving Germany. Margaret Thatcher's campaign to dismantle the edifice of socialism in Britain could never have happened if those policies had not failed miserably and if the population wasn't desperate for change. Andrew Grove arrived at Intel just as the high-tech revolution in the United States was gaining force in Silicon Valley, and as the United States was rapidly ramping up its investment in science and technology to compete with the post-Sputnik Soviet Union. And Deng Xiaoping's economic counterrevolution would not have generated so much momentum if his predecessor's policies had not decimated the Middle Kingdom at the same time that the West was embarking on an era of unprecedented prosperity.

They Were Accidental Globalists

When I started out to write this book, my assumption was that the ten people I selected were visionaries. After all, their accomplishments reshaped the world for many generations to come, sometimes for centuries. Having delved into their lives more deeply, however, I came to a different conclusion: they did not have grand strategies in mind, and they did not spend much time envisioning the major transformations for which they would be responsible. None set out to change the broader world, just the smaller, personal one they could see and understand. Accelerating the interconnectedness of nations was never their motivation;

instead, they were propelled by the urge to acquire power, or fortune, or fame, or something else that came with solving a concrete set of urgent problems in front of them. Indeed, their achievements were the result of putting one foot in front of another, dealing with one challenge at a time.

Thus, Genghis Khan was driven by power, greed, and lust for revenge. Prince Henry, being the ambitious third son of a nearly bankrupt monarch, had to figure out a role for himself and a way to make the kind of living appropriate for royalty. Conquest, religious conversion, and riches were his motivation, not the romance of discovering new worlds—even if that was what he often said. When Mayer Amschel Rothschild started to manage the finances of a wealthy German prince, his ambition was to become rich. One deal just led to another, one innovation to another, but he didn't set out to transform the nature of global banking. When John D. Rockefeller was conspiring with the railroads to build Standard Oil in America, it seems doubtful that this former accountant was thinking about creating a global behemoth that would play so monumental a role in propelling prosperity in the twentieth century and beyond. When Jean Monnet presented a plan to establish the European Coal and Steel Community to French finance minister Robert Schuman, he was solving an acute political problem of the moment—although I must admit that if there is one visionary among my ten, Monnet is that person, as he had envisioned something like a United Europe for at least a decade.

Serendipity was a guiding light for these accidental globalists. Would Genghis Khan have marched west if the governor of his Islamic neighbor had signed a trade agreement instead of murdering the Khan's commercial emissaries? Would Cyrus Field have gotten involved with the transatlantic cable if his brother had not had a chance encounter in a hotel lobby with a Canadian engineer trying to raise funds for a cable between Newfoundland and the United States? Suppose that Chancellor Adenauer had rejected Monnet's last-minute plan for the European Coal and Steel Community? Would we ever have heard of Andrew Grove if he had been captured by the Soviets while sneaking over the border from Hungary to Austria? Would it have been Deng Xiaoping who opened China to the world if, in his early days in Paris, the French police who were trying to arrest him had not arrived just a few days too late? There is a thin line between immortality and obscurity.

Details and Demons

To some extent, the very fact that I selected doers and not just thinkers could explain the common theme that all of them possessed a capacity to oversee complex projects. But leaders can be doers by being great strategists and not getting involved in the fine details of everyday management. What surprised me is just how deep in the nitty-gritty most of my subjects waded. Prince Henry had to lead and organize a vast industrial empire that ran the gamut from building ships to administering overseas territories. Mayer Amschel Rothschild's ability to establish a banking operation—in addition to a smuggling network—that was exquisitely coordinated across several countries was a key to his success. Cyrus Field oversaw a team of engineers, scientists, and financiers, all the while having to negotiate political support from the US and the UK governments. John D. Rockefeller managed a global corporate empire long before multinational companies had the experience, talent, and communications technology to do that easily. Margaret Thatcher's first speech to Parliament was packed with data, foreshadowing a command of policy details that would be one of her most powerful political weapons. Andrew Grove's most outstanding trait was managing every aspect of high-technology manufacturing, down to data on how quickly the janitors can clean the restrooms. Deng Xiaoping started out as an organizationally deft administrator at the end of the Long March, and he went on to bring invaluable management skills to the task of opening what is likely to become the largest country in the world.

None of our subjects accomplished what they did by acting alone; indeed, each surrounded himself or herself with considerable talent. Genghis Khan could preside over an empire in part because he could inspire a small band of soldiers to join him in the quest for treasure, and in part because he understood the need to recruit, capture, or strategically relocate the best engineers, doctors, and scholars from the lands he conquered. Prince Henry attracted to his castle in Sagres the best shipbuilding engineers, mapmakers, and experts on navigational instruments from all corners of the world. Cyrus Field brought together and coordinated dozens of preeminent technological and scientific experts, not to mention investors on two continents and two proud, assertive governments. You would expect John D. Rockefeller and Andrew Grove, both legendary businessmen, to understand the need for bringing in top people; but so did Deng Xiaoping, whose only real business experience took place in France as a lowly worker when he was eighteen, and who not only reversed the course of history with his ideas but proceeded to implement them, in large part, by refreshing China's bureaucratic talent pool with younger staffers and opening its doors to foreign expertise and thinking.

Finally, none of the ten was without character flaws, even streaks of evil. Think of Genghis Khan's brutality, Robert Clive's duplicity, Margaret Thatcher's callousness, John D. Rockefeller's and Andrew Grove's anything-goes competitiveness. Prince Henry not only kidnapped Africans to serve as his guides, he organized the African slave trade. Deng Xiaoping ordered Chinese troops to massacre their own citizens at Tiananmen. But it should not be surprising that in any group of highly ambitious men and women, most would have a dark side. As Lord Acton, the nineteenth-century British historian, is said to have stated, "Great men are almost always bad."

Echoes in the Daily News

Another big surprise that revealed itself to me in the writing was how regularly the lives of these historic figures echo in today's newspaper headlines. I heard the current president of China evoking not only the legacy and achievements of Deng Xiaoping but also, in almost poetic language, the age of Genghis Khan and his Silk Road. I saw modern explorers tipping their hat to Prince Henry, his swift caravels and his pioneering role in refining exploration into a science of discovery that remains a blueprint for travel to outer space and the ocean depths. I see no carbon copies of Robert Clive's East India Company but many distant echoes, particularly in the giant state-owned companies in Asia and the Persian Gulf. The trust built by the House of Rothschild is a striking counterpoint to the operations of global finance today, particularly after the complete collapse of trust between banks and their customers and regulators in the crisis of 2008–9. Cyrus Field may find a worthy successor as the world's master builder, given the dire needs for roads, ports, pipelines, transmission facilities, and especially new sources of energy and new environmental technologies to deal with climate change. John D. Rockefeller's greatest contribution, in my view, was the establishment of a philanthropy that was run with the efficiency of a business and had a major global focus, and the same basic strategy is visible in many modern philanthropies, led by the Bill and Melinda Gates Foundation. As Europe struggled in 2015 to find a way to bail out the debt-ridden Greeks in a manner acceptable to the austerity-minded Germans, it was hard not to think that Europe needed a latter-day Monnet and one of his simple plans to rally all sides and prevent the union from breaking up over this crisis. Margaret Thatcher's aggressive push for free markets continues to shape the mainstream consensus on how best to advance globalization, and to infuriate its critics and the many disaffected groups it has left behind. Andrew Grove's invention of new management systems for high-technology industries continues to drive the digital revolution that is reshaping modern industry and communications. And if you had to cite one leader behind the China we see in the news today—an ever more influential player in the global economy, ruled by a tightly controlled Communist cadre—it would be Deng Xiaoping.

Their Times and Ours

Though the ten leading characters in this book faced circumstances very different from the ones we face today, it strikes me that their challenges were at least as formidable as ours. The sense that the world has grown too wild and complicated for any one leader to control is no more frightening today than it was in the late twelfth and early thirteenth centuries, when much of the passage between China and the Middle East was a forbidden zone of brigands and bandits, yet Genghis Khan carved out a heavily guarded pony express route right through the chaos, and it lasted for generations. None of today's known unknowns are more formidable than Cape Bojador was in the fifteenth century, when Prince Henry finally broke its psychological hold on the world and opened the western coast of Africa—and much more—to Western explorers. In the nineteenth century, at a time when no one had successfully laid a cable across water to carry a telegraphic transmission, and at a time when no one had even mapped the floor of the Atlantic Ocean, Cyrus Field tried repeatedly to build the transatlantic telegraph, pushing not just the forefront of technology but securing the political backing of governments in support of this grandiose project despite his recurrent failures. The guns of World War II were barely silent when Jean Monnet successfully persuaded Germany and France to put past the blood rivalry and join in a cooperative government body. Soon after, Margaret Thatcher launched her campaign to roll back socialism, and won. Deng Xiaoping committed to prodding the world's most populous nation out of a seemingly hopeless epoch of poverty, insularity, and crippling ideology, and created a fast-rising middle-class giant. Each achieved results that would have seemed impossible when they began.

The accomplishments of the ten leaders here were made when the ground was shifting fundamentally under their feet, and they grasped opportunity amid periods of mind-boggling change. We, too, are living in such revolutionary times, as the challenges ahead amply illustrate.

A Leadership Deficit

While globalization started with the migration of Africans to other continents tens of thousands of years ago, and while it is certain to continue until or unless our planet experiences a massive catastrophe such as a nuclear war or an environmental collapse, it is also a sure thing that it will have its ups and downs, including interruptions and partial reversals. In fact, the tides of globalization may be receding right now. In the past ten years, for example, global economic growth has been much slower than the average of the previous four decades, and trade and capital flows across borders, measured against global GDP, show signs of decelerating, too. Trade and technology are posing serious threats to jobs and income distribution around the world. Immigration has become highly contentious. Terrorism, cyber threats, financial instability, and natural disasters have together made us all feel more vulnerable in a world with so many open channels that connect our societies. All of these factors are causing lots of citizens and their leaders to wonder whether there's been too much globalization, too fast, and leading many politicians to want to reclaim national control, reflected in the growth of stridently nationalist political parties in major countries such as France, the UK, Spain, and Germany.

Several longer-term challenges to globalization also promise to test governments, corporations, and nongovernmental organizations. With the opening of China and growing prosperity in so much of the developing world, for example, we have seen the rise of a global middle class of some two billion people, demanding all manner of goods and services and thereby creating huge opportunities for business but also straining production and delivery systems. Hyper-urbanization is under way around the world and will bring with it a new chance for poor people from the countryside but also a critical need for better water, food, health care, and public safety. Climate change, in particular, poses existential challenges for which we haven't yet found the answer. In the minds of many people, probably the silent global majority, the challenges of modern globalization have grown too large for any one government, much less any one person, to handle. There is widespread fear of a global leadership deficit.

Finally, we should not minimize the impact of changing geopolitics on globalization. In a nutshell, the world order in which globalization prospered in the last decades of the twentieth century is over. The United States no longer dominates as it did, and China, India, Brazil, Russia, and others are claiming more influence. It is no exaggeration to say that where the global economy is concerned a huge number of underpinnings are in flux: philosophy, ideals, rules, enforcement.

All this said, the trends of history lead to the conclusion that whatever setbacks occur in the next decade or even next generation, the long-term trend will be for globalization to regain momentum, and to widen and deepen. History does not move in a straight line, but the direction over time seems clear.

A World Too Complex for Transformational Leaders?

One of the questions that arose in my mind as I was writing this book was whether it would be possible for individuals to wield as much influence in the future as they once did, given all the complexities and all the crosscurrents of our era. A century from now, will we be able to point to distinct people as the architects of globalization, or will their contributions meld into the work of others? I would wager that we will see the emergence of many more courageous hedgehogs of the kind I have described.

The obvious counterargument is that I am simply underestimating how complex the world has become, how advances in globalization have made it ungovernable. The roadblocks to individual achievement on a grand scale could indeed be formidable. We have many more governments and levels of governance than ever before, all wanting to have their say. We have more laws and regulations, often conflicting with one another, and designed to protect various segments of the population who feel life is a zero-sum game. The financial costs of global projects could be massive, requiring multiple levels of participation from banks, governments, and wealthy individuals. More work is now being done collaboratively, requiring coordination among governments, businesses, and nongovernmental organizations. In light of all this, is it reasonable to expect a John D. Rockefeller, a Margaret Thatcher, or a Jean Monnet to emerge?

I genuinely think so. First of all, every age in history tends to look newly complex and scary from the narrow perspective of those living in it, but studying these ten lives compelled me to consider how the world must have looked to them. It is easy to forget now how isolated the Middle Kingdom was in the twelfth century, how impassable Africa appeared in the early fifteenth century, how ungovernable India appeared in the early nineteenth century, how intractably bitter French-German relations must have seemed after World War II, how hopeless China's poverty appeared in the 1970s, how permanent the communist world and the rise of the socialist welfare state looked in the 1980s—yet hedgehogs pushed ahead and changed all of that. There is no reason to believe that rising complexity has altered the basic process of forcing change: taking advantage of shifting circumstances, identifying a major problem, and attacking it at its weakest point with the clear strategic thinking and single-minded tenacity of a Prince Henry, a Cyrus Field, or an Andy Grove.

I began this book with an assertion that globalization has been advancing in waves since the dawn of mankind, with periods of advance and periods of retreat. Even the periods of progress have had their dark sides, as have the characters who shaped those epochs. Another striking storyline in the lives of the ten figures in this book was that it was often not until long after their deaths that history began to tell a balanced story, to weigh Genghis Khan the builder against Genghis Khan the destroyer, Prince Henry the explorer alongside the slave trader, Clive the reformer alongside the brigand, Rockefeller the philanthropist against the bare-knuckle monopolist. I have a similar sense of globalization today, when we tend to see the threats of global financial crises, terror, and overwhelming numbers of refugees in starker relief than all the advances of recent years.

Despite the tendency of the media to focus on bad news, the fundamental trends of recent decades have on the whole been highly positive. Due in part to improvements in medical technology, sanitation, vaccinations, and diet, the average citizen of the world could expect to live to seventy in 2010, up from just fifty-nine only forty years before. More children have enrolled in school in absolute numbers and as a percentage of world population than at any time in history. Civil and political rights exceed any other historical age. In 2015, in fact, the UN was able to report very favorably on efforts to achieve a set of "Millennium Development Goals" that it had set in 2000. Among the targets that were met or exceeded in this time span: the world reduced extreme poverty by half; major progress was made in fighting malaria and tuberculosis; access to improved drinking water was extended to 2.3 billion people; the political participation of women has been increasing steadily; hunger continues to decline; child mortality has been cut almost in half; 90 percent of children in developing countries are now attending primary schools. Despite the widespread perception that the world suffers from a leadership deficit, all this progress did not spring from nowhere. I suspect that in the future, historians will look back and find that there were many more heroic hedgehogs at work during this period than most of us understood.

There is also an easy counter to the argument that globalization has grown too big for any one leader to make a difference: globalization has given individuals powerful new avenues to make an impact. Thanks in great part to the foundations built by the ten people in this book, modern communications make it possible to instantly identify new threats, and pool the best available knowledge, talent, and experience in response. Because of global trade and transport networks, it's possible to move both people and material at a speed and scale previously impossible, whether to help out in a crisis or collaborate at a conference on curing cancer. Because of modern finance, never before have so many large pools of capital been available, ready to move across borders at lightning speed to fund big ideas. Because of the growth of global companies, never before has advanced research and development been conducted in so many locations. In short, riding the wave of global breakthroughs made by leaders like

Field and Rockefeller and Rothschild and Grove, the current and future hedgehogs have much more powerful tools to shape globalization than the past generations. This is a rapidly changing environment that can nurture new leaders at least as impressive as the ones I have written about. It just may be that the best is yet to come.

AUTHOR'S NOTE

Globalization has been woven into the fabric of my life from my very early days. Coming from a military family, half of my first fourteen years was spent in Germany and England. From high school through graduate school I was riveted by the subject of international relations and I wrote my PhD thesis on international economics, international law, and sovereign debt renegotiation. As an army officer in the late 1960s and early '70s, I was an adviser to the Royal Thai Army on the Burma border and in Vietnam. Later I held senior positions in the field of international economic policy in the Nixon, Ford, Carter, and Clinton administrations, focusing on trade, finance, energy, and economic development abroad. I spent twelve years on Wall Street doing big, complex deals in Asia, Latin America, and Africa. I have written extensively on global economics and politics, including four books and a monthly column for eight years in *BusinessWeek*. At Yale I have taught courses in global economics, finance, and business for over a decade.

I have also grown up with an intense interest in leadership. Perhaps this stems from living with a father who was a distinguished military leader in World War II, Korea, and Vietnam. Perhaps it's because I have been fascinated with different leadership styles in the arenas in which I have worked. What has impressed me the most is the marriage of big ideas and exquisite implementation. I have so often seen the first without the second and so admired men and women who could combine both.

Background to This Book

Several years ago, I was approached by a major publishing house—not HarperCollins—to write a book on the history of globalization. I was very excited by the prospect. But after a year of research, I went back to my editor and said that I didn't think I could write something that was differentiated enough from the many books on the topic. In reading as much as I had about the subject in preparation for a historical book, however, I found that one aspect of the literature was extremely interesting to me as well as being understudied: the life stories of the men and women who had so much to do with expanding globalization itself. It wasn't the book the publisher wanted, but it was the one that matched my interests and experiences and the one I wanted to write, and I was delighted that HarperCollins was glad to publish it. Thus, several years later, *From Silk to Silicon* emerged.

On Selecting Ten People to Write About

As I said in the introduction, I used several criteria to select the individuals I wrote about. They had to be as close to first movers as possible, recognizing that almost every high achiever builds on the work of others. They had to be doers and not just thinkers, people who rolled up their sleeves and implemented their ideas. They had to have accomplished something so transformative that the results not only changed their world but still reverberate today.

In addition, I focused on people who, on balance, made a positive contribution to the world. No Hitlers, no Osama bin Ladens.

Picking ten people is of course arbitrary and I am often asked who else I considered. In fact, if I had thought readers would have the endurance for more subjects, I would have added Johannes Gutenberg (approximately 1398–1468), who invented mechanical movable type printing that allowed the mass production of printed books, obviously with profound global consequences. A second person who would have deserved inclusion was Norman Borlaug (1914–2009), father of the "green revolution," who used modern agricultural technology to increase food production around the world, including Mexico, Pakistan, India, the Philippines, and throughout Africa. I could have added John Maynard Keynes (1883–1945), too, among history's great economic thinkers, to be sure, but also a chief architect and negotiator of the establishment of two pillars of the global economy—the International Monetary Fund and the World Bank.

I'm also frequently asked why I did not include Bill Gates and Steve Jobs. The answer is that my cutoff was the end of the twentieth century. Gates's lasting contribution, I believe, will be as a great global philanthropist, and that will be something he will achieve in this century. Jobs, too, hit his stride after the year 2000.

On Not Being a Historian

I have been conscious throughout this project that I am not a trained historian, and in delving so deeply into the past I have acquired enormous respect for the challenges of the craft. In *The Lessons of History*, a reflection written at the conclusion of their majestic multivolume series on the history of the world, authors Will and Ariel Durant write that "history plays havoc with our generalizations." Writing this book gave me a new appreciation for what they meant, since I have had to revise my own working assumptions repeatedly. It is also the case that history never repeats itself precisely and thus we should be skeptical of anyone trying to make tight analogies between periods, events, or individuals. Throughout my career, however, I have prided myself on being able to synthesize large amounts of information and to identify critical patterns and themes, and I will let the reader judge whether I have achieved that goal here. I've made no pretense of unearthing original documents or of devising novel theories in this book, but I have tried to create simple narratives that take account of many sources and many different perspectives. In the end, my goal was to tell a straightforward story of why the ten people I've written about really mattered in their time, and why they still do in ours.

ACKNOWLEDGMENTS

I have greatly benefited from the help of a number of people in writing this book.

I am deeply grateful to Tony Emerson, formerly editor of the international editions of *Newsweek*, who sharpened my thinking and writing over the last two years with many discussions and extensive editorial assistance. Tony helped me to condense the book by over half its original length by eliminating extensive history and shining a bright light on the characters themselves.

Special thanks to Abraham Parrish, who was head of Yale's map department until July 2015, and who very generously devoted his spare time to helping me. Many thanks also to Kai Alexis Smith for fact-checking and proofreading.

Over the years several students have contributed to my research, some extensively, some just for a few days. My thanks to: Katie Berk (Yale School of Management, 2012); Colleen Ferrand-Andrew (Yale SOM, 2012); Anna Grotberg (Yale SOM, 2013); Jessica Hanser (Yale PhD, 2012); Justin Kolbeck (Yale SOM, 2013); Robert Liu (Yale SOM, 2010); Serin Marshall ("significant other" of a Yale student); Socheata Poeuv (Yale SOM, 2012);

Aaron Schwirian (School of Advanced International Studies, Johns Hopkins University, 2007); Rakesh Shankar (Yale SOM, 2009); John Wang (Yale Law and Yale SOM, 2011); Lindsey White (Yale SOM, 2012); Jeffrey Woodward (Yale SOM, 2014); and Kimberly Yerino (Yale SOM, 2009).

In addition, a number of people have provided valuable comments on parts of the manuscript as it evolved: Nayan Chanda; Rita Chepulis; Kent Harrington; Reed Hundt; Yuxin Lin; Cait Murphy; David Rothkopf; Rafe Sagalyn; David Sanger; Julia Sweig; Steve Weisman; Walter Wells; and David Westin.

I have been blessed with wonderful assistants who tirelessly produced draft after draft of this book: Rhona Ceppos, Mary Ann Green, and Kelly Jessup. Kelly, in particular, was invaluable in the last stages of the manuscript that included constant rewrites, preparation of endnotes, and the all-important final editing. She was a critical confidante in judgment calls on a large number of the final edits.

And of course, many thanks to my agent, James Levine of Levine Greenberg Rostan, and to my editor at HarperCollins, Jonathan Jao, and to assistant editor Sofia Ergas Groopman.

NOTES

This book constitutes a synthesis of many, many books and articles. It is not possible to give all the credit that is due to the authors who have conducted so much original research and written comprehensively on the lives and times of my subjects, for I would have had to footnote almost every line. Instead, I have credited quotes and additional information where I felt it was particularly important, but not when I thought the information was generally known or when multiple authors described the same event in similar ways. At the beginning of the endnotes for each chapter I have identified the sources on which I relied most heavily. I have also included a full bibliography of all the sources I used and learned from, whether or not I cited them.

Introduction

xii It started about sixty thousand years ago: Nayan Chanda, *Bound Together: How Traders, Preachers, Adventurers, and Warriors Shaped Globalization* (New Haven, CT: Yale University Press, 2007), 5.

xiii had to be transformational leaders: James MacGregor Burns, *Leadership* (New York: Harper & Row, 1978), 455. The concept of transformational leadership is his and refers to political leaders, but I have interpreted it in the context of this book.

xv "The farther backward you can look": Stephen Mansfield, *Never Give In: The Extraordinary Character of Winston Churchill* (Nashville, TN: Cumberland House Publishing, 1995), 200, citing James C. Humes, *The Wit and Wisdom of Winston Churchill* (New York: HarperCollins, 1994), 44.

Chapter I: Genghis Khan

In writing this chapter I relied heavily on these sources: Jack Weatherford, *Genghis Khan and the Making of the Modern World* (New York: Crown, 2004); John Man, *Genghis Khan: Life, Death, and Resurrection* (New York: Bantam Press, 2004); Thomas T. Allsen, *Commodity and Exchange in the Mongol Empire: A Cultural History of Islamic Textiles* (New York: Cambridge University Press, 1997); Thomas T. Allsen, "Mongolian Princes and Their Merchant Partners, 1260," *Asia Major* Third Series 2, no. 2 (1989): 83–126; Thomas T. Allsen, "Mongols as Vectors for Cultural Transmission," in *The Cambridge History of Inner Asia: The Chinggisid Age*, eds. Nicola Di Cosmo, Frank J. Allen, and Peter B. Golden (New York: Cambridge University Press, 2009), 135–54; Irene M. Franck and David M. Brownstone, *The Silk Road: A History* (New York: Facts on File Publications, 1986); Janet Lippman Abu-Lughod, *The World System in the Thirteenth Century: Dead End or Precursor?* (Washington, DC: American Historical Association, 1993); and Janet Lippman Abu-Lughod, *Before European Hegemony: The World System A.D. 1250–1350* (New York: Oxford University Press, 1989).

10 "[You are] like an attacking panther": Weatherford, *Genghis Khan*, 24–25.
12 "attacking swarm of individuals": Ibid., 62.
12 delicacies such as boiled horse flesh: Michael Prawdin, *The Mongol Empire: Its Rise and Legacy* (New York: Free Press, 1967), 86.
12 "whatever Authority of Power he had given him": Weatherford, *Genghis Khan*, 66.
13 "If you wish me to be your ruler": Account of this exchange comes from Prawdin, *The Mongol Empire*, 85.
14 Genghis and his mounted army: The description of the geopolitical setting and the preparation of the Mongols for war draws heavily on Weatherford, *Genghis Khan*, 87–110.
19 Stories of devastation: Man, *Genghis Khan: Life, Death and Resurrection*, 136–42.
19 "I am not the author of this trouble": Ibid., 155.
21 Genghis reputedly put a platform over them: Ibid., 90.
21 not clear how many of these stories are true: Prawdin, *The Mongol Empire*, 48.
22 Genghis turned to foreign talent: Man, *Genghis Khan: Life, Death, and Resurrection*, 195–207.
26 paper money, primacy of state over church, and freedom of religion: Weatherford, *Genghis Khan*, 236.
26 In 1620, the English scientist Francis Bacon: Ibid.
26 "Empires played a key role": Nayan Chanda, "Runaway Globalization Without Governance," *Global Governance: A Review of Multilateralism and International Organizations* 14 (2008): 121.

28 Europeans were able to buy silk: Nayan Chanda, *Bound Together: How Traders, Preachers, Adventurers, and Warriors Shaped Globalization* (New Haven, CT: Yale University Press, 2007), 202–3.

28 In 1500, for example: Angus Maddison, *Contours of the World Economy 1–2030 AD: Essays in Macro-Economic History* (New York: Oxford University Press, 2007), 117. See also Joseph S. Nye, "Shaping the Future," in *What Matters: Ten Questions That Will Shape Our Future* (New York: McKinsey Publishing Group, 2009).

29 "I can almost hear the ring": Jane Perlez, "China Looks Westward as It Bolsters Ties," *New York Times*, September 8, 2013.

29 inject $62 billion: Gabriel Wildau, "China Backs Up Silk Road Ambitions with $62 Bn Capital Injection," *Financial Times*, April 20, 2015.

29 "One Belt, One Road": Jacob Stokes, "China's Road Rules: Beijing Looks West Toward Eurasian Integration," *Foreign Affairs*, April 19, 2015.

29 encompasses new rail freight routes: Stokes, "China's Road Rules"; Philip Stephens, "New China Starts to Make the Rules," *Financial Times*, May 28, 2015.

29 new international bank for infrastructure: Jonathan Wheatley, "Q&A: The Asia Infrastructure Investment Bank," *Financial Times*, March 12, 2015.

Chapter II: Prince Henry

I relied heavily on Peter Russell, *Prince Henry "the Navigator": A Life* (New Haven, CT: Yale University Press, 2001), by far the most comprehensive, thoughtful, and nuanced book on the prince; on Daniel J. Boorstin, *The Discoverers: A History of Man's Search to Know His World and Himself* (New York: Vintage Books, 1985); and on C. R. Boxer, *The Portuguese Seaborne Empire, 1415–1825* (New York: Alfred A. Knopf, 1969). I also used Ernle Bradford, *A Wind from the North: The Life of Henry the Navigator* (New York: Harcourt, Brace, 1960), especially for colorful detail at the beginning of the chapter and the portions dealing with the rounding of Cape Bojador, some of which Bradford almost certainly had to conjure up based on his own experiences as a sailor with direct experience in many of the waters that Prince Henry's ships traveled.

35 The ships were built: The description in this paragraph draws on Bradford, *Wind*, 29–30. I cannot be sure this picture is truly accurate in every detail, but I do believe it is broadly representative.

36 the armada included: Carlos B. Carreiro, *Portugal's Golden Years: The Life and Times of Prince Henry "The Navigator"* (Pittsburgh: Dorrance, 2005), 35; Russell, *Prince Henry*, 31.

40 "great and noble conquests": Felipe Fernández-Armesto, *Pathfinders: A Global History of Exploration* (New York: W. W. Norton, 2006), 131.

43 Bojador was cast as the point of no return: Boorstin, *The Discoverers*, 166.

43 Some historians say: Russell, *Prince Henry*, 113.
44 But one account gives us a picture: Bradford, *Wind*, 105–8.
44 It was a blistering summer day: Ibid., 106–8.
49 On the morning of August 8: Description of the beginning of slavery drawn from Anthony Pagden, *Peoples and Empires: A Short History of European Migration, Exploration, and Conquest, from Greece to the Present* (New York: Modern Library, 2003), 102.
49 The numbers grew rapidly: Russell, *Prince Henry*, 258.
49 Some of Henry's sympathetic biographers: Bradford, *Wind*, 169; Russell, *Prince Henry*, 249.
49 They hunted down men, women, and children: Russell, *Prince Henry*, 239–43.
49 "What heart could be so hard": Pagden, *Peoples and Empires*, 103.
52 retreat as a meeting place: Boorstin, *The Discoverers*, 162–64.
52 "To Sagres came sailors": Ibid., 162.
54 "The discovery of America": J. H. Elliott, *The Old World and the New 1492–1650* (New York: Cambridge University Press, 1992), 1, as quoted in Ronald Findlay and Kevin O'Rourke, *Power and Plenty: Trade, War, and the World Economy in the Second Millennium* (Princeton, NJ: Princeton University Press, 2007), 143.
55 "collaborative national adventure": Boorstin, *The Discoverers*, 157.
56 "what the space community is trying to accomplish": Wernher von Braun and Frederick I. Ordway III, *Space Travel: A History*, 4th ed. (New York: Harper and Row, 1975), 281, quoted in Stephen J. Pyne, *Voyager: Seeking Newer Worlds in the Third Great Age of Discovery* (New York: Viking Press, 2010), 9.
56 "lineal descendants": Carl Sagan, *Cosmos* (New York: Random House, 1980), quoted in Pyne, *Voyager*, 343.
56 and educated speculation: Chris Impy, *Beyond: Our Future in Space* (New York: W. W. Norton, 2015), 214–15.
56 "Our Manifest Destiny": "Our Manifest Destiny Is to Move Beyond Earth," *Financial Times*, December 23, 2014.

Chapter III: Robert Clive

I found Robert Harvey, *Clive: The Life and Death of a British Emperor* (London: Hodder & Stoughton, 1998) to be an exceptionally comprehensive narrative of Clive and his time and have relied heavily on it. I also drew substantially from Nirad C. Chaudhuri, *Clive of India: A Political and Psychological Essay* (London: Berrie & Jenkins, 1975); and Thomas Babington Macaulay, *Macaulay's Essay on Lord Clive*, ed. Vida D. Scudder (Boston: Leach, Shewell & Sanborn, 1889). The descriptions of battles in this chapter are usually summarized from much more elaborate and nuanced accounts by Harvey, *Clive: Life and*

Death; Mark Bence-Jones, *Clive of India* (New York: St. Martin's Press, 1974); and James P. Lawford, *Clive, Proconsul of India: A Biography* (London: Allen & Unwin, 1976). Among the many sources I used to describe the British Empire were Niall Ferguson, *Empire: The Rise and Demise of the British World Order and the Lessons for Global Power* (New York: Basic Books, 2002) and Lawrence James, *Raj: The Making and Unmaking of British India* (London: Little, Brown, 1997). For background on the East India Company I found Philip Lawson, *The East India Company: A History* (New York: Longman, 1993) very helpful, as well as Brian Gardner, *The East India Company: A History* (New York: McCall Publishing, 1971).

64 Warm air drifted off the land: Bence-Jones, *Clive of India*, 9.
64 They likely were carried by Indian boatmen: Norma Partington, *Master of Bengal* (New York: St. Martin's Press, 1974), 20.
66 born into a lower-middle-class family: The entire description of Clive's early days in Madras is summarized from accounts in Harvey, *Clive: Life and Death*; Bence-Jones, *Clive of India*; and Macaulay, *Essay on Lord Clive*.
66 "fierce and imperious": Lawford, *Clive, Proconsul of India*, 20.
67 "I have not enjoyed one happy day": Bence-Jones, *Clive of India*, 11.
67 Clive loaded a flintlock pistol: This account and the description of the work environment come from Bence-Jones, *Clive of India*, 15.
69 "bankers richer than the richest": Macaulay, *Essay on Lord Clive*, 14.
69 an incredible stretch of imagination: Ferguson, *Empire*, 29.
71 "Whosoever commands the trade of the world": Stephen R. Brown, *Merchant Kings: When Companies Ruled the World, 1600–1900* (New York: St. Martin's Press, 2009), 1.
71 between 1709 and 1748: Lawson, *East India Company*, 73.
72 The Company had established a gold-plated reputation: Bence-Jones, *Clive of India*, 5–6.
74 As quartermaster, Clive had to negotiate: Ibid., 37.
74 "This man, in the other parts of his life": Macaulay, *Essay on Lord Clive*, 61.
75 Clive's many big gambles: Lawson, *East India Company*, 98.
76 they had fled before Clive arrived: Harvey, *Clive: Life and Death*, 69.
77 the foundation from which Clive would ascend: Bence-Jones, *Clive of India*, 48; James, *Raj*, 25.
80 how this "grandest" of his undertakings: This and the following quotes from Clive's writings come from Gardner, *East India Company: A History*, 93–94.
81 "This was his Rubicon": Harvey, *Clive: Life and Death*, 209.
81 Clive reached Plassey in the dead of night: Bence-Jones, *Clive of India*, 139.
81 fate of India would be sealed: Macaulay, *Essay on Lord Clive*, 68.

83 "We had lost our glory": Bence-Jones, *Clive of India*, 169.
84 he boarded the *Royal George*: Ibid., 185.
84 It was now becoming clear: Details in this paragraph draw from Macaulay, *Essay on Lord Clive*, 95.
85 "See what an Augean Stable": Bence-Jones, *Clive of India*, 211.
87 "After securing a continent": Harvey, *Clive: Life and Death*, 6.
87 "one of the most extensive, difficult, and salutary reforms": Macaulay, *Essay on Lord Clive*, 98.
88 spinning out of control: James, *Raj*, 49.
88 the Company had become ever more controversial: Details in this paragraph drawn from Macaulay, *Essay on Lord Clive*, 116.
89 Consider the situation in which: Account by Clive quoted in Sir Colin Campbell, *Narrative of the Indian Revolt from Its Outbreak to the Capture of Lucknow* (London: G. Vickers, 1858), 27, quoted in Daniel Litvin, *Empires of Profit: Commerce, Conquest and Corporate Responsibility* (New York: Texere, 2003), 2.
89 PREMIS IN INDIS: William Dalrymple, "The East India Company: The Original Corporate Raiders," *Guardian*, March 4, 2015.
90 Oxford professor John Darwin writes: John Darwin, *The Empire Project: The Rise and Fall of the British World-System, 1830–1970* (New York: Cambridge University Press, 2009), 1–10.
90 Imperial India was more: Ibid., 10.
91 "[Arcot] was the first great triumph": Harvey, *Clive: Life and Death*, 80.
91 "It has to be noted that": Chaudhuri, *Clive of India*, 361.
91 would eventually build its own ships: Description of East India Company activities drawn from Antony Wild, *The East India Company: Trade and Conquest from 1600* (New York: HarperCollins Illustrated, 2000), 67.
91 control of over five hundred thousand square miles: H. V. Bowen, *The Business of Empire: The East India Company and Imperial Britain, 1756–1833* (New York: Cambridge University Press, 2006), 5.
91 Its army consisted of 250,000: Wild, *East India Company*, 168.
93 "an unstable sociopath": Dalrymple, "Corporate Raiders."
94 it was the British Empire: This is a central theme of Walter Russell Mead's *God and Gold: Britain, America and the Making of the Modern World* (New York: Alfred A. Knopf, 2007).

Chapter IV: Mayer Amschel Rothschild

I relied especially heavily on the works of Amos Elon, *Founder: A Portrait of the First Rothschild and His Time* (New York: Viking, 1996); Niall Ferguson, *The House of Rothschild*, vol. 1: *Money's Prophets, 1798–1848* (New York: Viking, 1998); and Niall Ferguson, *The House of Rothschild*, vol. 2: *The World's Banker, 1849–1999* (New York: Viking, 1999). The latter's two-volume history of the Rothschilds is a tour de force that contains a sensational amount of information and insight into the family and eighteenth- and nineteenth-century finance.

98 The ghetto contained not one tree: The description of the ghetto comes in large part from Elon, *Founder*, 33–53, and Ferguson, *Money's Prophets*, 36–40.
98 An anonymous traveler: Elon, *Founder*, 40.
99 fat sow holding up its tail: Ibid., 33.
100 Germany as a nation: The description of historic Germany is drawn from Elon, *Founder*, 21–24.
101 rulers often hired Jews to advise them: This section on the Court Jew draws heavily from Michael Graetz, "Court Jews in Economics and Politics," in *From Court Jews to the Rothschilds: Art, Patronage, and Power: 1600–1800*, ed. Vivian B. Mann and Richard I. Cohen (New York: Prestel Publishing, 1996), 29–33, 36.
102 "Lengthy terms of imprisonment": Ibid., 34.
102 In the broader region: David S. Landes, *Bankers and Pashas: International Finance and Economic Imperialism in Egypt* (Cambridge, MA: Harvard University Press, 1958), 17.
107 a nine-foot-square room: Egon Caesar Corti, *The Rise of the House of Rothschild*, trans. Brian Lunn and Beatrix Lunn (New York: Grosset & Dunlap, 1928), 18.
109 Rothschild purchased all of the available Danish bonds: Fritz Backhaus, "The Last of the Court Jews—Mayer Amschel Rothschild and His Sons," in *From Court Jews to the Rothschilds: Art, Patronage, Power, 1600–1800*, ed. Vivian B. Mann and Richard I. Cohen (New York: Prestel Publishing, 1996), 90.
114 It would provide the template: Ferguson, *Money's Prophets*, 85–89; Elon, *Founder*, 166–70.
114 idea of an integrated international market: Ferguson, *Money's Prophets*, 4.
115 a clearinghouse for international finance: Corti, *Rise*, 147.
117 "The significance of this system": Ferguson, *Money's Prophets*, 4.
117 "this growing international bond market brought together Europe's true 'capitalists' ": Ibid, 4–5.
118 constantly upgrading its internal courier service: The description of the development of the courier service is drawn from Ferguson, *Money's Prophets*, 233.
119 "Keep your brothers together": Elon, *Founder*, 174.
120 the London operation rescued the brothers: Ferguson, *Money's Prophets*, 465.
120 an increasingly dominant force: Ibid., 162–63.
120 three years after Mayer died: Ibid., 162.

120 By the 1820s, the Rothschild sons were expanding: The history of the Rothschilds after the death of Mayer Amschel Rothschild is drawn heavily from Ferguson, *World's Banker.*
120 loosen the repayment terms on Austria's debts to Britain: Corti, *Rise*, 300.
121 Rothschild mythology took on such dimensions: Ibid., 323–24.
121 His funeral was attended by: Ferguson, *Money's Prophets*, 299.
121 "more like that of an emperor": Ferguson, *World's Banker*, 155.
122 "The emergence of global capitalism": Fritz Stern, *Gold and Iron: Bismarck, Bleichröder, and the Building of the German Empire* (New York: Vintage Books, 1979), xv.

Chapter V: Cyrus Field

I have drawn heavily on Samuel Carter III, *Cyrus Field: Man of Two Worlds* (New York: G. P. Putnam's Sons, 1968), a complete biography of Cyrus Field and one of the few such accounts in print, Henry M. Field, *The Story of the Atlantic Telegraph* (New York: Charles Scribner's Sons, 1892), and on historians John Steele Gordon, *A Thread Across the Ocean: The Historical Story of the Transatlantic Cable* (New York: Walker, 2002); Chester G. Hearn, *Circuits in the Sea: The Men, the Ships, and the Atlantic Cable* (Westport, CT: Praeger, 2004); and Tom Standage, *The Victorian Internet: The Remarkable Story of the Telegraph and the Nineteenth Century's On-Line Pioneers* (New York: Walker, 1998).

132 "manned expedition to Mars": Gordon, *Thread*, 12.
133 commitments that totaled $1.5 million: Carter, *Man of Two Worlds*, 103.
133 about $40 million in today's dollars: Calculations assume average annual inflation rate of 2.04 percent. Data from Federal Reserve Bank of Minneapolis, "Consumer Price Index (Estimate) 1800– ," accessed April 15, 2015, https://www.minneapolisfed.org/community/teaching-aids/cpi-calculator-information/consumer-price-index-1800.
134 Cyrus Field was born: This section is drawn heavily from Carter, *Man of Two Worlds*; Isabella Field Judson, ed., *Cyrus W. Field: His Life and Work* (New York: Harper & Brothers, 1896); and Field, *Story*.
134 "To *know* is a thing which pleaseth": Carter, *Man of Two Worlds*, 30.
138 one working telegraph line in 1846: The information in this paragraph is drawn from Standage, *Victorian Internet*, 58–69.
138 "No one who knew anything about telegraphy": Ibid., 75.
140 breeder of large show dogs: Gordon, *Thread*, 68.
140 Opinions differed on whether the splice: Carter, *Man of Two Worlds*, 119–20.
141 "grandest work which has ever been attempted": Gordon, *Thread*, 75.
141 risk with Field's extreme haste: Ibid.
142 "I cannot bind myself to more": Carter, *Man of Two Worlds*, 138.

142 "There was a strange unnatural silence": Ibid., 139.
143 "Do not think I am discouraged": Field, *Story*, 139.
144 "The masts were rapidly getting worse": Gordon, *Thread*, 116.
144 "Of all the many mishaps": Field, *Story*, 159.
144 "I think there is nothing to be done": Gordon, *Thread*, 121–22.
145 "that the Company was possessed": Field, *Story*, 165–66.
146 "Arrived here yesterday": Gordon, *Thread*, 131.
146 "Since the discovery of Columbus" and "The Atlantic is dried up": Ibid., 134.
146 "All Wall Street stirred up": Ibid., 133.
146 "I most heartedly rejoice with you": Hearn, *Circuits*, 121.
147 mutiny against British rule: Carter, *Man of Two Worlds*, 172; Gillian Cookson, *The Cable: The Wire That Changed the World* (Stroud, Gloucestershire: Tempus Publishing, 2003), 102.
147 open the Middle Kingdom to foreign trade: Hearn, *Circuits*, 123.
147 "were used to having the Atlantic Ocean": Gordon quoted in David Axelrod, "The Great Transatlantic Cable," Transcript of PBS special, *American Experience*, November 30, 2004.
149 "in the first instance": Hearn, *Circuits*, 155.
150 "cross-examination" and "was in a witness box": Carter, *Man of Two Worlds*, 211.
150 gargantuan ship called the SS *Great Eastern*: Hearn, *Circuits*, 175–77; Gordon, *Thread*, 146, 155.
150 he was ready to load the *Great Eastern*: Hearn, *Circuits*, 191.
151 "It is all over": Field, *Story*, 284.
151 "Well, it's so": Carter, *Man of Two Worlds*, 236.
152 The crew was riveted: Field, *Story*, 329–31.
152 "Heart's Content, July 27": Hearn, *Circuits*, 229.
152 "On this day": Ibid., 230.
153 "I went to my cabin": Carter, *Man of Two Worlds*, 256.
153 It has been a long hard struggle: Cyrus Field quoted in Carter, *Man of Two Worlds*, 258.
153 In 1858, it had taken sixteen hours: Hearn, *Circuits*, 229–31.
154 "The Atlantic Cable will tend to equalize prices": Kenneth D. Garbade and William L. Silber, "Technology, Communication and the Performance of Financial Markets: 1840–1975," *Journal of Finance* 33, no. 3 (June 1978): 827.
154 Soon every major nation: Gordon, *Thread*, 212.
155 "If any of the foremost men of our time": Hearn, *Circuits*, 241.
156 "In 1830 a message from London to New York or Bombay": Debora L. Spar, *Ruling the Waves: Cycles of Discovery, Chaos, and Wealth from the Compass to the Internet* (New York: Harcourt, 2001), 60.
156 it took between fourteen days and four months: Gordon, *Thread*, 2.

157 twenty-three thousand miles of telegraph wires: Ibid., 61.
159 "Victorian Internet": Standage, *Victorian Internet*, title.
159 nearly 95 percent of communications traffic: Robert Martinage, "Under the Sea," *Foreign Affairs* 94, no. 1 (January/February 2015): 117; Stuart Taylor, "The Next Generation of the Internet: Revolutionizing the Way We Work, Live, Play, and Learn," Cisco Systems., Inc., April 2013, 2, http://www.cisco.com/web/about/ac79/docs/sp/Next-Generation-of-the-Internet.pdf.
159 a consortium of six global companies: "A Global Consortium to Build New Trans-Pacific Cable System 'FASTER,'" NEC Corporation, news release, August 11, 2014, http://www.nec.com/en/press/201408/global_20140811_01.html.

Chapter VI: John D. Rockefeller

I relied heavily on five sources: Ron Chernow, *Titan: The Life of John D. Rockefeller, Sr.* (New York: Random House, 1998); Allan Nevins, *John D. Rockefeller: The Heroic Age of American Enterprise*, vol. 1 (New York: Scribner's, 1940); Allan Nevins, *John D. Rockefeller: The Heroic Age of American Enterprise*, vol. 2 (New York: Scribner's, 1940); Daniel Yergin, *The Prize: The Epic Quest for Oil, Money and Power* (New York: Free Press, 2008); and John D. Rockefeller, *Random Reminiscences of Men and Events* (New York: Doubleday, Page, 1909; repr., Creative English Publishing, 2013).

166 Rockefeller learned to handle: Rockefeller, *Random Reminiscences*, 30.
166 "fallible emotion" and "a solid empirical reality": Chernow, *Titan*, 46.
170 "The drillers started shouting": Nevins, *Heroic* Age, 1:165.
171 prices of crude: Chernow, *Titan*, 85.
172 "I ever point to the day": Nevins, *Heroic* Age, 1:191.
172 "The only time I saw John Rockefeller enthusiastic": Chernow, *Titan*, 79.
172 producing 103,691 gallons of refined products: Nevins, *Heroic* Age, 1:177–78.
173 In 1860, for instance: Ibid., 1:194.
174 Businessmen enjoyed the benefit: Ibid., 1:193–94.
174 "Institutions were still sufficiently plastic": Ibid., 1:7.
174 "The United States more than any other": Ibid., 1:604.
174 2,000 barrels of oil: Ibid., 1:277.
174 "A commodity that had been a curiosity": Ibid., 1:207.
175 "The proliferation of rebates": Chernow, *Titan*, 115.
177 "You are responsible": Nevins, *Heroic Age*, 1:626.
177 "[He] was not a dictator": Ibid., 1:512.
178 "The day of individual competition": Rockefeller, *Reminiscences*, 47–48.
179 Rockefeller's oil empire encompassed: Chernow, *Titan*, 249.

180 "Shorthand for the trust itself": Ibid., 222.
180 America's most valuable non-farm export: Yergin, *The Prize*, 41.
180 Standard Oil received extensive assistance: Ibid., 40, 45; Chernow, *Titan*, 246.
180 its own importing and distribution companies: Chernow, *Titan*, 246.
181 exporting 50,000 barrels of oil: Yergin, *The Prize*, 41.
181 In England, for example: Nevins, *Heroic* Age, 1:663–64.
182 to cut foreign prices to well below: Chernow, *Titan*, 259.
183 "one of the most famous polemics": Nevins, *Heroic* Age, 2:331.
183 investigative reporter Ida Tarbell: Chernow, *Titan*, 438–39.
184 "It is your own fault, Mr. Rockefeller": This exchange is taken from *John D. Rockefeller Interview, 1917–1920*, conducted by William O. Inglis (Westport, CT: Meckler Publishing, in association with the Rockefeller Center Archive, 1984), 13.
185 "No disinterested person can survey": Chief Justice Edward White quoted in Yergin, *The Prize*, 93.
185 "Father Lennon, have you some money": Elizabeth Deane and Adriana Bosch, "The Rockefellers," *American Experience*, PBS, PDF transcript, http://www.pbs.org/wgbh/americanexperience/features/transcript/rockefellers-transcript/.
185 a vast interconnected empire: Yergin, *The Prize*, 93–94.
186 No one came out better than John Rockefeller: Chernow, *Titan*, 556–57.
186 In 2014, *Forbes* estimated: Sam Roberts, "Why Are Rockefellers Moving from 30 Rock?" *New York Times*, November 23, 2014.
187 "the daily vocation of life is one thing": Rockefeller, *Reminiscences*, 101.
187 "we can well afford to ask the ablest": Ibid., 99.
188 the world's first global foundation: Judith Rodin, preface to Eric Abrahamson, Sam Hurst, and Barbara Shubinski, *Democracy and Philanthropy: The Rockefeller Foundation and the American Experiment* (New York: Rockefeller Foundation, 2013), 14–17.
190 international grant making by American foundations: Anne Petersen and Gail D. McClure, "Trends in Global Philanthropy Among U.S. Foundations: A Brief Review of Data and Issues," *Foundation Review* 2, no. 4 (2010), http://scholarworks.gvsu.edu/tfr/vol2/iss4/8/.
190 "An individual institution of learning": Rockefeller, *Reminiscences*, 116.
191 "the art of laying out roads": Rockefeller, *Reminiscences*, 21.
191 "the pioneer of efficient business organization": "Financier's Fortune in Oil Amassed in Industrial Era of 'Rugged Individualism,'" *New York Times*, May 24, 1937.

Chapter VII: Jean Monnet

I relied heavily on the following sources: Douglas Brinkley and Clifford Hackett, eds., *Jean Monnet: The Path to European Unity* (New York: St. Martin's

Press, 1991); François Duchêne, *Jean Monnet: The First Statesman of Interdependence* (New York: Norton, 1994); Richard Mayne, *The Community of Europe* (New York: Norton, 1963); and Jean Monnet, *Memoirs*, trans. Richard Mayne (Garden City, NY: Doubleday, 1978).

200 he wanted to speak to the prime minister: Monnet, *Memoirs*, 48–49.
200 "It was not some conceited need": Ibid., 50.
201 "The European Union is an unparalleled historical experiment": Kathleen R. McNamara, "A Less Perfect Union," *Foreign Affairs*, July 19, 2015, https://www.foreignaffairs.com/articles/western-europe/2015-07-19/less-perfect-union.
201 "Don't take any books": Monnet, *Memoirs*, 44.
202 "In China you have to know": Jean Monnet quoted in Duchêne, *First Statesman*, 30–31.
202 listening to tales of adventure: Duchêne, *First Statesman*, 28–29.
202 "One did one thing": Monnet, *Memoirs*, 38.
202 "Unlike my father": Ibid., 37.
203 had to act as one entity: Duchêne, *First Statesman*, 32; Monnet, *Memoirs*, 53–54.
203 This buyers' cooperative: Duchêne, *First Statesman*, 36.
204 British diplomatic historian Harold Nicolson: Ibid., 37.
204 "He realized over the Allied arguments": James Reston, "Monnet's Noble Life," *New York Times*, March 18, 1979.
205 league guaranteed that the self-interest of member states: Monnet, *Memoirs*, 95.
205 "Cooperation between nations": Ibid., 83.
205 parable of a man who scales a mountain: John Brooks, "The Common Market—I," *New Yorker*, September 22, 1962.
207 meet with President Franklin D. Roosevelt: Duchêne, *First Statesman*, 64, 67; Monnet, *Memoirs*, 117.
208 ceaseless discussions with Roosevelt's inner circle: Monnet, *Memoirs*, 155.
208 Monnet's efforts shortened the war: Duchêne, *First Statesman*, 93.
209 "Rather than make a frontal attack": Monnet, *Memoirs*, 219.
209 national income, adjusted for inflation: Ibid., 233–34.
210 "You speak of greatness": Duchêne, *First Statesman*, 145.
210 He understood that while political leaders: Monnet, *Memoirs*, 230.
212 "Everything I have seen": Ibid., 272–73.
214 Acheson told French foreign minister Robert Schuman: Duchêne, *First Statesman*, 190.
215 "Monnet thought most effectively": Ibid., 13.
215 "electrically charged with priorities": Ibid., 348.
215 His genius was in honing big ideas: Multiple sources describe his work style in nearly identical terms. See, for example, François Duchêne, "Jean

Monnet's Methods," in *Jean Monnet: The Path to European Unity*, ed. Douglas Brinkley and Clifford Hackett (New York: St. Martin's Press, 1991).

216 "Experience had taught me": Monnet, *Memoirs*, 286.

218 "no longer the moment for vain words": Mayne, *Community of Europe*, 85.

218 "a first step in the federation of Europe": Ibid., 88.

218 "In other words": Monnet, *Memoirs*, 305.

218 "By the signature of this Treaty": Strobe Talbott, "Monnet's Brandy & Europe's Fate," Brookings Essay (Washington, DC: Brookings Institution Press, 2014), last modified February 11, 2014, http://www.brookings.edu/research/essays/2014/monnets-brandy-and-europes-fate.

218 "Never before have states undertaken": Monnet, *Memoirs*, 318.

219 "we are here to build a European Community": Ibid., 328.

220 "All these institutions can be modified": Jean Monnet quoted in Duchêne, *First Statesman*, 235.

220 "Each man begins the world afresh": Jean Monnet quoted in ibid., 401.

220 He wanted to apply this concept: Talbott, "Monnet's Brandy."

220 "If we succeeded in proving": Duchêne, *First Statesman*, 386.

221 Monnet's strength was his vision: Description of managerial style from John Gillingham, *Coal, Steel, and the Rebirth of Europe 1945–1955: The Germans and French from Ruhr Conflict to Economic Community* (New York: Cambridge University Press, 1991), 314–15.

221 Europe's first experiment in supranationalism: Ibid., 363.

223 twelve of the twenty-eight European Union members: European Central Bank, "Initial Changeover (2002)," accessed August 27, 2015, http://www.ecb.europa.eu/euro/changeover/2002/html/index.en.html.

224 "Men in power are short of ideas": Monnet, *Memoirs*, 84.

224 when politicians are desperate for ideas: George Ball, introduction to Monnet, *Memoirs*, 13.

224 "In crises, most people don't know what to do": Duchêne, *First Statesman*, 347.

225 yet he had a great gift of persuasion: Ibid., 353.

225 "the crushing simplicity and elemental clarity": John Brooks, "The Common Market—II," *New Yorker*, September 29, 1962.

225 specific to time and place: Duchêne, *First Statesman*, 23.

225 "There are two kinds of people": Monnet, *Memoirs*, 519.

225 He was a master at building: Duchêne, "Jean Monnet's Methods," 206.

226 "circumspect manner of inspecting": Duchêne, *First Statesman*, 27.

226 "I have never met a man before or since": Ibid., 21.

228 the EU had by 2014 grown: European Union, "Countries," accessed May 5, 2015, http://europa.eu/about-eu/countries/index_en.htm.

229 an analysis by Deutsche Bank: Stefan Vetter, *The Single European Market 20 Years On: Achievements, Unfulfilled Expectations & Further*

Potential (Frankfurt: Deutsche Bank Research, October 31, 2013), www.dbresearch.com.

229 one of the more credible analyses: Ibid., citing Andrea Boltho and Barry Eichengreen, "The Economic Impact of European Integration," Centre for Economic Policy Research Discussion Paper No. 6820, May 2008.

229 intra-European trade increased from 12 to 22 percent of GDP: Vetter, *The Single European Market*.

229 One analysis showed: Ibid.

229 Ten years after joining: Daniel Gros, "Europe's Big Bang at Ten," Project Syndicate, last modified May 8, 2014, http://www.project-syndicate.org/commentary/daniel-gros-assesses-the-surprises--both-good-and-bad--in-the-decade-since-the-eu-admitted-ten-new-members.

230 engineered one of the greatest advances: Talbott, "Monnet's Brandy."

230 In the dynamic region of Southeast Asia: Aladdin Rillo, Pitchaya Sirivunnabood, Ahmad Ahsan, and Jean-Christophe Maur, *Association of Southeast Asian Nations (ASEAN) Integration Monitoring Report: A Joint Report by the ASEAN Secretariat and the World Bank* (Washington, DC: World Bank Group, 2013), http://documents.worldbank.org/curated/en/2013/01/18780456/association-southeast-asian-nations-asean-integration-monitoring-report-joint-report-asean-secretariat-world-bank.

230 the African Union has launched negotiations: Andrew England, "Africa Sets Deadline for Free Trade Area as Regional Blocs Come Together," *Financial Times*, June 15, 2015.

230 a deeply intertwined North American market: David H. Petraeus and Robert B. Zoellick, chairs, *North America: Time for a New Focus*, Independent Task Force Report No. 71 (New York: Council on Foreign Relations, 2014).

231 oversight of the resource-rich Arctic Circle: Jeffrey E. Garten, "We Need a Global Authority for Arctic Oil and Gas," *Financial Times*, August 15, 2008.

231 For centuries, emperors, kings and dictators: Monnet, *Memoirs*, 472.

231 "In fact, he was something far rarer": Talbott, "Monnet's Brandy."

232 the European electorate has been deeply disillusioned: Timothy Garton Ash, "Let the Next Generation Speak Up for Europe," *Guardian*, last modified December 8, 2014, http://www.theguardian.com/commentisfree/2014/dec/07/europe-brussels-european-eu.

232 It took centuries before the peoples and nations of Europe found their way: Angela Merkel, "The 2014 Lowy Lecture" (Lecture, Lowy Institute for International Policy, Sydney, Australia, November 17, 2014).

233 a report signed by the president of the European Commission: Jean-Claude Juncker, with Donald Tusk, Jeroen Dijsselbloem, Mario Draghi, and Martin Schulz, *Completing Europe's Economic and Monetary Union* (Brussels: European Commission, June 2015).

233 "Europe's Economic and Monetary Union (EMU) today": Ibid., 4.
234 "that Europe would be established through crisis": Jean Monnet quoted in Talbott, "Monnet's Brandy."
234 "The sovereign nations of the past": Monnet, *Memoirs*, 524.

Chapter VIII: Margaret Thatcher

I have relied heavily on the following sources: the two-volume biography of Margaret Thatcher by John Campbell, *Margaret Thatcher*, vol. 1: *The Grocer's Daughter* (London: Pimlico, 2001) and *Margaret Thatcher*, vol. 2: *The Iron Lady* (London: Pimlico, 2003); Mrs. Thatcher's two-volume memoirs, *The Downing Street Years* (New York: HarperCollins, 1993) and *The Path to Power* (New York: HarperCollins, 1995); and on Charles Moore, *Margaret Thatcher: The Authorized Biography, from Grantham to the Falklands* (New York: Alfred A. Knopf, 2013). I have also drawn significantly on Daniel Yergin and Joseph Stanislaw, *The Commanding Heights: The Battle for the World Economy* (New York: Simon & Schuster, 2002); and Hugo Young, *The Iron Lady: A Biography of Margaret Thatcher* (New York: Farrar, Straus & Giroux, 1989).

237 Margaret was helping her father: Young, *Iron Lady*, 8.
238 her father's corner store in Grantham: Campbell, *Grocer's Daughter*, 1.
238 "[My father] told me": Thatcher, *Path to Power*, 6.
239 "There were certain things you just didn't do": Young, *Iron Lady*, 6.
240 At school she read voraciously: Many of the books she read are described in Thatcher, *Path to Power*, 50–57.
241 dressing up as a barmaid: Campbell, *Grocer's Daughter*, 76.
241 The themes in the stump speeches: Ibid., 83.
241 We are going into one of the biggest battles: Thatcher, *Path to Power*, 73.
242 "Marriages are made in heaven": Richard Vinen, *Thatcher's Britain: The Politics and Social Upheaval of the Thatcher Era* (New York: Simon & Schuster, 2009), 23.
243 "The force which transformed British politics": Campbell, *Grocer's Daughter*, 187.
243 Heath seemed more comfortable managing the system: Yergin and Stanislaw, *Commanding Heights*, 78.
243 She deeply resented Heath: Young, *Iron Lady*, 75; Thatcher, *Path to Power*, 161.
244 "We [in Britain] are now more socialist": Young, *Iron Lady*, 84.
244 the first overt sign: Ibid., 86.
244 driving subordinates crazy: Descriptions in this paragraph are drawn from Campbell, *Grocer's Daughter*, 215–17.
245 "[Thatcher] was to British education": Ibid., 231.
245 the most unpopular woman in Britain: Ibid., 232.

245 "Iron entered my soul": Young, *Iron Lady*, 73.
246 "You must be out of your mind": Thatcher, *Path to Power*, 266.
247 "I saw how they destroyed Keith [Joseph]": Ibid., 269.
247 "too many Conservatives have become socialists": Campbell, *Grocer's Daughter*, 295.
247 What kind of leadership Mrs. Thatcher will provide: Ibid., 311.
248 "This is what we [Conservatives] believe": Yergin and Stanislaw, *Commanding Heights*, 89.
248 forceful on the world stage: Campbell, *Grocer's Daughter*, 338.
248 "the Iron Lady": Young, *Iron Lady*, ix.
248 "never did me a greater favor": Thatcher, *Path to Power*, 362.
249 "It goes wherever I go": Campbell, *Grocer's Daughter*, 349.
249 choreographing her public appearances: Ibid., 408–10.
250 By late 1978: Yergin and Stanislaw, *Commanding Heights*, 90; Thatcher, *Downing Street*, 93.
250 "there are times, perhaps every thirty years": Earl A. Reitan, *The Thatcher Revolution: Margaret Thatcher, John Major, Tony Blair, and the Transformation of Modern Britain, 1979–2001* (Lanham, MD: Rowman & Littlefield, 2003), 241.
251 "Having previously exaggerated our power": Thatcher, *Downing Street*, 8.
251 welfare expenditures were growing: Dennis Kavanagh, *Thatcherism and British Politics: The End of Consensus?* (New York: Oxford University Press, 1987), 304.
251 In these nations: Ajay Chhibber, "We Need Our States to Be Smarter, Not Bigger," *Financial Times*, August 25, 2010.
252 "Don't tell me *what* to do": Campbell, *Grocer's Daughter*, 366.
252 "Chatham famously remarked": Thatcher, *Downing Street*, 10.
253 "the process of abandoning all beliefs": Moore, *Grantham*, 646.
254 Public approval for the Conservative Party: Yergin and Stanislaw, *Commanding Heights*, 92.
254 "To those waiting with bated breath": Young, *Iron Lady*, 209.
254 "deepen the depression": Nigel Lawson, *The View from No. 11: Memoirs of a Tory Radical* (New York: Bantam Press, 1992), 97.
254 "the guts to see it through": Campbell, *Margaret Thatcher: The Iron Lady*, 87.
255 "She has reasserted her political dominance": Thatcher, *Downing Street*, 152.
255 unemployment stuck at 3 million: Details on unemployment, economic growth and taxes are drawn from Young, *Iron Lady*, 316–17.
255 "The humiliation of Britain was sudden": Moore, *Grantham*, 656.
256 By the end of the ordeal: Ibid., 749; Campbell, *Margaret Thatcher: The Iron Lady*, 157.
256 "It may well have been the happiest moment": Moore, *Grantham*, 758.

257 "No transformation in modern British history": Ibid., 755.

257 "I had a very tough time the first three years": Yergin and Stanislaw, *Commanding Heights*, 94.

257 Whereas between 1980 and 1981: Moore, *Grantham*, 631.

258 "the moment when the hundred-year-old political argument": Campbell, *Margaret Thatcher: The Iron Lady*, 205.

258 "striking and internationally influential": Peter Riddell, *The Thatcher Era and Its Legacy* (Cambridge, MA: Blackwell, 1991), 206.

258 "the most decisive element of Thatcherism": Yergin and Stanislaw, *Commanding Heights*, 96.

259 "crucial to the survival of freedom and democracy": These quotes come from Lawson, *View from No. 11*, 206.

259 "no industry should remain under State ownership": Ibid., 211.

259 sold two out of every five companies: Riddell, *The Thatcher Era*, 87.

259 adult population owning stock shares: Lawson, *View from No. 11*, 207.

260 "In advance of every significant privatization": Ibid., 201.

261 In her memoirs: Thatcher, *Downing Street*, 371.

262 "At the other end are the hard left": Young, *Iron Lady*, 373.

262 The outcome shattered the labor movement: Riddell, *The Thatcher Era*, 227.

262 "What the strike's defeat established": Thatcher, *Downing Street*, 378.

263 financial services accounting for 10 percent of GDP: Henry Laurence, *Money Rules: The New Politics of Finance in Britain and Japan* (Ithaca, NY: Cornell University Press, 2001), 69.

264 Her approval ratings sank: Peter Jenkins, "Mrs. Thatcher's Last Stand?" review of *The Iron Lady: A Biography of Margaret Thatcher*, by Hugo Young, and of *The Thatcher Decade*, by Peter Riddell, *New York Review of Books*, April 12, 1990.

265 "All Britons remember where they were": Daniel Hannan, "The Woman Who Saved Britain," *Wall Street Journal*, May 17, 2013.

265 from 23 percent to 20 percent: Intelligence Unit, *Economist*, www.eiu.com/.

265 public debt declined: Ibid.

265 a dramatic decrease in inflation: Ibid.

266 Between 1955 and 1979: John Cassidy, "The Economic Case For and Against Thatcherism," *New Yorker*, April 10, 2013.

266 Tax rates on income were dramatically simplified: Paul Johnson, "The World-Changing Margaret Thatcher," *Wall Street Journal*, April 8, 2013.

266 all shock and no therapy: Cassidy, "The Economic Case."

266 too many children and people: "Obituary: Margaret Thatcher," BBC News, April 8, 2013, http://www.bbc.com/news/uk-politics-10364876.

266 "most popular and successful way of running a country": Johnson, "The World-Changing Margaret Thatcher."

267 "cult of greed": Alan Cowell, "Britons Reflect on Divided Views of Thatcher Legacy," *New York Times*, April 9, 2013.

267 Whatever the verdict: Janan Ganesh, "The Iron Lady Towers over Modern Britain," *Financial Times*, April 8, 2013.

267 "Many think she saved Britain": Moore, *Grantham*, xviii.

268 One study of forty-three major developing nations: Chhibber, "Smarter, Not Bigger."

268 To describe it another way: Campbell, *Margaret Thatcher: The Iron Lady*, 625.

268 "Whether they managed well or badly": "The Lady Who Changed the World," *Economist*, April 8, 2013.

268 "Washington Consensus": Larry Neal and Jeffrey G. Williamson, "The Future of Capitalism," in *The Cambridge History of Capitalism*, ed. Larry Neal and Jeffrey G. Williamson, vol. 2 (New York: Cambridge University Press, 2014), 534.

270 a fundamentally new long-term direction: Fareed Zakaria, "It's the Age of Bloomberg, Not Thatcher," *Newsweek*, October 13, 2008, http://www.newsweek.com/zakaria-its-age-bloomberg-not-thatcher-91827.

271 a major debate is now taking place: John Micklethwait and Adrian Wooldridge, "The Age of the Unthinkable," *Wall Street Journal*, September 10, 2014.

Chapter IX: Andrew Grove

I am deeply indebted to Professor Richard S. Tedlow, whose exceptionally comprehensive biography, *Andy Grove: The Life and Times of an American* (New York: Portfolio, 2006), was the overarching framework for my story, not to mention the source for so many of the ideas I wrote about and the facts I used.

I also drew extensively from Tim Jackson, *Inside Intel: Andy Grove and the Rise of the World's Most Powerful Chip Company* (New York: Dutton, 1997), and, of course, on Andy Grove's own writings, particularly *Swimming Across: A Memoir* (New York: Warner Books, 2001); *High Output Management* (New York: Vintage Books, 1995); and *Only the Paranoid Survive: How to Exploit the Crisis Points That Challenge Every Company and Career* (New York: Currency Doubleday, 1996). Michael S. Malone, *The Intel Trinity: How Robert Noyce, Gordon Moore, and Andy Grove Built the World's Most Important Company* (New York: Harper Business, 2014; Kindle edition) is a source of many of my observations, too. Finally, a wonderful article on the early days of Silicon Valley, Tom Wolfe, "How the Sun Rose on the Silicon Valley: The Tinkerings of Robert Noyce," *Esquire*, December 1983, was a critical source.

276 McKinsey & Company says: Richard Dobbs, James Manyika, and Jonathan Woetzel, "The Four Global Forces Breaking All the Trends," *McKinsey Insights*, April 2015.

276 "Ideas are a commodity": Stratford Sherman et al., "The New Computer Revolution," *Fortune*, June 14, 1993.

277 He was born András Gróf: The first section of this chapter on Grove's life in Hungary is drawn largely from Grove, *Swimming Across*.

277 "Things were gloomy": Ibid., 47.

278 "my rich and multifaceted self in writing": Ibid., 124.

279 I lost track of time: Ibid., 235.

280 By the time I was twenty: Ibid., 1.

284 "The research lab and the manufacturing location": Randall MacLowry and Tracy Heather Strain, "Silicon Valley," *American Experience*, PBS, 2013, http://www.pbs.org/wgbh/americanexperience/films/silicon/.

284 It pointed to the opportunity: Thomas L. Friedman, "Moore's Law Turns 50," *New York Times*, May 13, 2015.

287 If Noyce and Moore led the way: Walter Isaacson, *The Innovators* (New York: Simon & Schuster, 2014), 190.

287 "this was Andy's unwritten role": Malone, *Trinity*, chap. 12, "The Wild West."

287 "unwittingly influenced by the totalitarian regime": Jackson, *Inside Intel*, 33.

288 "It was terrifying": Tedlow, *Life and Times*, 118.

288 the 1103 was the biggest-selling semiconductor: Ibid., 138.

290 one of which went like this: Wolfe, "Tinkerings."

290 "Occasionally we . . . suggest": Elizabeth Corcoran, "Intel CEO Andy Grove Steps Aside: A Founding Father of Silicon Valley," *Washington Post*, March 27, 1998.

291 "Andy, I've got bigger problems": Jackson, *Inside Intel*, 114.

292 Busicom had just made one of the worst: Michael S. Malone, *The Microprocessor: A Biography* (New York: Telos/Springer-Verlag, 1995), 12.

293 "the most important single product": Ibid., 19.

293 "What's it good for?": Jackson, *Inside Intel*, 201.

295 "You can never come out of a downturn": Jeffrey E. Garten, "Listen Up, Execs: Playing It Safe Won't Cut It," *BusinessWeek*, March 3, 2003.

295 the company saw yearly revenues grow: Tedlow, *Life and Times*, 138–39.

295 "the perfect expression of [Grove's] conception": Ibid., 176.

297 By the early 1980s, Japan was better: Jackson, *Inside Intel*, 244.

297 As the company's profits fell: Richard Tedlow, "The Education of Andy Grove," *Fortune*, December 12, 2005.

297 "If we got kicked out": Grove, *Paranoid*, 89.

298 "For more than thirty years": Malone, *Trinity*, chap. 21, "Silicon Valley Aristocracy."

299 "It's like sailing a boat": Grove, *Paranoid*, 19.

299 inflection point compares to ordinary change: Ibid., 27.

300 "Allow chaos": Grove, *High Output Management*, xv.

301 "never been a company to be altruistic or even fair": David Kirkpatrick and John Wyatt, "Intel Goes for Broke," *Fortune*, May 16, 1994.

301 Grove would open his customer presentations: Andrew Pollack, "An 'Awesome' Intel Corners Its Markets," *New York Times*, April 3, 1988.

302 "It is the paranoia that is ingrained": Tedlow, *Life and Times*, 370.

302 "I worry": Grove, *Paranoid*, 3.

303 Intel's stock market share had grown: Tedlow, *Life and Times*, 385.

303 Sales grew from $1.9 billion: Ibid., xv.

303 the most valuable manufacturing company: Malone, *Trinity*, chap. 56, "Running the Asylum."

304 By 2008, it had added dozens more: Information gathered from various information formally submitted to the Securities and Exchange Commission, "SEC Filings," Intel Corporation, http://www.intc.com/sec.cfm.

304 employed approximately eighty-two thousand people: "Backgrounder: 40th Anniversary," Intel Corporation, news release, July 2008, http://www.intel.com/pressroom/enhanced/40th_Anniversary/40th_anniversary_backgrounder.pdf?iid=pr_smrelease_40th_addlmat1.

307 Between 1971 and 2011: Information about transistors is drawn from Robert Bryce, *Smaller Faster Lighter Denser Cheaper: How Innovation Keeps Proving the Catastrophists Wrong* (New York: PublicAffairs, 2014), 121–24.

307 the computing power of a 1975 supercomputer: James Manyika et al., *Disruptive Technologies: Advances That Will Transform Life, Business, and the Global Economy*, McKinsey Global Institute, May 2013, http://www.mckinsey.com/insights/business_technology/disruptive_technologies.

307 Fearful that they will soon have crammed: John Pavlus, "The Search for a New Machine," *Scientific American* 312, no. 5 (2015): 60.

308 took off in early–twentieth-century America: "The Third Industrial Revolution," *Economist*, April 21, 2012.

308 "The [previous] Industrial Revolution": Andrew McAfee and Erik Brynjolfsson, "Artificial Intelligence Meets the C-Suite," in "Management: The Next 50 Years," *McKinsey Quarterly*, special issue, September 2014, 67.

309 "aren't just traditional devices": John Chambers and Wim Elfrink, "The Future of Cities: The Internet of Everything Will Change How We Live," *Foreign Affairs*, October 31, 2014.

309 proposes this thought experiment: John Steele Gordon, "The Little Miracle Spurring Inequality," *Wall Street Journal*, June 3, 2014.

Chapter X: Deng Xiaoping

I am especially indebted to a series of articles that appeared in *The China Quarterly* special issue entitled "Deng Xiaoping: An Assessment," *China Quarterly*, no. 135 (1993): 409–572; Harrison E. Salisbury, *The New Emperors: China in the Era of Mao and Deng* (New York: HarperPerennial, 1993); John King Fairbank, *The Great Chinese Revolution: 1800–1985*, Perennial Library edition (New York: Harper & Row, 1987); the definitive work on Deng Xiaoping, Ezra

F. Vogel, *Deng Xiaoping and the Transformation of China* (Cambridge, MA: Belknap Press of Harvard University Press, 2011); Jonathan Fenby, *Modern China: The Rise and Fall of a Great Power, 1850 to the Present* (New York: Ecco, 2008); and Orville Schell and John Delury, *Wealth and Power: China's Long March to the Twenty-First Century* (New York: Random House, 2013).

314 Deng left for Paris: Benjamin Yang, *Deng: A Political Biography* (Armonk, NY: M. E. Sharpe, 1998), 29.
314 Socialist Youth League of China: David S. G. Goodman, *Deng Xiaoping and the Chinese Revolution: A Political Biography* (New York: Routledge, 1994), 26–27.
317 encouraged peasants to kill landlords: Salisbury, *The New Emperors*, 38.
320 showing his "capitalist leanings": Goodman, *Deng Xiaoping and the Chinese Revolution*, 43.
320 "the four hundred and seventy-five million people": Salisbury, *The New Emperors*, 55.
322 "Let one hundred flowers bloom": Ibid., 136.
322 he was a zealous supporter of the crackdown: Richard Evans, *Deng Xiaoping and the Making of Modern China* (New York: Penguin Books, 1997), 137–38.
323 "doesn't matter if the cat is black or white": Schell and Delury, *Wealth and Power*, 268.
323 pushed to dismantle large communes: Summary of Deng's actions is drawn from Salisbury, *The New Emperors*, 209; Yang, *Political Biography*, 151.
324 "Whenever I think of the damages caused": Lucian W. Pye, "An Introductory Profile: Deng Xiaoping and China's Political Culture," *China Quarterly*, no. 135 (1993): 438.
324 campaign to tear down bourgeois authority: This summary of the Red Guards and their actions is drawn from Fairbank, *Chinese Revolution*, 316–20.
325 Deng had to chop his own wood: Description of Deng's home life is drawn from Vogel, *Transformation*, 52–53.
327 "Deng is a rare talent": Salisbury, *The New Emperors*, 328.
327 "You will see": Yang, *Political Biography*, 180.
328 He was silent on several Maoist themes: Ibid., 185.
328 Deng defied the Gang: Evans, *Making of Modern China*, 206–8.
329 a deep base of support for more pragmatic policies: Vogel, *Transformation*, 183.
329 "large numbers of pathbreakers": All quotations in this paragraph come from Schell and Delury, *Wealth and Power*, 290.
329 volunteered to take responsibility for science: Vogel, *Transformation*, 197.
330 "In the annals of world political history": Ibid., 247.
331 omnipotence lies in the mystery: Pye, "Introductory Profile," 415–17.
331 "discouraged any suggestion of a cult of personality": Ibid., 418.

331 "accustomed to Mao's philosophical disquisitions": Henry Kissinger, *On China* (New York: Penguin Press, 2011), 323–24.

331 Deng came to exactly the opposite conclusion: Evans, *Making of Modern China*, 23.

331 To Deng the message was clear: Roderick MacFarquhar and Michael Schoenhals, *Mao's Last Revolution* (Cambridge, MA: Belknap Press of Harvard University Press, 2006), 2.

332 As a result of these reforms: Evans, *Making of Modern China*, 254.

332 In 1981, China was the world's fourth-largest importer of cotton: Vogel, *Transformation*, 444.

333 "When you open the windows": Schell and Delury, *Wealth and Power*, 295.

333 "I am just a layman": Evans, *Making of Modern China*, 256.

333 many of his specific economic reforms: Pye, "Introductory Profile," 424.

333 "In order to cross a large river": This expression is attributed to Deng in countless sources, sometimes as his repeating an old Chinese proverb, sometimes as his original formulation. See, for example, Patrick E. Tyler, "Deng Xiaoping: A Political Wizard Who Put China on the Capitalist Road," *New York Times*, February 20, 1997.

334 everyone was an entrepreneur: Fenby, *Rise and Fall*, 567.

335 In 1975, he called in the People's Liberation Army: MacFarquhar and Schoenhals, *Mao's Last Revolution*, 388.

335 Deng especially viewed the Cultural Revolution: Schell and Delury, *Wealth and Power*, 303.

335 Deng's reforms were a mixture: Ibid., 261.

335 In the first six months of 1988: Fenby, *Rise and Fall*, 581.

337 "If I have any useful ideas": Ibid., 642.

337 "Whoever is opposed to reform": Vogel, *Transformation*, 677.

338 the editorial in the *People's Daily*: Ibid., 680.

338 As the sun rose on Beijing: Seth Faison, "Deng Xiaoping Is Dead at 92; Architect of Modern China," *New York Times*, February 20, 1997.

339 "The structural changes that took place": Vogel, *Transformation*, 693.

339 "150 years of modern China's history": Tyler, "Political Wizard."

339 "Perhaps never in human history": Pye, "Introductory Profile," 412.

340 "The scale and speed of the transformation": Gideon Rose, "China Now," *Foreign Affairs* 94, no. 3 (2015), https://www.foreignaffairs.com/articles/china/2015-04-20/china-now.

340 some 1.4 million Chinese students: Vogel, *Transformation*, 697.

341 "Great times make great men": Russell Leigh Moses, "Deng Redux? Parsing the Grand Ambitions of China's President," ChinaRealTime, *Wall Street Journal*, August 22, 2014.

341 the Communist Party has adhered to his principles: "Xi Speech Celebrates Deng Xiaoping's Legacy," China.org.cn, http://www.china.org.cn/opinion/2014-08/21/content_33299018.htm.

342 President Xi is strengthening the authoritarian nature of the regime: Evan Osnos, "Born Red," *New Yorker*, April 6, 2014. See also: Edward Wong, "Security Law Suggests a Broadening of China's 'Core Interests,'" *New York Times*, July 2, 2015; and Sarah Cook, *The Politburo's Predicament: Confronting the Limitations of the Chinese Communist Party Repression*, Special Report (Washington, DC: Freedom House, January 2015).

343 When it comes to its ties to the United States: Andrew Browne, "Can China Be Contained?" *Wall Street Journal*, June 12, 2015.

Chapter XI: The Best Is Yet to Come

345 "The fox knows many things": Isaiah Berlin, *The Hedgehog and the Fox*, Elephant Paperback edition (Chicago: Ivan R. Dee, 1993; first published in 1953 by Weidenfeld & Nicolson), 3.

348 "The history of the world is but the biography of great men": Thomas Carlyle, *On Heroes, Hero-Worship, and the Heroic in History* (1841). Project Gutenberg, last updated November 30, 2012, http://www.gutenberg.org/files/1091/1091-h/1091-h.htm.

348 "Before he can remake his society": Robert L. Carneiro, *The Muse of History and the Science of Culture* (New York: Kluwer Academic/Plenum Publishers, 2000), 115.

348 "There is a tide in the affairs of men": William Shakespeare, *Julius Caesar*, Act 4, Scene 3.

352 "Great men are almost always bad": Lord Acton Quote Archive, Acton Institute, http://www.acton.org/research/lord-acton-quote-archive.

358 the fundamental trends of recent decades: For extensive documentation on these developments, see: Charles Kenny, *Getting Better: Why Global Development Is Succeeding—and How We Can Improve the World Even More* (New York: Basic Books, 2011); Joshua S. Goldstein, *Winning the War on War: The Decline of Armed Conflict Worldwide* (New York: Dutton, 2011); Peter H. Diamondis and Steven Kotler, *Abundance: The Future Is Better Than You Think* (New York: Free Press, 2012); Steven Pinker, *The Better Angels of Our Nature: Why Violence Has Declined* (New York: Viking Penguin, 2011).

359 the UN was able to report: *The Millennium Development Goals Report 2015* (New York: United Nations Department of Economic and Social Affairs, July 2015). http://www.un.org/en/development/desa/publications/mdg-report-2015.html.

Author's Note

363 "history plays havoc": Will and Ariel Durant, *The Lessons of History* (New York: Simon & Schuster, 1968), 13.

BIBLIOGRAPHY

Books

Abrahamson, Eric, Sam Hurst, and Barbara Shubinski. *Democracy and Philanthropy: The Rockefeller Foundation and the American Experiment.* New York: Rockefeller Foundation, 2013.

Abu-Lughod, Janet Lippman. *Before European Hegemony: The World System A.D. 1250–1350.* New York: Oxford University Press, 1989.

———. *The World System in the Thirteenth Century: Dead-End or Precursor?* Washington, DC: American Historical Association, 1993.

Adenauer, Konrad. *Memoirs, 1945–53.* London: Weidenfeld & Nicolson, 1965.

Adshead, S. A. M. *Central Asia in World History.* London: Macmillan, 1993.

Al-Khalili, Jim. *The House of Wisdom: How Arab Science Saved Ancient Knowledge and Gave Us the Renaissance.* New York: Penguin Press, 2011.

Allsen, Thomas T. *Commodity and Exchange in the Mongol Empire: A Cultural History of Islamic Textiles.* New York: Cambridge University Press, 1997.

Barber, Benjamin. *Jihad vs. McWorld: Terrorism's Challenge to Democracy.* New York: Ballantine Books, 1996.

Beazley, C. Raymond. *Prince Henry the Navigator: The Hero of Portugal and of Modern Discovery, 1394–1460 A.D.* New York: G. P. Putnam's Sons, 1895.

Beckett, Andy. *When the Lights Went Out: Britain in the Seventies.* London: Faber and Faber, 2009.

Bence-Jones, Mark. *Clive of India*. New York: St. Martin's Press, 1974.

Bergsten, C. Fred, Bates Gill, Nicholas R. Lardy, and Derek J. Mitchell. *China: The Balance Sheet: What the World Needs to Know Now About the Emerging Superpower*. New York: PublicAffairs, 2006.

Bergsten, C. Fred, Charles Freeman, Nicholas R. Lardy, and Derek J. Mitchell. *China's Rise: Challenges and Opportunities.* Washington, DC: Peterson Institute for International Economics and Center for Strategic and International Studies, 2008.

Berlin, Isaiah. *The Hedgehog and the Fox.* Elephant Paperback edition. Chicago: Ivan R. Dee, 1993. First published 1953 by Weidenfeld & Nicolson.

Berlin, Leslie. *The Man Behind the Microchip: Robert Noyce and the Invention of Silicon Valley*. New York: Oxford University Press, 2005.

Berlinski, Claire. *"There Is No Alternative": Why Margaret Thatcher Matters.* New York: Basic Books, 2011.

Bernstein, William J. *A Splendid Exchange: How Trade Shaped the World.* New York: Atlantic Monthly Press, 2008.

Bhagwati, Jagdish. *In Defense of Globalization*. New York: Oxford University Press, 2004.

Blum, Andrew. *Tubes: A Journey to the Center of the Internet*. New York: Ecco, 2012.

Boorstin, Daniel J. *The Discoverers: A History of Man's Search to Know His World and Himself*. New York: Vintage Books, 1985.

———. *The Discoverers.* Vol. 1. New York: Harry N. Abrams, 1991.

Bowen, H. V. *The Business of Empire: The East India Company and Imperial Britain, 1756–1833*. New York: Cambridge University Press, 2006.

Boxer, C. R. *The Portuguese Seaborne Empire, 1415–1825*. New York: Alfred A. Knopf, 1969.

Bradford, Ernle. *A Wind from the North: The Life of Henry the Navigator*. New York: Harcourt, Brace, 1960.

Braun, Ernest, and Stuart Macdonald. *Revolution in Miniature: The History and Impact of Semiconductor Electronics*. New York: Cambridge University Press, 1978.

Brinkley, Douglas, and Clifford Hackett, eds. *Jean Monnet: The Path to European Unity*. New York: St. Martin's Press, 1991.

Brown, Stephen R. *Merchant Kings: When Companies Ruled the World, 1600–1900*. New York: St. Martin's Press, 2009.

Bryce, Robert. *Smaller Faster Lighter Denser Cheaper: How Innovation Keeps Proving the Catastrophists Wrong*. New York: PublicAffairs, 2014.

Brynjolfsson, Erik, and Andrew McAfee. *The Second Machine Age: Work, Progress, and Prosperity in a Time of Brilliant Technologies.* New York: W. W. Norton, 2014.

Burgelman, Robert A. *Strategy Is Destiny: How Strategy-Making Shapes a Company's Future*. New York: Free Press, 2002.
Burns, James MacGregor. *Leadership*. New York: Harper & Row, 1978.
Calder, Kent E. *The New Continentalism: Energy and Twenty-First-Century Eurasian Geopolitics.* New Haven, CT: Yale University Press, 2012.
The Cambridge History of Capitalism. 2 vols. Edited by Larry Neal and Jeffrey G. Williams. New York: Cambridge University Press, 2014.
Campbell, Colin. *Narrative of the Indian Revolt from Its Outbreak to the Capture of Lucknow.* London: G. Vickers, 1858. Quoted in Daniel Litvin, *Empires of Profit: Commerce, Conquest and Corporate Responsibility.* New York: Texere, 2003.
Campbell, John. *Margaret Thatcher*. Vol. 1: *The Grocer's Daughter*. London: Pimlico, 2001.
———. *Margaret Thatcher*. Vol. 2: *The Iron Lady*. London: Pimlico, 2003.
Carlyle, Thomas. *On Heroes, Hero-Worship, and the Heroic in History* (1841). Project Gutenberg. Last updated November 30, 2012. http://www.gutenberg.org/files/1091/1091-h/1091-h.htm.
Carneiro, Robert L. *The Muse of History and the Science of Culture.* New York: Kluwer Academic/Plenum Publishers, 2000.
Carreiro, Carlos B. *Portugal's Golden Years: The Life and Times of Prince Henry "The Navigator."* Pittsburgh: Dorrance, 2005.
Carter, Samuel, III. *Cyrus Field: Man of Two Worlds*. New York: G. P. Putnam's Sons, 1968.
Caryl, Christian. *Strange Rebels: 1979 and the Birth of the 21st Century*. New York: Basic Books, 2013.
Chanda, Nayan. *Bound Together: How Traders, Preachers, Adventurers, and Warriors Shaped Globalization*. New Haven, CT: Yale University Press, 2007.
Chang, David W. *China Under Deng Xiaoping: Political and Economic Reform.* New York: St. Martin's Press, 1988.
Chaudhuri, K. N. *The Trading World of Asia and the English East India Company 1660–1760.* New York: Cambridge University Press, 1978.
Chaudhuri, Nirad C. *Clive of India: A Political and Psychological Essay.* London: Berrie & Jenkins, 1975.
Cheng, Pei-kai, and Michael Lestz, eds. *The Search for Modern China: A Documentary Collection*. With Jonathan D. Spence. New York: W. W. Norton, 1999.
Chernow, Ron. *The Death of the Banker: The Decline and Fall of the Great Financial Dynasties and the Triumph of the Small Investor*. New York: Vintage Books, 1997.
———. *Titan: The Life of John D. Rockefeller, Sr.* New York: Random House, 1998.
———. *The Warburgs: The Twentieth-Century Odyssey of a Remarkable Jewish Family.* New York: Random House, 1993.

Chua, Amy. *Day of Empire: How Hyperpowers Rise to Global Dominance—and Why They Fall.* New York: Doubleday, 2007.

Clark, Gregory. *A Farewell to Alms: A Brief Economic History of the World.* Princeton, NJ: Princeton University Press, 2007.

Clark, Ronald W. *Works of Man.* New York: Viking, 1985.

Clarke, Arthur C. *Voice Across the Sea.* New York: Harper, 1958.

Cole, Juan. *The New Arabs: How the Millennial Generation Is Changing the Middle East.* New York: Simon & Schuster, 2014.

Coll, Steve. *Private Empire: ExxonMobil and American Power.* New York: Penguin Press, 2012.

Cookson, Gillian. *The Cable: The Wire That Changed the World.* Stroud, Gloucestershire, UK: Tempus Publishing, 2003.

Corti, Egon Caesar. *The Rise of the House of Rothschild.* Translated by Brian Lunn and Beatrix Lunn. New York: Grosset & Dunlap, 1928.

Craughwell, Thomas J. *How the Barbarian Invasions Shaped the Modern World.* Beverly, MA: Fair Winds Press, 2008.

Crosby, Alfred W. *The Columbian Exchange: Biological and Cultural Consequences of 1492.* Westport, CT: Greenwood Press, 1972.

Darwin, John. *The Empire Project: The Rise and Fall of the British World-System, 1830–1970.* New York: Cambridge University Press, 2009.

———. *Unfinished Empire: The Global Expansion of Britain.* London: Allen Lane, 2012.

Dawson, Christopher. *Mission to Asia.* Toronto: University of Toronto Press, 1980.

Delpar, Helen, ed. *The Discoverers: An Encyclopedia of Explorers and Exploration.* New York: McGraw-Hill, 1980.

Deng, Maomao. *Deng Xiaoping: My Father.* New York: Basic Books, 1995.

Deng, Rong. *Deng Xiaoping and the Cultural Revolution: A Daughter Recalls the Critical Years.* Translated by Sidney Shapiro. Beijing: Foreign Language Press, 2002.

De Soto, Hernando. *The Mystery of Capital: Why Capitalism Triumphs in the West and Fails Everywhere Else.* New York: Basic Books, 2000.

Diamandis, Peter H., and Steven Kotler. *Abundance: The Future Is Better Than You Think.* New York: Free Press, 2012.

Di Cosmo, Nicola, Frank J. Allen, and Peter B. Golden, eds. *The Cambridge History of Inner Asia: The Chinggisid Age.* New York: Cambridge University Press, 2009.

Diebold, William, Jr. *The Schuman Plan: A Study in Economic Cooperation, 1950–1959.* New York: Praeger for the Council on Foreign Relations, 1959.

Dirks, Nicholas B. *The Scandal of Empire: India and the Creation of Imperial Britain.* Cambridge, MA: Belknap Press of Harvard University Press, 2006.

Duchêne, François. *Jean Monnet: The First Statesman of Interdependence.* New York: Norton, 1994.

Durant, Will, and Ariel Durant. *The Lessons of History.* New York: Simon & Schuster, 1968.

Economy, Elizabeth C., and Michael Levi. *By All Means Necessary: How China's Resource Quest Is Changing the World.* New York: Oxford University Press, 2014.

Edwardes, Michael. *The Battle of Plassey and the Conquest of Bengal.* London: B. T. Batsford, 1963.

Elliott, J. H. *The Old World and the New 1492–1650.* New York: Cambridge University Press, 1992. Quoted in Ronald Findlay and Kevin O'Rourke, *Power and Plenty: Trade, War, and the World Economy in the Second Millennium.* Princeton, NJ: Princeton University Press, 2007.

Elon, Amos. *Founder: A Portrait of the First Rothschild and His Time.* New York: Viking, 1996.

Evans, Richard. *Deng Xiaoping and the Making of Modern China.* New York: Penguin Books, 1997.

Fairbank, John King. *The Great Chinese Revolution, 1800–1985.* Perennial Library edition. New York: Harper & Row, 1987.

Fenby, Jonathan. *Modern China: The Fall and Rise of a Great Power, 1850 to the Present.* New York: Ecco, 2008.

Ferguson, Niall. *Empire: The Rise and Demise of the British World Order and the Lessons for Global Power.* New York: Basic Books, 2002.

———. *The House of Rothschild.* Vol. 1: *Money's Prophets, 1798–1848.* New York: Viking, 1998.

———. *The House of Rothschild.* Vol. 2: *The World's Banker, 1849–1999.* New York: Viking, 1999.

Fernañdez-Armesto, Felipe. *Pathfinders: A Global History of Exploration.* New York: W. W. Norton, 2006.

Field, Henry M. *The Story of the Atlantic Telegraph.* New York: Charles Scribner's Sons, 1892.

Findlay, Ronald, and Kevin H. O'Rourke. *Power and Plenty: Trade, War, and the World Economy in the Second Millennium.* Princeton, NJ: Princeton University Press, 2007.

Finkbeiner, Ann. *A Grand and Bold Thing: An Extraordinary New Map of the Universe Ushering in a New Era of Discovery.* New York: Free Press, 2010.

Foltz, Richard C. *Religions of the Silk Road: Overland Trade and Cultural Exchange from Antiquity to the Fifteenth Century.* New York: St. Martin's Press, 1999.

Franck, Irene M., and David M. Brownstone. *The Silk Road: A History.* New York: Facts on File Publications, 1986.

Franklin, Daniel, and John Andrews. *Megachange: The World in 2050.* Hoboken, NJ: John Wiley & Sons, 2012.

Friedman, Thomas L. *The Lexus and the Olive Tree: Understanding Globalization.* New York: Farrar, Straus & Giroux, 1999.

———. *The World Is Flat: A Brief History of the Twenty-First Century.* New York: Farrar, Straus & Giroux, 2005.

Gardner, Brian. *The East India Company: A History.* New York: McCall Publishing, 1971.

Garten, Jeffrey E. *The Mind of the CEO.* New York: Perseus Publishing, 2002.

Ghemawat, Pankaj. *World 3.0: Global Prosperity and How to Achieve It.* Boston: Harvard Business Review Press, 2011.

Gillingham, John. *Coal, Steel, and the Rebirth of Europe, 1945–1955: The Germans and French from Ruhr Conflict to Economic Community.* New York: Cambridge University Press, 1991.

Goetzmann, William M., and K. Geert Rouwenhorst, eds. *The Origins of Value: The Financial Innovations That Created Modern Capital Markets.* New York: Oxford University Press, 2005.

Golden, Peter B. *Central Asia in World History.* New York: Oxford University Press, 2011.

Goldin, Ian, and Mike Mariathasan. *The Butterfly Defect: How Globalization Creates Systemic Risks, and What to Do About It.* Princeton, NJ: Princeton University Press, 2014.

Goldstein, Joshua S. *Winning the War on War: The Decline of Armed Conflict Worldwide.* New York: Dutton, 2011.

Goodman, David S. G. *Deng Xiaoping and the Chinese Revolution: A Political Biography.* New York: Routledge, 1994.

Gordon, John Steele. *A Thread Across the Ocean: The Heroic Story of the Transatlantic Cable.* New York: Walker, 2002.

Gordon, Stewart. *When Asia Was the World: Traveling Merchants, Scholars, Warriors, and Monks Who Created the "Riches of the East."* Cambridge, MA: Da Capo Press, 2008.

Green, Edwin. *Banking: An Illustrated History.* Oxford: Phaidon, 1989.

Gross, Donald. *The China Fallacy: How the U.S. Can Benefit from China's Rise and Avoid Another Cold War.* New York: Bloomsbury, 2012.

Grove, Andrew S. *High Output Management.* New York: Vintage Books, 1995.

———. *Only the Paranoid Survive: How to Exploit the Crisis Points That Challenge Every Company and Career.* New York: Currency Doubleday, 1996.

———. *Swimming Across: A Memoir.* New York: Warner Books, 2001.

Hansen, Valerie. *The Silk Road: A New History.* New York: Oxford University Press, 2012.

Hanson, Dirk. *The New Alchemists: Silicon Valley and the Microelectronics Revolution.* Boston: Little, Brown, 1982.

Hanson, Elizabeth. *The Rockefeller University Achievements.* New York: Rockefeller University Press, 2000.

Harding, Harry. *China's Second Revolution: Reform After Mao.* Washington, DC: Brookings Institution, 1987.

Harvey, Robert. *Clive: The Life and Death of a British Emperor.* London: Hodder & Stoughton, 1998.

Hearn, Chester G. *Circuits in the Sea: The Men, the Ships, and the Atlantic Cable.* Westport, CT: Praeger, 2004.

Houser, Trevor, and Shashank Mohan. *Fueling Up: The Economic Implications of America's Oil and Gas Boom.* Washington, DC: Peterson Institute for International Economics, 2014.

Howe, Daniel Walker. *What Hath God Wrought: The Transformation of America, 1815–1848.* New York: Oxford University Press, 2007.

Hughes, Thomas P. *Rescuing Prometheus.* New York: Pantheon Books, 1998.

Hunt, Lynn. *Writing History in the Global Era.* New York: W. W. Norton, 2014.

Iggulden, Conn. *Genghis: Birth of an Empire.* New York: Delacorte Press, 2007. (Historical fiction)

———. *Genghis: Lords of the Bow.* New York: Delacorte Press, 2008. (Historical fiction)

Impy, Chris. *Beyond: Our Future in Space.* New York: W. W. Norton, 2015.

Isaacson, Walter. *The Innovators: How a Group of Hackers, Geniuses, and Geeks Created the Digital Revolution.* New York: Simon & Schuster, 2014.

———. *Steve Jobs.* New York: Simon & Schuster, 2011.

Jackson, Ashley. *The British Empire: A Very Short Introduction.* Oxford: Oxford University Press, 2013.

Jackson, Tim. *Inside Intel: Andy Grove and the Rise of the World's Most Powerful Chip Company.* New York: Dutton, 1997.

James, Lawrence. *Raj: The Making and Unmaking of British India.* London: Little, Brown, 1997.

Jay, Peter. *The Wealth of Man.* New York: PublicAffairs, 2000.

Jebb, Gladwyn. *The European Idea.* London: Weidenfeld & Nicolson, 1966.

Judson, Isabella Field, ed. *Cyrus W. Field: His Life and Work.* New York: Harper & Brothers, 1896.

Kahn, Paul. *The Secret History of the Mongols: The Origins of Chinghis Khan.* Based primarily on the English translation by Francis Woodman Cleaves. San Francisco: North Point Press, 1984.

Kaplan, Herbert H. *Nathan Mayer Rothschild and the Creation of a Dynasty: The Critical Years 1806–1816.* Stanford, CA: Stanford University Press, 2006.

Kavanagh, Dennis. *Thatcherism and British Politics: The End of Consensus?* New York: Oxford University Press, 1987.

Keegan, John. *The Mask of Command.* New York: Viking, 1987.

Kenny, Charles. *Getting Better: Why Global Development Is Succeeding—and How We Can Improve the World Even More.* New York: Basic Books, 2011.

Kierzkowski, H., ed. *Europe and Globalization.* New York: Palgrave, 2002.

Kindleberger, Charles Poor. *A Financial History of Western Europe.* 2nd ed. New York: Oxford University Press, 1993.

Kirkpatrick, David. *The Facebook Effect: The Inside Story of the Company That Is Connecting the World*. New York: Simon & Schuster, 2010.

Kissinger, Henry. *On China*. New York: Penguin Press, 2011.

Kristof, Nicholas D., and Sheryl WuDunn. *China Wakes: The Struggle for the Soul of a Rising Power*. New York: Times Books, 1994.

Kynge, James. *China Shakes the World: The Rise of a Hungry Nation*. London: Weidenfeld & Nicolson, 2006.

Landes, David S. *Bankers and Pashas: International Finance and Economic Imperialism in Egypt*. Cambridge, MA: Harvard University Press, 1958.

———. *Dynasties: Fortunes and Misfortunes of the World's Great Family Businesses*. New York: Viking, 2006.

———. *The Wealth and Poverty of Nations: Why Some Are So Rich and Some So Poor*. New York: W. W. Norton, 1998.

Lane, George. *Daily Life in the Mongol Empire*. Westport, CT: Greenwood Press, 2006.

———. *Genghis Khan and the Mongol Rule*. Westport, CT: Greenwood Press, 2004.

Langford, Paul. *Eighteenth-Century Britain: A Very Short Introduction*. Oxford: Oxford University Press, 1984.

Lardy, Nicholas R. *China's Unfinished Economic Revolution*. Washington, DC: Brookings Institution, 1998.

———. *Foreign Trade and Economic Reform in China, 1978–1990*. New York: Cambridge University Press, 1992.

Lasch, Christopher. *The True and Only Heaven: Progress and Its Critics*. New York: W. W. Norton, 1991.

Laurence, Henry. *Money Rules: The New Politics of Finance in Britain and Japan*. Ithaca, NY: Cornell University Press, 2001.

Lawford, James P. *Clive, Proconsul of India: A Biography*. London: Allen & Unwin, 1976.

Lawson, Nigel. *The View from No. 11: Memoirs of a Tory Radical*. New York: Bantam Press, 1992.

Lawson, Philip. *The East India Company: A History*. New York: Addison-Wesley Longman, 1993.

Litvin, Daniel. *Empires of Profit: Commerce, Conquest and Corporate Responsibility*. New York: Texere, 2003.

Liu, Xinru. *The Silk Road in World History*. New York: Oxford University Press, 2010.

Livermore, H. V. *A New History of Portugal*. New York: Cambridge University Press, 1976.

Lowe, Keith. *Savage Continent: Europe in the Aftermath of World War II*. New York: St. Martin's Press, 2012.

Macaulay, Thomas Babington. *Macaulay's Essay on Lord Clive*. Edited by Vida D. Scudder. Boston: Leach, Shewell & Sanborn, 1889.

MacFarquhar, Roderick, and Michael Schoenhals. *Mao's Last Revolution.* Cambridge, MA: Belknap Press of Harvard University Press, 2006.

MacGillivray, Alex. *A Brief History of Globalization: The Untold Story of Our Incredible Shrinking Planet*. New York: Carroll & Graf, 2006.

MacMillan, Margaret. *The Uses and Abuses of History*. London: Profile Books, 2009.

Maddison, Angus. *Contours of the World Economy, 1–2030 AD: Essays in Macro-Economic History*. New York: Oxford University Press, 2007.

Mahbubani, Kishore. *The New Asian Hemisphere: The Irresistible Shift of Global Power to the East*. New York: PublicAffairs, 2008.

Malone, Michael S. *The Intel Trinity: How Robert Noyce, Gordon Moore, and Andy Grove Built the World's Most Important Company.* New York: Harper Business, 2014. Kindle edition.

———. *The Microprocessor: A Biography*. New York: Telos/Springer-Verlag, 1995.

Man, John. *Genghis Khan: Life, Death, and Resurrection*. New York: Bantam Press, 2004.

———. *Kublai Khan: From Xanadu to Superpower*. London: Bantam Press, 2006.

Mann, Charles C. *1493: Uncovering the New World Columbus Created*. New York: Knopf, 2011.

Mann, Vivian B., and Richard I. Cohen, eds. *From Court Jews to the Rothschilds: Art, Patronage, and Power, 1600–1800*. New York: Prestel Publishing, 1996.

Mansfield, Stephen. *Never Give In: The Extraordinary Character of Winston Churchill.* Nashville, TN: Cumberland House Publishing, 1995.

Marjolin, Robert. *Architect of European Unity: Memoirs, 1911–1986*. Translated by William Hall. London: Weidenfeld & Nicolson, 1989.

Mayne, Richard. *The Community of Europe*. New York: Norton, 1963.

McNeill, William H. *The Rise of the West: A History of Human Community.* Chicago: University of Chicago Press, 1963.

Mead, Walter Russell. *God and Gold: Britain, America, and the Making of the Modern World.* New York: Alfred A. Knopf, 2007.

Meyer, Christopher. *Standing on the Sun: How the Explosion of Capitalism Abroad Will Change Business Everywhere.* With Julia Kirby. Boston: Harvard Business Review Press, 2012.

Micklethwait, John, and Adrian Wooldridge. *The Company: A Short History of a Revolutionary Idea.* New York: Modern Library, 2005.

———. *A Future Perfect: The Challenge and Hidden Promise of Globalization.* London: Heinemann, 2000.

Mitter, Rana. *Modern China: A Very Short Introduction.* New York: Oxford University Press, 2008.

Monnet, Jean. *Memoirs*. Translated by Richard Mayne. Garden City, NY: Doubleday, 1978.

Moore, Charles. *Margaret Thatcher: The Authorized Biography, from Grantham to the Falklands*. New York: Alfred A. Knopf, 2013.
Moore, Karl, and David Lewis. *The Origins of Globalization*. New York: Routledge, 2009.
Morris, Ian. *Why the West Rules—for Now: The Patterns of History, and What They Reveal About the Future*. New York: Farrar, Straus & Giroux, 2010.
Muasher, Marwan. *The Second Arab Awakening: And the Battle for Pluralism*. New Haven, CT: Yale University Press, 2014.
Nevins, Allan. *John D. Rockefeller: The Heroic Age of American Enterprise*. 2 vols. New York: Scribner's, 1940.
Nisbet, Robert A. *History of the Idea of Progress*. New York: Basic Books, 1980.
O'Rourke, Kevin H., and Jeffrey G. Williamson. *Globalization and History: The Evolution of a Nineteenth-Century Atlantic Economy*. Cambridge, MA: MIT Press, 1999.
Osterhammel, Jürgen, and Niels Peterson. *Globalization: A Short History*. Translated by Dona Geyer. Princeton, NJ: Princeton University Press, 2005.
Pagden, Anthony. *Peoples and Empires: A Short History of European Migration, Exploration, and Conquest, from Greece to the Present*. New York: Modern Library, 2003.
Pan, Philip P. *Out of Mao's Shadow: The Struggle for the Soul of a New China*. New York: Simon & Schuster, 2008.
Panitchpakdi, Supachai, and Mark L. Clifford. *China and the WTO: Changing China, Changing World Trade*. Singapore: John Wiley & Sons (Asia), 2002.
Pantsov, Alexander V., with Steven I. Levine. *Deng Xiaoping: A Revolutionary Life*. New York: Oxford University Press, 2015.
Parker, David. *The Official History of Privatization*. 2 vols. New York: Routledge, 2009 and 2012.
Partington, Norma. *Master of Bengal*. New York: St. Martin's Press, 1974. (Historical novel)
Pinker, Steven. *The Better Angels of Our Nature: Why Violence Has Declined*. New York: Viking, 2011.
Prawdin, Michael. *The Mongol Empire: Its Rise and Legacy*. New York: Free Press, 1967.
Prestage, Edgar. *The Portuguese Pioneers*. London: A. & C. Black, 1933.
Price-Groff, Claire. *Great Conquerors*. San Diego, CA: Lucent Books, 1999.
Pyne, Stephen J. *Voyager: Seeking Newer Worlds in the Third Great Age of Discovery*. New York: Viking, 2010.
Rapport, Mike. *1848, Year of Revolution*. New York: Basic Books, 2009.
Ratchnevsky, Paul. *Genghis Khan: His Life and Legacy*. Translated by Thomas Nivison Haining. Cambridge, MA: Blackwell, 1992.
Reeves, John. *The Rothschilds: The Financial Rulers of Nations*. Chicago: A. C. McClurg, 1887.

Reinhart, Carmen M., and Kenneth S. Rogoff. *This Time Is Different: Eight Centuries of Financial Folly.* Princeton, NJ: Princeton University Press, 2009.

Reitan, Earl A. *The Thatcher Revolution: Margaret Thatcher, John Major, Tony Blair, and the Transformation of Modern Britain, 1979–2001.* Lanham, MD: Rowman & Littlefield, 2003.

Riddell, Peter. *The Thatcher Era and Its Legacy.* Cambridge, MA: Blackwell, 1991.

Ringrose, David R. *Expansion and Global Interaction, 1200–1700.* New York: Longman, 2001.

Roach, Stephen S. *Unbalanced: The Codependency of America and China.* New Haven, CT: Yale University Press, 2014.

Rockefeller, John D. *John D. Rockefeller Interview, 1917–1920.* Conducted by William O. Inglis. Westport, CT: Meckler Publishing in association with the Rockefeller Archive Center, 1984.

———. *Random Reminiscences of Men and Events.* New York: Doubleday, Page, 1909. Republished by Creative English Publishing, 2013.

Rodrik, Dani. *The Globalization Paradox: Democracy and the Future of the World Economy.* New York: W. W. Norton, 2011.

Rossabi, Morris. *The Mongols: A Very Short Introduction.* New York: Oxford University Press, 2012.

Rothkopf, David. *Power, Inc.: The Epic Rivalry Between Big Business and Government—and the Reckoning That Lies Ahead.* New York: Farrar, Straus & Giroux, 2012.

Roussel, Eric. *Jean Monnet, 1888–1979.* Paris: Fayard, 1996.

Russell, Peter. *Prince Henry "the Navigator": A Life.* New Haven, CT: Yale University Press, 2001.

Sagan, Carl. *Cosmos.* New York: Random House, 1980. Quoted in Stephen J. Pyne, *Voyager: Seeking Newer Worlds in the Third Great Age of Discovery.* New York: Viking Press, 2010.

Salisbury, Harrison E. *The Long March: The Untold Story.* New York: Harper & Row, 1985.

———. *The New Emperors: China in the Era of Mao and Deng.* New York: HarperPerennial, 1993.

Sampson, Anthony. *The Seven Sisters: The Great Oil Companies and the World They Shaped.* New York: Bantam Books, 1976.

Schell, Orville, and John Delury. *Wealth and Power: China's Long March to the Twenty-First Century.* New York: Random House, 2013.

Schuman, Robert. *For Europe.* Luxembourg: Robert Schuman Foundation, 2010.

Shambaugh, David L., ed. *Deng Xiaoping: Portrait of a Chinese Statesman.* New York: Oxford University Press, 1995.

Shirk, Susan L. *How China Opened Its Door: The Political Success of the PRC's Foreign Trade and Investment Reforms.* Washington, DC: Brookings Institution, 1994.

Simpfendorfer, Ben. *The New Silk Road: How a Rising Arab World Is Turning*

Away from the West and Rediscovering China. New York: Palgrave Macmillan, 2009.

Snow, Edgar. *Red Star over China.* New York: Grove Press, 1968.

Spar, Debora L. *Ruling the Waves: Cycles of Discovery, Chaos, and Wealth from the Compass to the Internet.* New York: Harcourt, 2001.

Spence, Jonathan D. *Mao Zedong.* New York: Lipper/Viking, 1999.

Standage, Tom. *The Victorian Internet: The Remarkable Story of the Telegraph and the Nineteenth Century's On-Line Pioneers.* New York: Walker, 1998.

Stapleton, Darwin H., ed. *Creating a Tradition of Biomedical Research: Contributions to the History of the Rockefeller University.* New York: Rockefeller University Press, 2004.

Stearns, Peter N. *Interpreting the Industrial Revolution.* Washington, DC: American Historical Association, 1991.

Stern, Fritz. *Gold and Iron: Bismarck, Bleichröder, and the Building of the German Empire.* New York: Vintage Books, 1979.

Stiglitz, Joseph E. *Making Globalization Work.* New York: W. W. Norton, 2006.

Subramanian, Arvind. *Eclipse: Living in the Shadow of China's Economic Dominance.* Washington, DC: Peterson Institute for International Economics, 2011.

Tarbell, Ida M. *The History of the Standard Oil Company.* 2 vols. New York: McClure, Phillips, 1904 (serialized by *McClure's* magazine in 1902 and 1903).

Taylor, Robert. *Life in Genghis Khan's Mongolia.* San Diego: Lucent, 2001.

Tedlow, Richard S. *Andy Grove: The Life and Times of an American.* New York: Portfolio, 2006.

Thackray, Arnold, David Brock, and Rachel Jones. *Moore's Law: The Life of Gordon Moore, Silicon Valley's Quiet Revolutionary.* New York: Basic Books, 2015.

Thatcher, Margaret. *The Downing Street Years.* New York: HarperCollins, 1993.

———. *The Path to Power.* New York: HarperCollins, 1995.

Vinen, Richard. *Thatcher's Britain: The Politics and Social Upheaval of the Thatcher Era.* New York: Simon & Schuster, 2009.

Vogel, Ezra F. *Deng Xiaoping and the Transformation of China.* Cambridge, MA: Belknap Press of Harvard University Press, 2011.

Von Braun, Wernher, and Frederick I. Ordway III. *Space Travel: A History.* 4th ed. New York: Harper and Row, 1985. Quoted in Stephen J. Pyne, *Voyager: Seeking Newer Worlds in the Third Great Age of Discovery.* New York: Viking Press, 2010.

Waley, Arthur. *The Secret History of the Mongols: And Other Pieces.* London: Allen & Unwin, 1963.

Weatherford, Jack. *Genghis Khan and the Making of the Modern World.* New York: Crown, 2004.

———. *The History of Money: From Sandstone to Cyberspace.* New York: Crown, 1997.

Wechsberg, Joseph. *The Merchant Bankers.* New York: Pocket Books, 1966.

Weightman, Gavin. *The Industrial Revolutionaries: The Making of the Modern World, 1776–1914.* New York: Grove Press, 2007.

Weinberg, Steve. *Taking On the Trust: The Epic Battle of Ida Tarbell and John D. Rockefeller.* New York: W. W. Norton, 2008.

Wende, Peter. *A History of Germany.* New York: Palgrave Macmillan, 2005.

Werth, Alexander. *France, 1940–1955.* New York: Henry Holt, 1956.

White, Jerry. *London in the Nineteenth Century: "A Human Awful Wonder of God."* London: Jonathan Cape, 2007.

Wild, Antony. *The East India Company: Trade and Conquest from 1600.* New York: HarperCollins Illustrated, 2000.

Wildavsky, Ben. *The Great Brain Race: How Global Universities Are Reshaping the World.* Princeton, NJ: Princeton University Press, 2010.

Winchester, Simon. *Atlantic: Great Sea Battles, Heroic Discoveries, Titanic Storms, and a Vast Ocean of a Million Stories.* New York: HarperCollins, 2010.

Winseck, David R., and Robert M. Pike. *Communication and Empire: Media, Markets, and Globalization, 1860–1930.* Durham, NC: Duke University Press, 2007.

Wolf, Martin. *Why Globalization Works.* New Haven, CT: Yale University Press, 2004.

Wood, Frances. *The Silk Road: Two Thousand Years in the Heart of Asia.* Berkeley: University of California Press, 2002.

Wright, Robert. *Nonzero: The Logic of Human Destiny.* New York: Vintage Books, 2001.

Wright, Robin. *Rock the Casbah: Rage and Rebellion Across the Islamic World.* New York: Simon & Schuster, 2011.

Yang, Benjamin. *Deng: A Political Biography.* Armonk, NY: M. E. Sharpe, 1998.

Yergin, Daniel. *The Prize: The Epic Quest for Oil, Money & Power.* New York: Free Press, 2008.

———. *The Quest: Energy, Security, and the Remaking of the Modern World.* New York: Penguin Press, 2011.

Yergin, Daniel, and Joseph Stanislaw. *The Commanding Heights: The Battle for the World Economy.* New York: Simon & Schuster, 2002.

Young, Hugo. *The Iron Lady: A Biography of Margaret Thatcher.* New York: Farrar, Straus & Giroux, 1989.

Yunus, Muhammad. *Creating a World Without Poverty: Social Business and the Future of Capitalism.* New York: PublicAffairs, 2007.

Ziegler, Philip. *The Sixth Great Power: A History of One of the Greatest of All Banking Families, the House of Barings, 1762–1929.* New York: Alfred A. Knopf, 1988.

Zhang, Liang. *The Tiananmen Papers.* Edited by Andrew J. Nathan and Perry Link. New York: PublicAffairs, 2001.

Zhao, Ziyang. *Prisoner of the State: The Secret Journal of Premier Zhao Ziyang.* Translated and edited by Bao Pu, Renee Chiang, and Adi Ignatius. New York: Simon & Schuster, 2009.

Articles, Research Reports, and Other Source Materials

Achenbach, Joel. "The Era of His Ways: In Which We Choose the Most Important Man in the Last Thousand Years." *Washington Post*, Style Section, December 31, 1995.

Acton Institute, Lord Acton Quote Archive. http://www.acton.org/research/lord-acton-quote-archive.

Aguiar, Marcos, Vladislav Boutenko, David Michael, Vaishali Rastogi, Arvind Subramanian, and Yvonne Zhou. *The Internet's New Billion: Digital Consumers in Brazil, Russia, India, China and Indonesia.* Boston: Boston Consulting Group, September 2010.

Allsen, Thomas T. "Mongolian Princes and Their Merchant Partners, 1200–1260." *Asia Major* Third Series 2, no. 2 (1989): 83–126.

———. "Mongols as Vectors for Cultural Transmission." In *The Cambridge History of Inner Asia: The Chinggisid Age*, edited by Nicola Di Cosmo, Frank J. Allen, and Peter B. Golden, 46–66. New York: Cambridge University Press, 2009.

American Meteorology Society. *Earth Observations, Science and Services for the 21st Century.* Boston: American Meteorological Society, 2012. http://www.ametsoc.org/oss.

American Society of Civil Engineers. *2009 Report Card for America's Infrastructure.* Reston, VA: Society of Civil Engineers, 2009.

Amos, Jonathan. "Voyager Probe 'Leaves Solar System.'" BBC News. September 12, 2013. http://www.bbc.com/news/science-environment-24026153.

Anderlini, Jamil. "China Flexes Its Economic Muscle with Soaring Overseas Investments." *Financial Times*, October 23, 2014.

Antholis, William. *Brookings Essay: New Players on the World Stage: Chinese Provinces and Indian States.* Washington, DC: Brookings Institution Press, October 2013.

"The Arab Spring: Has It Failed?" *Economist*, July 13, 2013.

"Are Telegrams Dead?" *Economist*, July 8, 2013.

Arnold, Martin, and Camilla Hall. "Big Banks Losing Interest in Global Ambitions." *Financial Times*, October 20, 2014.

Arrison, Sonia. "Living to 100 and Beyond." *Wall Street Journal*, August 27, 2011.

Ash, Timothy Garton. "Let the Next Generation Speak Up for Europe." *Guardian*, December 8, 2014. Last modified December 8, 2014. http://www.theguardian.com/commentisfree/2014/dec/07/europe-brussels-european-eu.

Axelrod, David. "The Great Transatlantic Cable." Transcript of PBS special, *American Experience*, November 30, 2004.

"Backgrounder: 40th Anniversary." Intel Corporation. News release, July 2008. http://www.intel.com/pressroom/enhanced/40th_Anniversary/40th_anniversary_backgrounder.pdf?iid=pr_smrelease_40th_addlmat1.

Backhaus, Fritz. "The Last of the Court Jews—Mayer Amschel Rothschild and His Sons." In *From Court Jews to the Rothschilds: Art, Patronage, and Power, 1600–1800*, edited by Vivian B. Mann and Richard I. Cohen, 79–95. Munich and New York: Prestel Publishing, 1996.

Ball, George. "Jean Monnet: He Altered the History of His Time." *Chicago Tribune*, March 23, 1979.

Barboza, David. "Berkeley Reveals Plan for Academic Center in China." *New York Times*, November 16, 2011.

———. "Despite Law, Job Conditions Worsen in China." *New York Times*, June 22, 2009.

Barboza, David, and John Markoff. "Power in Numbers: China Aims for High-Tech Primacy." *New York Times*, December 5, 2011.

Barton, Dominic. "Age of Disruption." Project Syndicate. January 7, 2014. www.project-syndicate.org.

Beatty, Jack. "A Capital Life." *New York Times*, May 17, 1998.

Beddoes, Zanny Minton. "Disruption from Above and Below." In "The World in 2015." Special issue, *Economist*, December 2014.

Beech, Hannah. "How China Sees the World." *Time* 181, no. 23 (June 17, 2013): 26.

Bell, David A. "Inglorious Revolutions." *National Interest* 129 (January–February 2014): 31–38. http://nationalinterest.org.

Berg, Maxine. "In Pursuit of Luxury: Global History and British Consumer Goods in the Eighteenth Century." *Past & Present* 182, no. 1 (February 2004): 85–142.

Bergsten, C. Fred. "A Partnership of Equals." *Foreign Affairs* 87, no. 4 (July–August 2008): 57–69.

Bergsten, C. Fred, and Jacob Funk Kirkegaard. *The Coming Resolution of the European Crisis: An Update*. Policy Briefs in International Economics PB 12–18. Washington, DC: Peterson Institute for International Economics, June 2012. http://www.iie.com/publications/pb/pb12-18.pdf.

Bisson, Peter, Elizabeth Stephenson, and S. Patrick Viguerie. "The Global Grid." *McKinsey Quarterly*, June 2010.

———. "How to Compete in a Rebalanced Global Economy." *McKinsey Quarterly*, no. 3 (September 2010): 31–38.

Blair, Tony. "We Can Help China Embrace the Future." *Wall Street Journal*, August 26, 2008.

Bloom, David. "The Challenging Billions." Project Syndicate. August 17, 2011. www.project-syndicate.org.

The Boeing Company. www.Boeing.com.

Boxwell, James. "Rothschilds Eye Cross-Channel Unity." *Financial Times*, April 4, 2012.

Bradshaw, Tim, Andrew Ward, and Richard Waters. "Wear Your Medicine." *Financial Times*, July 19–20, 2014.

Bradsher, Keith. "China Builds High Wall to Guard Energy Industry." *New York Times*, July 13, 2009.

———. "Hauling New Treasure on the Silk Road." *New York Times*, Sunday Business, July 21, 2013.

———. "U.S. and China Agree to Study Ways to Make Buildings More Efficient." *New York Times*, July 16, 2009.

Brenner, Brian, ed. "Trading at the Speed of Light." *BusinessWeek*, April 2–8, 2012.

Broad, William J. "Billionaires with Big Ideas Are Privatizing American Science." *New York Times*, March 15, 2014.

———. "Deep, Dark, Cold and Mostly Unknown." *New York Times*, May 30, 2010.

Brooks, John. "The Common Market—I." *New Yorker*, September 22, 1962.

———. "The Common Market—II." *New Yorker*, September 29, 1962.

Brown, Kevin. "The Asian Century Calls for a Rethink on Growth." *Financial Times*, June 30, 2010.

Browne, Andrew. "Can China Be Contained?" *Wall Street Journal*, June 12, 2015.

Bughin, Jacques, Michael Chui, and James Manyika. "Ten IT-Enabled Business Trends for the Decade Ahead." *McKinsey Quarterly*, May 2013.

Buruma, Ian. "Fear and Loathing in Europe." *New York Review of Books*, October 17, 1996.

———. "Mrs. Thatcher's Revenge." Review of *The Path to Power*, by Margaret Thatcher, *Letters from London*, by Julian Barnes, and *The Disenchanted Isle*, by Charles Dellheim, in *New York Review of Books*, March 21, 1996.

Butler, Nick. "Climate Change and the New Congress: Bill Gates to the Rescue?" *Globalist*, November 11, 2014.

Cassidy, John. "The Economic Case For and Against Thatcherism." *New Yorker*, April 10, 2013.

Central Intelligence Agency. *The World Factbook 2014*. Washington, DC: Central Intelligence Agency; Office of Public Affairs, 2014. https://www.cia.gov/library/publications/the-world-factbook/index.html.

Chamber of Commerce of New York. *The Atlantic Telegraph: Report of the Proceedings at a Banquet Given to Mr. Cyrus Field*. New York: J. W. Amerman, 1866.

Chambers, John, and Wim Elfrink. "The Future of Cities: The Internet of Everything Will Change How We Live." *Foreign Affairs*, October 31, 2014. https://www.foreignaffairs.com/articles/2014-10-31/future-cities.

Chanda, Nayan. "Runaway Globalization Without Governance." *Global Governance: A Review of Multilateralism and International Organization* 14 (2008): 119–25.

Chang, Kenneth. "Machinery of an Energy Dream." *New York Times*, March 17, 2014.

———. "Obama Views Renewed Space Program." *New York Times*, April 15, 2010.

Chao, Loretta. "Intel, Tencent to Form Tablet Center in China." *Wall Street Journal*, April 13, 2011.

Chhibber, Ajay. "We Need Our States to Be Smarter, Not Bigger." *Financial Times*, August 25, 2010.

"China: The Paradox of Prosperity." *Economist*, January 28, 2012.

China and the U.S. Economy: Advancing a Winning Trade Agenda. Washington, DC: US-China Business Council, January 2013.

"China to Establish $40 billion Silk Road Infrastructure Fund." Reuters, November 8, 2014.

The Churchill Centre. www.winstonchurchill.org.

Cingano, Federico, and Michael Forster. "Trends in Income Inequality and Its Impact on Economic Growth." Presentation at OECD Conference in Florence, Italy, November 22, 2014.

Citigroup. "Treasury and Trade Solutions." Citigroup.com. http://citigroup.com/transactionservices/home/trade_svcs/financing/global.jsp.

Clark, Pilita. "Energy Groups Face 'Existential' Climate Threat, Says Ex-BP Chief." *Financial Times*, November 19, 2014.

Cohen, Joel E. "Human Population Grows Up." In "Crossroad for Planet Earth." Special issue, *Scientific American* 293, no. 3 (September 2005): 48–55.

"The Company That Ruled the Waves: The East India Company." *Economist*, December 17, 2011.

Cook, Sarah. *The Politburo's Predicament: Confronting the Limitations of the Chinese Communist Party Repression*. Special Report. Freedom House: Washington, DC, January 2015.

Cookson, Clive. "Let There Be Life." *Financial Times*, May 21, 2010.

Corcoran, Elizabeth. "Grove Took Intel to the Top." *Washington Post*, March 27, 1998.

———. "Intel CEO Andy Grove Steps Aside: A Founding Father of Silicon Valley." *Washington Post*, March 27, 1998.

Cowell, Alan. "Britons Reflect on Divided Views of Thatcher Legacy." *New York Times*, April 9, 2013.

"Crossroad for Planet Earth." Special issue, *Scientific American* 293, no. 3 (September 2005).

Crowell, Paul. "John D. Rockefeller Dies at 97 in His Florida Home; Funeral to Be Held There." *New York Times*, May 24, 1937.

Cyr, Arthur. Review of *The Downing Street Years*, by Margaret Thatcher, and *The Anatomy of Thatcherism*, by Shirley Robin Letwin. *Society* 31, no. 6 (1994).

Dalrymple, William. "The East India Company: The Original Corporate Raiders." *Guardian*, March 4, 2015.

Das, Saswato R. "The Chip That Changed the World." *International Herald Tribune*, September 20, 2008.

Daudin, Guillaume, Matthias Morys, and Kevin H. O'Rourke. "Globalization, 1870–1914." The Institute for International Integration Studies Discussion Paper No. 250, May 2008. https://www.tcd.ie/iiis/documents/discussion/pdfs/iiisdp250.pdf.

Davenport, Coral. "Nations Plod Forward on Climate Change Accord." *New York Times*, December 13, 2014.

Davenport, John. "M. Monnet of Cognac." *Fortune*, August 14, 1944.

Davis, Ralph. "English Foreign Trade, 1660–1700." *Economic History Review* 7, no. 2 (1954): 150–66.

Deane, Elizabeth, and Adriana Bosch. "The Rockefellers." *American Experience.* PBS.

"Deng Xiaoping: An Assessment," *China Quarterly*, no. 135 (1993): 409–572.

Denver Museum of Nature and Science. Documents accompanying exhibit on Genghis Khan, on display October 16, 2009, to February 7, 2010.

Dobbs, Richard, Anu Madgavkar, Dominic Barton, Eric Labaye, James Manyika, Charles Roxburgh, Susan Lund, and Siddarth Madhav. *The World at Work: Jobs, Pay and Skills for 3.5 Billion People.* McKinsey Global Institute, June 2012.

Dobbs, Richard, James Manyika, and Jonathan Woetzel. "The Four Global Forces Breaking All the Trends." *McKinsey Insights*, April 2015.

Dobbs, Richard, Sree Ramaswamy, Elizabeth Stephenson, and S. Patrick Viguerie. "Management Intuition for the Next 50 Years." *McKinsey Quarterly*, September 14, 2014.

Dobbs, Richard, Sven Smit, Jaana Remes, James Manyika, Charles Roxburgh, and Alejandra Restrepo. *Urban World: Mapping the Economic Power of Cities.* McKinsey Global Institute, 2011.

Dolan, Kerry A. "Andy Grove's Last Stand." *Forbes*, January 11, 2008.

Donnan, Shawn. "WTO Head Roberto Azevêdo Urges Shape-Up of Organization." *Financial Times*, October 16, 2014.

Duchêne, François. "Jean Monnet's Methods." In *Jean Monnet: The Path to European Unity*, edited by Douglas Brinkley and Clifford Hackett, 184–209. New York: St. Martin's Press, 1991.

Duggett, Michael. "Jean Monnet and Project Management." *Public Policy and Administration* 15, no. 1 (March 2000): 77–83.

Dyer, Geoff. "Little Leaps Forward?" *Financial Times*, May 27, 2009.

Eckholm, Eric. "A Secret Memoir Offers Look Inside China's Politics," *New York Times*, May 14, 2009.

"The Eclipse of the Public Company." *Economist*, August 19, 2010.

Economy, Elizabeth C. "The Game Changer: Coping with China's Foreign Policy Revolution." *Foreign Affairs* 89, no. 6 (November/December 2010): 142–52.

Economy, Elizabeth C., and Adam Segal. "China's Olympic Nightmare." *Foreign Affairs* 87, no. 4 (July/August, 2008): 47–56.

England, Andrew. "Africa Sets Deadline for Free Trade Area as Regional Blocs Come Together." *Financial Times*, June 15, 2015.

Esty, Daniel C. "Bottom-Up Climate Fix." *New York Times*, September 21, 2014.

European Central Bank. "Initial Changeover (2002)." http://www.ecb.europa.eu/euro/changeover/2002/html/index.en.html.

European Union. "Countries." http://europa.eu/about-eu/countries/index_en.htm (accessed May 5, 2015).

"Europe: Then Will It Live . . ." *Time* 78, no.14 (October 6, 1961): 28.

Fairless, Tom. "U.S. Tech Giants Battle Europe's Sovereign States." *Wall Street Journal*, December 8, 2014.

Faison, Seth. "Deng Xiaoping Is Dead at 92; Architect of Modern China." *New York Times*, February 20, 1997.

Fallaci, Oriana. "Deng: Cleaning Up Mao's 'Feudal Mistakes.'" *Washington Post*, August 31, 1980.

———. "Deng: A Third World War Is Inevitable." *Washington Post*, September 1, 1980.

Fallows, James. "China's Way Forward." *Atlantic*, April 2009.

Federal Reserve Bank of Minneapolis. "Consumer Price Index (Estimate) 1800– ." https://www.minneapolisfed.org/community/teaching-aids/cpi-calculator-information/consumer-price-index-1800. An annual inflation rate of 2.06% was assumed.

Ferguson, Niall. "The First Eurobonds." In *The Origins of Value: Financial Innovations That Created Modern Capital Markets*, edited by William M. Goetzmann and K. Geert Rouwenhorst, 311–23. New York: Oxford University Press, 2005.

"Financier's Fortune in Oil Amassed in Industrial Era of 'Rugged Individualism.'" *New York Times*, May 24, 1937.

Forney, Matthew. "What Really Happened?" *Time*, January 15, 2001.

Foroohar, Rana. "A New Age of Capitalism Starts Now." *Newsweek*, October 13, 2008.

———. "Why China Works." *Newsweek*, January 19, 2009.

Freed, Josh. *Brookings Essay: Back to the Future*. Washington, DC: Brookings Institution Press, December 2014.

Freeman, Chas. W. "From Mao to Now: A Reevaluation." *Globalist*, August 28, 2008.

French, Henry. "The New Empire." *Atlantic*, May 2010.

French, Howard W. "China's Dangerous Game." *Atlantic*, October 13, 2014.

Friedman, Thomas L. "Foreign Policy by Whisper and Nudge." *New York Times*, August 24, 2013.

———. "Moore's Law Turns 50." *New York Times*, May 13, 2015.

Galston, William A. "Countering Tech's Damaging Effect on Jobs." *Wall Street Journal*, October 14, 2014.

———. *The New Challenges to Market Democracies*. Washington, DC: Brookings Institution, October 2014.

———. " 'Secular Stagnation' May Be for Real." *Wall Street Journal*, August 27, 2014.

Ganesh, Janan. "The Iron Lady Towers over Modern Britain." *Financial Times*, April 8, 2013.

Gapper, John. "Expect Dark Sequels to the Sony Hack." *Financial Times*, December 10, 2014.

Garbade, Kenneth D., and William L. Silber. "Technology, Communication and the Performance of Financial Markets: 1840–1975." *Journal of Finance* 33, no. 3 (1978): 819–32.

Garten, Jeffrey E. "Andy Grove Made the Elephant Dance." *BusinessWeek*, April 11, 2005.

———. "Listen Up, Execs: Playing It Safe Won't Cut It." *BusinessWeek*, March 3, 2003.

———. "We Need a Global Authority for Arctic Oil and Gas." *Financial Times*, August 14, 2008.

———. "We Need a New Global Monetary Authority." *Financial Times*, September 25, 2008.

Gates, Bill. "Great Expectations." In "The World in 2015." Special issue, *Economist,* December 2014.

Giles, Chris. "A Road Less Travelled." *Financial Times*, June 22, 2011.

Gladney, Dru C. "China's Ethnic Fault Lines." *Wall Street Journal*, July 16, 2009.

Gladstone, Rick. "Rights Groups Accuse Turkish Police of Excessive Force Against Protesters." *New York Times*, July 13, 2013.

Gladwell, Malcolm. "The Tweaker: The Real Genius of Steve Jobs." *New Yorker*, November 14, 2011.

"A Global Consortium to Build New Trans-Pacific Cable System 'FASTER.' " NEC Corporation, news release, August 11, 2014. http://www.nec.com/en/press/201408/global_20140811_01.html.

Global Construction Outlook. IHS, April 2009. www.ihs.com.

"The Global Oil Industry." Survey. *Economist*, August 3, 2013.

Golden, Frederic. "Big Dimwits and Little Geniuses." *Time*, January 3, 1983.

Golden, Frederic, Michael D. Lemonick, and Dick Thompson. "The Race Is Over." (Cover story). *Time* 156, no. 1 (July 3, 2000): 18–23.

Gopnik, Adam. "Vive La France." *New Yorker*, May 7, 2012.

Gordon, John Steele. "The Little Miracle Spurring Inequality." *Wall Street Journal*, June 3, 2014.

Gordon, Stewart. "The Dawn of Global Trade." *Globalist*, August 5, 2008.

Gorman, Siobhan. "NSA Director Warns of 'Dramatic' Counterattack in Next Decade." *Wall Street Journal*, November 20, 2014.

Graetz, Michael. "Court Jews in Economics and Politics." In *From Court Jews to the Rothschilds: Art, Patronage, and Power: 1600–1800*, edited by Vivian B. Mann and Richard I. Cohen, 27–43. Munich and New York: Prestel Publishing, 1996.

Grindell, Julia, and Robert Armstrong. "CNOOC—Gutsy Call." *Financial Times*, November 14, 2012.

Gros, Daniel. "Europe's Big Bang at Ten." Project Syndicate. Last modified May 8, 2014. www.project-syndicate.org.

Grove, Andy. "Taking On Prostate Cancer." *Fortune*, May 13, 1996.

Gutterman, Steve. "Russia Blocks Internet Sites of Putin Critics." Reuters, March 13, 2014.

Hagerty, James R. "U.S. Loses High-Tech Jobs as R&D Shifts Toward Asia." *Wall Street Journal*, January 18, 2012.

Hamm, Steve. "Is Silicon Valley Losing Its Magic?" *BusinessWeek*, January 12, 2009.

Hannan, Daniel. "The Woman Who Saved Britain." *Wall Street Journal*, May 17, 2013.

Hansakul, Syetarn, and Willie Keng. *A Potential Game Changer for ASEAN Countries*. Frankfurt: Deutsche Bank Research, June 14, 2013. www.dbresearch.com.

Hanser, Jessica. "Teatime in the North Country: Consumption of Chinese Imports in North-East England." *Northern History* 49, no. 1 (March 2012): 51–74.

Headrick, Daniel R., and Pascal Griset. "Submarine Telegraph Cables: Business and Politics, 1838–1939." *Business History Review* 75, no. 3 (Autumn 2001): 543–78.

"Heal Thyself." *Economist*, December 13, 2014.

Hoffman, Stanley. "Goodbye to a United Europe?" *New York Review of Books*, May 27, 1993.

Holbrooke, Richard. "The Day the Door to China Opened Wide." *Washington Post*, December 15, 2008.

Hollingsworth, J. Rogers. "Institutionalizing Excellence in Biomedical Research: The Case of Rockefeller University." In *Creating a Tradition of Biomedical Research: Contributions to the History of the Rockefeller University*, edited by Darwin H. Stapleton, 1–49. New York: Rockefeller University Press, 2004.

"How Does China Censor the Internet?" *Economist*, April 21, 2013.

Hubbard, Ben, and Rick Gladstone. "Arab Spring Countries Find Peace Is Harder Than Revolution." *New York Times*, August 14, 2013.

Hurd, Douglas. "No Reverse Gear." Review of *Margaret Thatcher: The Iron Lady*, by John Campbell. *New Statesman*, October 20, 2003.

Ignatieff, Michael. "The Post-Ukraine World Order." Lecture, The Ditchley Foundation 50th Annual Lecture. Oxfordshire, UK, July 12, 2014.

Ilzkovitz, Fabienne, Adriaan Dierx, Viktoria Kovacs, and Nuno Sousa. *Steps Toward Deeper Economic Integration: The Internal Market in the 21st Century*. Brussels: European Commission Directorate-General for Economic and Financial Affairs, January 2007.

Immelt, Jeffrey R., Vijay Govindarajan, and Chris Trimble. "How GE Is Disrupting Itself." *Harvard Business Review*, October 2009.

Institute for the Future. "The Future of Coordination." November 2012. http://www.IFTF.org.

Intelligence Unit. *Economist*. www.eiu.com.

International Monetary Fund. *People's Republic of China: 2014 Article IV Consultation*. IMF Country Report No. 14/235. Washington, DC: International Monetary Fund Publication Services, July 2014.

Isaacson, Andy. "The Challenger's Deep Sea Brethren." *New York Times*, March 27, 2012.

Isaacson, Walter. "Andrew Grove: Man of the Year." *Time* 150, no. 27 (December 29, 1997): 46.

"Jean Monnet: A Man with a Vision." *Los Angeles Times*, March 18, 1979.

"Jean Monnet, 1888–1979." *New York Times*, March 18, 1979.

Jenkins, Peter. "Mrs. Thatcher's Last Stand?" Review of *The Iron Lady: A Biography of Margaret Thatcher*, by Hugo Young, and of *The Thatcher Decade*, by Peter Riddell. *New York Review of Books*, April 12, 1990.

Jenkins, Roy. "A Great European." *Economist*, March 24, 1979.

Johnson, Ian. "Beijing's Ethnic Policy Faces Data Challenge." *Wall Street Journal*, July 22, 2009.

———. "Uighurs Lose Economic Ground to Han." *Wall Street Journal*, July 21, 2009.

Johnson, Paul. "The World-Changing Margaret Thatcher." *Wall Street Journal*, April 8, 2013.

Jolliffe, Dean, Peter Lanjouw, Shaohua Chen, Aart Kraay, Christian Meyer, Mario Negre, Espen Prydz, Renos Vakis, and Kyla Wethli. *A Measured Approach to Ending Poverty and Boosting Shared Prosperity: Concepts, Data, and Twin Goals*. Washington, DC: World Bank Group, 2015.

Juncker, Jean-Claude, with Donald Tusk, Jeroen Dijsselbloem, Mario Draghi, and Martin Schulz. *Completing Europe's Economic and Monetary Union*. Brussels: European Commission, June 2015.

Kehoe, Louise. "Not a Grove in the Valley." *Financial Times*, March 28, 1998.

Kerr, Richard. "It's Official—Voyager Has Left the Solar System." *Science* 341, no. 6151 (September 13, 2013): 1158–59.

Kharif, Olga, Peter Burrows, and Cliff Edwards. "Windows and Intel's Digital Divide." *BusinessWeek*, February 23, 2009.

Kirkpatrick, David, and John Wyatt. "Intel Goes for Broke." *Fortune*, May 16, 1994.

Korn, Melissa. "Chinese Students Flood U.S. Grad Schools." *Wall Street Journal*, August 16, 2011.

Körner, Kevin, and Oliver Masetti. *The GCC Going East*. Frankfurt: Deutsche Bank Research, February 18, 2014. www.dbresearch.com.

"The Lady Who Changed the World." *Economist*, April 8, 2013.

LaFraniere, Sharon. "China's Students Feel a Faint Tug from the Ghosts of Tiananmen." *New York Times*, May 22, 2009.

Lagarde, Christine. "A New Multilateralism for the 21st Century." Lecture, International Monetary Fund Richard Dimbleby Lecture. London, February 3, 2014.

Lawson, Nigel. "Mrs. Thatcher's Lasting Legacy." *Standpoint Magazine*, May 2009.

Ledgard, J. M. "A Radical But Possible Plan to Connect African Nations with Cargo Drones." *Wired*, September 22, 2014.

Lex Column. *Financial Times*, September 2, 2010.

Lohr, Steve. "Intel Chief Steps Down After 11 Years." *New York Times*, March 27, 1998.

"London as a Financial Centre: Capital City." *Economist*, October 19, 2006.

Lund, Susan, Toos Daruvala, Richard Dobbs, Philipp Härle, Ju-Hon Kwek, and Ricardo Falcón. *Financial Globalization: Retreat or Reset?* McKinsey Global Institute, March 2013.

Lyons, Daniel. "Short-Circuiting Malaria." *Newsweek*, April 18, 2010.

MacFarquhar, Neil. "Human Development Report Shows Great Gains, and Some Slides." *New York Times*, November 4, 2010.

MacLowry, Randall, and Tracy Heather Strain. "Silicon Valley." *American Experience*. PBS, 2013.

Mahbubani, Kishore, and William Weld. "Asia Keeps the West's Betrayed Faith." *Financial Times*, July 21, 2009.

Mallaby, Sebastian. "Inequality May Lead to Rage Against the Machines." *Financial Times*, March 9, 2012.

———. "The Politically Incorrect Guide to Ending Poverty." *Atlantic*, July, 2010.

Manyika, James, Jacques Bughin, Susan Lund, Olivia Nottebohm, David Poulter, Sebastian Jauch, and Sree Ramaswamy. *Global Flows in a Digital Age: How Trade, Finance, People and Data Connect the World Economy*. McKinsey Global Institute, 2014.

Manyika, James, Michael Chui, Jacques Bughin, Richard Dobbs, Peter Bisson, and Alex Marrs. *Disruptive Technologies: Advances That Will Transform Life, Business, and the Global Economy.* McKinsey Global Institute, 2013. http://www.mckinsey.com/insights/business_technology/disruptive_technologies.

Marcus, Gary, and Christof Koch. "The Plug-and-Play Brain." *Wall Street Journal*, March 15–16, 2014.

"Margaret Thatcher, Prime Minister." *New York Times*, April 8, 2013.

Markoff, John. "Burned Once, Intel Prepares New Chip Fortified by Constant Test." *New York Times*, November 16, 2008.

Marsh, Peter. "Future Factories." *Financial Times*, June 11, 2012.

Martinage, Robert. "Under the Sea: The Vulnerability of the Commons." *Foreign Affairs* 94, no. 1 (January/February 2015): 117–26.

Masood, Salman, and Declan Walsh. "Xi Jinping Plans to Fund Pakistan." *New York Times*, April 21, 2015.

Maude, Francis. "No Surrender." Review of *Statecraft: Strategies for a Changing World*, by Margaret Thatcher. *New Statesman*, April 15, 2002.

Max, D. T. "The End of the Book?" *Atlantic Monthly*, September 1994.

McAfee, Andrew, and Erik Brynjolfsson. "Artificial Intelligence Meets the C-Suite." Interview by Rik Kirkland. In "Management: The Next 50 Years." Special issue, *McKinsey Quarterly*, September 2014.

McGillem, Clare D. "Telegraph." In *Encyclopedia Britannica Online*. http://www.britannica.com/technology/telegraph.

McKay, Betsy, and Ellen Knickmeyer. "Saudi Researchers Mount Genome-Sequencing Push." *Wall Street Journal*, February 5, 2014.

McKibbon, Ross. "Why Did It End So Badly?" Review of *Margaret Thatcher: The Iron Lady*, by John Campbell. *London Review of Books* 26, no. 6 (March 18, 2004).

McNamara, Kathleen R. "A Less Perfect Union." *Foreign Affairs*, July 19, 2015. https://www.foreignaffairs.com/articles/western-europe/2015-07-19/less-perfect-union.

Merkel, Angela. "The 2014 Lowy Lecture." Lecture, Lowy Institute for International Policy. Sydney, Australia, November 17, 2014.

Metzl, Jamie F. "The Genetics Epidemic." *Foreign Affairs*, October 10, 2014. https://www.foreignaffairs.com/articles/united-states/2014-10-10/genetics-epidemic.

Micklethwait, John. "The West's Malaise." In "The World in 2015." Special issue, *Economist*, December 2014.

Micklethwait, John, and Adrian Wooldridge. "The Age of the Unthinkable." *Wall Street Journal*, September 10, 2014.

The Millennium Development Goals Report 2015. United Nations Department of Economic and Social Affairs. http://mdgs.un.org/unsd/mdg/Resources/Static/Products/Progress2015/English2015.pdf.

Miller, Greg. "Intel Chief Andrew Grove Steps Down." *Los Angeles Times*, March 27, 1998.

Miller, John W. "Exhausting the Earth's Resources? Not So Fast." *Wall Street Journal*, June 4, 2010.

Millius, Susan. "An Oceanic Endeavor." *Science News* 178, no. 9 (October 23, 2010): 22–27.

Mishra, Pankaj. "Guilt and Glory." *Financial Times*, October 21, 2011.

Moffett, Matt. "Beyond Petrosaurus: How a Sleepy Oil Giant Became a World Player." *Wall Street Journal*, August 30, 2007.

Morse, Edward L. "Welcome to the Revolution." *Foreign Affairs* 93, no. 3 (May/June 2014): 3–7.

Moses, Russell Leigh. "Deng Redux? Parsing the Grand Ambitions of China's President." ChinaRealTime, *Wall Street Journal*, August 22, 2014.

Mozur, Paul. "In China, Internet Czar Is Taking a Blunt Tone." *New York Times*, November 3, 2014.

Mr. Europe and the Common Market. CBS Reports, March 16, 1962. (David Schoenbrun, reporting.)

Myerson, Harold. "Building a Better Capitalism." *Washington Post*, March 12, 2009.

Naik, Gautam. "Census Shows Ocean's Vast Diversity." *Wall Street Journal Europe*, October 5, 2010.

Nathan, Andrew J. "Modern China's Original Sin." *Foreign Affairs*, June 3, 2014. https://www.foreignaffairs.com/articles/china/2014-06-03/modern-chinas-original-sin.

National Geographic Society. "The New Universe: Here, Now and Beyond." Special issue, *National Geographic*, November 5, 2010.

National Research Council. *Pathways to Exploration: Rationales and Approaches for a U.S. Program of Human Space Exploration*. Washington, DC: National Academies Press, 2014.

Neal, Larry, and Jeffrey G. Williamson. "The Future of Capitalism." In *The Cambridge History of Capitalism*, edited by Larry Neal and Jeffrey G. Williamson, vol. 2, 547–67. Cambridge, UK, and New York: Cambridge University Press, 2014.

"The New Special Relationship: Emerging-Market Firms and Britain." *Economist*, September 10, 2011.

Noble, John. "Hong Kong Protests Turn Spotlight on Income Split." *Financial Times*, October 22, 2014.

Norman, Laurence, and Yeliz Candemir. "E.U. Urges Members to Stop Stalling with Turkey." *Wall Street Journal*, October 8, 2014.

Nye, Joseph S., contributor. "The Asian Century: Reality or Hype?" A Symposium of Views. *International Economy* 27, no. 3 (Summer 2013): 8–31. See especially his comments, 12–13.

Nye, Joseph S. "Shaping the Future." In *What Matters: Ten Questions That Will Shape Our Future.* New York: McKinsey Publishing Group, 2009.

"Obituary: Margaret Thatcher." BBC News. April 8, 2013. http://www.bbc.com/news/uk-politics-10364876.

The OECD Development Centre: 50 Years of Sharing Knowledge. OECD Development Centre, 2013.

O'Rourke, Kevin H. "Europe and the Causes of Globalization, 1790–2000." In *From Europeanization of the Globe to the Globalization of Europe*, edited by H. Kierzkowski, 64–86. Houndmills, Basingstoke, UK, and New York: Palgrave, 2002.

Osnos, Evan. "Born Red." *New Yorker*, April 6, 2015.

"Our Manifest Destiny Is to Move Beyond Earth." *Financial Times,* December 23, 2014.

Packer, George. "The Birth of a New Century: What the World Lost in 2014." In "A World Disrupted: Leading Global Thinkers of 2014." Special issue, *Foreign Policy*, November/December 2014. http://globalthinkers.foreignpolicy.com.

Parris, Matthew. "The Global Spread of English Is a Seismic Event in Man's History." *Times* (London), January 15, 2005.

Pasztor, Andy. "China Sets Ambitious Space Goals." *Wall Street Journal*, April 15, 2010.

Petraeus, David H., and Robert B. Zoellick, Chairs. *North America: Time for a New Focus*. Independent Task Force Report No. 71. New York: Council on Foreign Relations, 2014.

Pavlus, John. "The Search for a New Machine." *Scientific American* 312, no. 5 (2015): 58–63.

Paxton, Robert O. "Papa Europe." *New York Review of Books*, November 23, 1978.

Pearce, Robert. Review of *Margaret Thatcher: Iron Lady*, by John Campbell. *History Today*, February 2004.

Pearlstine, Norman. "The Man and the Magic." *Time*, December 29, 1997.

Peel, Quentin. "Germany & Europe: A Very Federal Formula." *Financial Times*, February 9, 2012.

Pei, Minxin. "Down with Dengism." Project Syndicate. September 1, 2014. www.project-syndicate.org.

———. "How Beijing Kept Its Grip on Power." *Financial Times*, June 2, 2009.

Perlez, Jane. "China Looks Westward as It Bolsters Ties." *New York Times*, September 8, 2013.

Perry, Alex. "China's New Continent." *Time*, July 5, 2010.

Petersen, Anne, and Gail D. McClure. "Trends in Global Philanthropy Among U.S. Foundations: A Brief Review of Data and Issues." *Foundation Review* 2, no. 4 (2010). http://scholarworks.gvsu.edu/tfr/vol2/iss4/8/.

Pfaff, William. "She Is Real and She Is Serious." Review of *The Iron Lady*, by Hugo Young. *New York Times*, November 12, 1989.

Pisano, Gary P., David J. Collis, and Peter K. Botticelli. "Intel Corporation: 1968–1997." Harvard Business School Case 797-137, May 1997. (Revised May 2008.)

Plender, John. "London's Big Bang in International Context." *International Affairs* (Royal Institute of International Affairs) 63, no. 1 (Winter 1986–1987): 39–48.

Pollack, Andrew. "An 'Awesome' Intel Corners Its Markets." *New York Times*, April 3, 1988.

Pye, Lucian W. "An Introductory Profile: Deng Xiaoping and China's Political Culture." *China Quarterly*, no. 135 (1993): 412–43.

Rachman, Gideon. "China or the U.S.? Make Your Choice." *Financial Times*, October 3, 2011.

———. "Zero-Sum World." *Financial Times*, October 22, 2010.

Ramo, Joshua Cooper. "Surviving in Digital Times." *Time* 148, no. 19 (October 21, 1996): 86.

Ramo, Joshua Cooper, and Daniel Eisenberg. "Andrew Grove: A Survivor's Tale." *Time* 150, no. 27 (December 29, 1997): 54.

Ramo, Joshua Cooper, and David S. Jackson. "Winner Take All: Microsoft v. Netscape." *Time* 148, no. 13 (September 16, 1996): 56.

Randles, W. G. L. "The Alleged Nautical School Founded in the Fifteenth Century at Sagres by Prince Henry of Portugal, called the 'Navigator.'" *Imago Mundi: The International Journal for the History of Cartography* 45, no. 1 (1993): 20–28.

Rashid, Ahmed. "Why, and What, You Should Know About Central Asia." *New York Review of Books*, August 15, 2013.

Restall, Hugo. "The Lessons from 30 Years of Chinese Reform." *Wall Street Journal*, December 16, 2008.

Reston, James. "Monnet's Noble Life." *New York Times*, March 18, 1979.

Rillo, Aladdin, Pitchaya Sirivunnabood, Ahmad Ahsan, and Jean-Christophe Maur. *Association of Southeast Asian Nations (ASEAN) Integration Monitoring Report: A Joint Report by the ASEAN Secretariat and the World Bank*. Washington, DC: World Bank Group, 2013. http://documents.worldbank.org/curated/en/2013/01/18780456/association-southeast-asian-nations-asean-integration-monitoring-report-joint-report-asean-secretariat-world-bank.

Roach, Stephen S. "China's Turning Point." Project Syndicate. February 4, 2011. www.project-syndicate.org.

———. "Ten Reasons Why China Is Different." Project Syndicate. May 27, 2011. www.project-syndicate.org.

Roberts, Andrew. "Europe's 'Proud Empire' Is Entering a Cul de Sac of History." *Financial Times*, February 17, 2012.

Roberts, Sam. "Why Are Rockefellers Moving from 30 Rock?" *New York Times*, November 23, 2014.

Rockefeller Foundation website. http://www.rockefellerfoundation.org.

Rockefeller Foundation. "Our History." http://www.rockefellerfoundation.org/about-us/our-history.

Rockefeller University website. http://www.rockefeller.edu.

Romero, Simon. "Brazil on Edge as World Cup Exposes Rifts." *New York Times*, June 9, 2014.

Roosevelt, Theodore. "Citizenship in a Republic." Speech delivered at the Sorbonne, Paris, April 23, 1910.

Rose, Gideon. "China Now." *Foreign Affairs* 94, no. 3 (2015). https://www.foreignaffairs.com/articles/china/2015-04-20/china-now.

Rosen, Daniel H. *Avoiding a Blind Alley: China's Economic Overhaul and Its Global Implications*. New York: Asia Society Policy Institute in Collaboration with Rhodium Group, October 2014. http://asiasociety.org.

Rothstein, Edward. "Marveling at Wonders out of This World." *New York Times*, July 29, 2010.

Roxburgh, Charles, Susan Lund, Charles Atkins, Stanlislas Belot, Wayne W. Hu, and Moira S. Pierce. *Global Capital Markets: Entering a New Era.* McKinsey Global Institute, September 2009.

Roxburgh, Charles, Susan Lund, and John Piotrowski. *Mapping Global Capital Markets 2011.* McKinsey Global Institute, August 2011.

Ryan, Alan. "After the Fall." *New York Review of Books*, November 3, 2005.

———. "What Happened to the American Empire." *New York Review of Books*, October 23, 2008.

———. "Where Did She Go Wrong?" Review of *The Disenchanted Isle: Mrs. Thatcher's Capitalist Revolution*, by Charles Dellheim. *New York Review of Books*, August 20, 1995.

———. "Yes, Minister." Review of *The Downing Street Years*, by Margaret Thatcher. *New York Review of Books*, December 2, 1993.

Sachs, Jeffrey D. "Writing the Future." Project Syndicate. January 24, 2013. www.project-syndicate.org.

Schäfer, Daniel. "Rockefeller and Rothschilds Unite." *Financial Times*, May 29, 2012.

Schama, Simon, and Simon Kuper. "Margaret Thatcher 1925–2013." *Financial Times*, April 12, 2013.

Schechter, Erik, and Dave Majumdar. "Lazers Are No Longer a 'Star Wars' Fantasy." *Wall Street Journal*, January 20, 2014.

Schüssel, Wolfgang. "Why Europe Needs to Put Privatization Back on the Agenda." *Globalist*, November 21, 2014.

Science News 178, no. 9 (October 23, 2010).

"Setting Out the Store." *Economist*, January 11, 2014.

Shambaugh, David, ed. "Deng Xiaoping: An Assessment." Special issue, *China Quarterly*, no. 135 (1993): 409–652.

Sherman, Stratford, Jane Furth, Alicia Jills Moore, and Joshua Mendes. "The New Computer Revolution." *Fortune*, June 14, 1993.

Shevory, Kristina. "With Squeeze on Credit, Microlending Blossoms." *New York Times*, July 28, 2010.

Shu-Ling Tan, Catherine, Steffen Dyck, and Syetarn Hansakul. *China's Financial Integration into the World Economy*. Frankfurt: Deutsche Bank Research, November 23, 2011. www.dbresearch.com.

Silver-Greenberg, Jessica, Matthew Goldstein, and Nicole Perlroth. "JPMorgan Chase Hacking Affects 76 Million Households." *New York Times*, October 2, 2014.

Specter, Michael. "The Gene Factory." *New Yorker*, January 6, 2014.

Spero, Joan. *Charity and Philanthropy in Russia, China, India, and Brazil.* New York: Foundation Center, 2014. http://foundationcenter.org.

Spiegel, Peter. "José Manuel Barroso: 'Not Everything I Did Was Right.' " *Financial Times*, November 4, 2014.

Stephens, Philip. "All That Needs to Be Said About Europe's Next Half Century." *Financial Times*, March 7, 2007.

———. "New China Starts to Make the Rules." *Financial Times*, May 28, 2015.

———. "Riches and Risk: Welcome to the World of Tomorrow." *Financial Times*, January 9, 2014.

Stephenson, Hugh. Review of *Margaret Thatcher: The Grocer's Daughter*, by John Campbell. *History Today* 50, no. 10 (October 2000): 57.

Stokes, Bruce. "Cooling Dragon." *National Journal*, August 2, 2008.

Stokes, Jacob. "China's Road Rules: Beijing Looks West Toward Eurasian Integration." *Foreign Affairs*, April 19, 2015. https://www.foreignaffairs.com/articles/asia/2015-04-19/chinas-road-rules.

Subramanian, Arvind. "The Inevitable Superpower." *Foreign Affairs* 90, no. 5 (September/October 2011): 66–78.

Swanström, Niklas. "China and Greater Central Asia: New Frontiers?" Silk Road Papers Series. Washington, DC, and Stockholm: Central Asia–Caucasus Institute and Silk Road Studies Program, December 2011. www.silkroadstudies.org.

Talbott, Strobe. "Monnet's Brandy and Europe's Fate." Brookings Essay. Washington, DC: Brookings Institution Press, 2014. Last modified February 11, 2014. http://www.brookings.edu/research/essays/2014/monnets-brandy-and-europes-fate.

"Talking to the Neighbors: A Modest Proposal for an Interstellar Communications Network." *Economist*, April 7, 2011.

Taylor, Stuart. "The Next Generation of the Internet: Revolutionizing the Way We Work, Live, Play, and Learn." Cisco Systems, Inc., April 2013. http://www.cisco.com/web/about/ac79/docs/sp/Next-Generation-of-the-Internet.pdf.

Tedlow, Richard S. "The Education of Andy Grove." *Fortune*, December 12, 2005.

"The Third Industrial Revolution." *Economist*, April 21, 2012.

"This Time It's Serious." *Economist*, February 18, 2012.

Thomas, Helen, and Megan Murphy. "Rothschild to Bolster US Presence." *Financial Times*, June 3, 2010.

Times (London). "M. Jean Monnet." Obituary, March 16, 1979.

Tita, Bob. "3-D Printer Makers Get Reality Check." *Wall Street Journal*, April 4, 2014.

Town Meeting of the World. CBS Reports, July 10, 1963. (Walter Cronkite, reporting.)

"The Tragedy of the Arabs." *Economist*, July 5, 2014.

"A Troubling Trajectory." *Economist*, December 13, 2014.

"20,000 Colleagues Under the Sea." *Economist*, June 9, 2012.

Tyler, Patrick E. "Deng Xiaoping: A Political Wizard Who Put China on the Capitalist Road." *New York Times*, February 20, 1997.

University of Chicago website. http://www.uchicago.edu.

Vance, Ashlee. "Forecast for Microsoft: Partly Cloudy." *New York Times*, October 17, 2009.

———. "Intel Looks to Invest Despite a Drop-Off." *New York Times*, January 16, 2009.

Van Rompuy, Herman. *Towards a Genuine Economic and Monetary Union.* Report presented by the President of the European Council at the European Commission Meeting, Brussels, June 26, 2012.

Vetter, Stefan. *The Single European Market 20 Years On: Achievements, Unfulfilled Expectations & Further Potential.* Frankfurt: Deutsche Bank Research, October 31, 2013. www.dbresearch.com.

Victor, David, M. Granger Morgan, Jay Apt, John Steinbrune, and Katharine Ricke. "The Geoengineering Option." *Foreign Affairs* 88, no. 2 (March/April 2009): 64–76.

"VW Conquers the World." *Economist*, July 7, 2012.

Wagstyl, Stefan, and Alex Barker. "Schäuble Revives Push for European Interpretation." *Financial Times*, March 27, 2014.

Wald, Matthew L. "Capturing Carbon as a Byproduct of Running a Fuel Cell." *New York Times*, December 15, 2014.

Waters, Richard. "California Scheming." *Financial Times*, January 15, 2009.

———. "Technology: Vanity or Visionary?" *Financial Times*, August 13, 2013.

Watson, James. *Hot Spots: Benchmarking Global City Competitiveness.* Edited by Sudhir Vadaketh. London: Economist Intelligence Unit, January 2012. http://www.economistinsights.com/sites/default/files/downloads/Hot%20Spots.pdf.

"Welcome to a (Rather Dour) Party." *Economist*, July 31, 2008.

Wheatley, Jonathan. "Q&A: The Asia Infrastructure Investment Bank." *Financial Times*, March 12, 2015.

"When Did Globalization Start?" *Economist*, September 23, 2013.

White, Theodore H. "China: Burnout of a Revolution." *Time* 122, no. 13 (September 26, 1983): 40.

Whitman, Alden. "Jean Monnet, 90, Architect of European Unity, Dies." *New York Times*, March 17, 1979.

Wildau, Gabriel. "China Backs Up Silk Road Ambitions with $62 Bn Capital Injection." *Financial Times*, April 20, 2015.

Wilford, John Noble. "3, 2, 1 and the Last Shuttle Leaves an Era Behind." *New York Times*, July 8, 2011.

Wilson, Dominic, and Raluca Dragusanu. *The Expanding Middle Class.* Global Economics Paper no. 170. Goldman Sachs Economic Research, July 7, 2008.

Wines, Michael, and Edward Wong. "China Takes Stage as World Economic Power, But Its Transformation Is Incomplete." *New York Times*, April 2, 2009.

Wolf, Martin. "A Climate Fix Would Ruin Investors." *Financial Times*, June 18, 2014.

Wolfe, Alexandra. "Weekend Confidential: Ray Kurzweil." *Wall Street Journal*, March 30, 2014.

Wolfe, Tom. "How the Sun Rose on the Silicon Valley: The Tinkerings of Robert Noyce." *Esquire*, December 1983.

Wong, Edward. "Security Law Suggests a Broadening of China's 'Core Interests.'" *New York Times*, July 2, 2015.

"Work Until You Drop." *Economist*, October 11, 2014.

"World Refugee Day: Global Forced Displacement Tops 50 Million for First Time in Post–World War II Era." UNHCR, June 20, 2014. http://www.unhcr.org/53a155bc6.html.

World Trade Organization. *World Trade Report: 2013*. Geneva: World Trade Organization, 2013.

"Xi Speech Celebrates Deng Xiaoping's Legacy," China.org.cn. http://www.china.org.cn/opinion/2014-08/21/content_33299018.htm.

Yergin, Daniel. "It's Still the One." *Foreign Policy*, no. 174 (September/October 2009): 88–95.

———. "There Will Be Oil." *Wall Street Journal*, September 17, 2011.

Yousafzai, Malala. "Nobel Lecture." Lecture, The Nobel Foundation, Stockholm, 2014.

Zakaria, Fareed. "The Future of American Power." *Foreign Affairs* 87, no. 3 (May/June 2008): 18–43.

———. "Greed Is Good (to a Point)." *Newsweek*, June 22, 2009.

———. "It's the Age of Bloomberg, Not Thatcher." *Newsweek*, October 13, 2008. http://www.newsweek.com/zakaria-its-age-bloomberg-not-thatcher-91827.

———. "The Return of Capitalism." *Washington Post*, June 15, 2009.

———. "This Is Brand New." *Newsweek*, April 23, 2010.

INDEX

Page numbers in *italics* refer to maps and illustrations.

ABOUT THE AUTHOR

JEFFREY E. GARTEN teaches courses on the global economy at the Yale School of Management, where he was formerly the dean. He has held senior positions in the Nixon, Ford, Carter, and Clinton administrations and was a managing director of Lehman Brothers and the Blackstone Group on Wall Street.

His articles have appeared in the *New York Times*, *Wall Street Journal*, *Financial Times*, *Foreign Affairs*, *Foreign Policy*, *BusinessWeek*, and the *Harvard Business Review*, and he is the author of four previous books on global economics and politics. He has been a frequent speaker around the world on global politics, global economics, and global leadership.